Mission-Critical Network Planning

For a listing of recent titles in the *Artech House Telecommunications Library*, turn to the back of this book.

Mission-Critical Network Planning

Matthew Liotine

Artech House, Inc.
Boston • London
www.artechhouse.com

Library of Congress Cataloging-in-Publication Data

Library of Congress CIP information is available on request.

British Library Cataloguing in Publication Data
Liotine, Matthew
   Mission-critical network planning. —(Artech House telecommunications library)
   1. Computer networks—Design and construction   2. Business enterprises—Computer
   networks
   I. Title
   004.6

   ISBN   1-58053-516-X

Cover design by Igor Valdman

© 2003 ARTECH HOUSE, INC.
685 Canton Street
Norwood, MA 02062

International Standard Book Number: 1-58053-516-X
A Library of Congress Catalog Card number is available for this book.

10 9 8 7 6 5 4 3 2 1

To Camille and Joseph—this is for them to remember me by.

Contents

CHAPTER 7

Network Access Continuity   153

CHAPTER 8

Mission-Critical Platforms   177

Foreword

September 11, 2001, is a defining date in U.S. and world history. September 11, or 911, may also prove to be a defining event for business networking, as 911 brought attention to the challenging practices and processes of maintaining states of preparedness in critical infrastructures. Among these were the communications infrastructure built around and upon Internet technology. While the news coverage in the days and weeks following 911 justly focused on the human tragedy, many individuals and organizations affected by the terrorist acts struggled to resume business as usual. Those for whom the Internet was critical to their business processes were either devastated by the extent to which they were isolated from customers, partners, and even their colleagues, or they were surprised (if not astonished) that their businesses processes were rapidly restored, and in some cases, remained nearly intact.

The first group undoubtedly struggled through a slow, painful, disruptive, and expensive process we call business resumption and disaster recovery. The latter group included a handful of "lucky ducks" but many more organizations for which business and network continuity were carefully planned and implemented. When needed, these continuity measures executed as intended. For these organizations, 911 was *not* a defining event in a business sense, but a catastrophic event for which they had prepared their business and network operations.

A second and equally compelling reason that 911 was not a defining event to the latter group is that maintaining business and network continuity is as much about maintaining good performance when confronted with incidental events and temporary outages as it is when confronted with catastrophic ones.

In this unique work, Matthew Liotine presents strategies, best practices, processes, and techniques to prepare networks that are survivable and have stable behavior. He shows us how to design survivable networks that perform well and can quickly recover from a variety of problems to a well-performing state using commercial, off-the-shelf equipment and public services. The proactive measures and anticipatory planning Matthew presents here are immensely more useful lessons to learn and apply than resumption and recovery measures.

This book discusses problems and offers recommendations and solutions in each of the IT disciplines most often seen in large enterprises today. All of the major functional areas—networking (wide area and local area networks); hardware and operating system platforms, applications, and services; facilities, recovery and mirroring sites, and storage; and management and testing—are examined.

Dr. Liotine provides the most comprehensive analysis I have seen thus far in our industry. He establishes a wonderful balance between technology and business, complementing useful technical insight for the IT manager and staff with equally useful insight into planning, practice, process, manpower, and capital expenses for the business development folks. If your network is truly something you view as *mission critical*, you cannot afford to overlook this work.

David Piscitello
Core Competence, Inc.
September 2003

Preface

I wrote this book in the aftermath of the terrorist attacks that took place on September 11, 2001. Until then, I was planning to write this book at a much later time. But the events of that day compelled me to immediately begin this work. As we have seen, those events have had far-reaching repercussions in the information technology (IT) industry. Now more than ever, private and public organizations are assessing their vulnerability to unexpected adverse events. I felt that this was the right time to develop a work that would serve as a reference for them to utilize in their efforts.

This book represents wisdom distilled from over 25 years of industry experience, the bulk of which was realized during my years at AT&T. I also researched many books and countless articles that were directly or peripherally related to this subject. I conducted conversations with colleagues and experts on the many topics within to obtain their viewpoints. I discovered that there wasn't any one book that embodied the many facets of IT continuity. I thus saw the need for a work that assembled and consolidated much of this conventional wisdom into an organized knowledge base for continuity practitioners.

This book is intended for IT managers, service providers, business continuity planners, educators, and consultants representing corporations, public agencies, or nonprofit organizations. It is assumed that the reader has some basic familiarity with data or voice networking. Even those seasoned continuity professionals, I hope, will find this book a valuable resource with a different perspective. All along, I stress the understanding of basic concepts and principles to maximize retention. The book is quite comprehensive. Many topics are covered at a level that will most likely require follow up by the reader—this is intentional. Under each topic, I have tried to flag those issues of importance and relevance to continuity.

Size and time constraints prevented me from including threads on several important subjects. One of them is security, which is fast becoming a critical issue for many IT organizations. Security is definitely tied to continuity—we touch upon it from time to time in this book. But security is a topic that requires an entire book for proper treatment. Likewise with business impact analysis—the process of identifying those elements of an IT infrastructure that require protection and justifying and allocating the related costs. Although I had originally prepared a chapter on this topic to include in this book, I left it out because I could not give the subject due justice in only a single chapter. Last, I did not include discussion on how to audit network environments for continuity and apply the many types of solutions that we discuss herein. Again, this requires another book.

Today's IT industry is staffed by many bright young individuals. They do not have the luxury of enduring many years to acquire the wherewithall on this topic, much of which is not readily obtained through academic training and books. (I had to learn about this stuff the hard way—spending many years in field practice making mistakes and learning from them.) I hope I have done them a favor by encapsulating this information so that those interested individuals can quickly get up to speed on this subject.

Acknowledgments

I am indebted to several individuals for helping me accomplish this work. First and foremost, special thanks go to Dave Piscitello at Core Competence, Inc., for being the technical reviewer of the manuscript. It gave me great comfort knowing that, with his vast information technology knowledge and experience, Dave was carefully evaluating the manuscript for technical accuracy. He provided invaluable comments, suggestions, and corrections that I incorporated wherever possible. Reviewing a manuscript of this magnitude is no easy job—for that I am forever grateful. I also thank him for writing the Foreword.

I would also like to thank my editor at Artech House Publishers, Barbara Lovenvirth, for guiding me through the authoring process and keeping the project on track. In addition, I would like to thank Mark Walsh at Artech House for recognizing the importance of this topic and having the vision to take this project on. My final thanks goes to my family—my wife, Billie, and my children, Camille and Joseph. I had to spend many hours at the keyboard creating this work. They gave me the love and support that I needed throughout the many months it took to complete this book.

CHAPTER 1

Introduction

A decade ago, a book on network continuity would have presented a different light on survivability and performance for information technology (IT) environments. Although enterprises possessed and operated their own IT infrastructure, end users were typically internal to the company. The advent of the Internet as an enabler for electronic commerce (e-commerce) eventually forced IT departments to take a somewhat different stand regarding their operating environments. IT environments now have greater exposure and interaction with entities and users external to the enterprise. Heavy reliance on the Internet and distributed processing has created an environment that is more susceptible to forces outside the company. Consequently, customer expectations regarding survivability and performance have risen far beyond the tolerances of the internal user. Processing availability and speed have become new requirements for competition. After all, there is not much an internal user can do when service is problematic except complain or file a trouble ticket. Customers, on the other hand, can take their business elsewhere.

As systems become increasingly distributed, cheaper, and innovative, greater focus has been placed on strategically arranging them using effective network design practices to ensure survivability and performance. Strategies and practices used in the past to deliver survivable and well-performing networks are being revised in light of more recent IT trends and world events. The practice of purchasing fault-tolerant hardware as a sole means of achieving service availability is no longer sound. The practice of simply backing up and restoring data can no longer ensure the tight recovery intervals required for continuous processing. A recent trend towards consolidating many operation centers into larger, more efficient centers is being revisited in light of recent threats to national security.

1.1 What Is Mission Critical?

Mission critical refers to infrastructure and operations that are absolutely necessary for an organization to carry out its mission. Each organization must define the meaning of mission critical based on its need. For a private enterprise, mission critical may be synonymous with business goals. For a public agency, mission critical may take on various contexts, depending on the nature of the agency. A law enforcement agency, for example, might associate it with public-safety goals. The meaning can also differ with scope. While for a large private enterprise it can be tied to broad strategic goals, a manufacturing operation might associate it with plant-production goals.

Each organization must determine what aspects of their IT network are mission critical to them. This includes resources such as switches, routers, gateways, service platforms, security devices, applications, storage facilities, and transmission facilities. Mission-critical resources are those that are absolutely necessary to achieve the given mission. This can constitute those resources that are essential to critical components and processes to perform their intended function. Because any of these elements can fail due to improper design, environmental factors, physical defects, or operator error, countermeasures should be devised that continue operation when key resources become unavailable. Highly visible front-end resources may have to take precedence over less critical back-office resources that have less pronounced influence on mission success.

1.2 Purpose of the Book

As organizations grow increasingly dependent on IT, they also grow more dependent on immunity to outages and service disruptions. This book presents strategies, practices, and techniques to plan networks that are survivable and have stable behavior. Although the practice of disaster recovery emphasizes restoration from outages and disruptions, this book is not intended to be a book on disaster recovery. Instead, we discuss how to design survivability and performance into a network, using conventional networking technologies and practices, and how to create the ability to recover from a variety of problems. We tell you what to look out for and what to keep in mind. Wherever possible, we try to discuss the benefits and caveats in doing things a certain way.

There is often a temptation to become too obsessed with individual technologies versus looking at the big picture. To this end, this book emphasizes higher-level architectural strategies that utilize and leverage the features of various technologies. As many of these strategies can be turned around and applied elsewhere, even at different network levels, one will find that the practice of network continuity planning is influenced more on how various capabilities are arranged together and less on a sole reliance on the capability of one technology.

Organizations with infinite money and resources and no competitors can certainly eliminate most disruptions and outages, but of course such firms do not exist. Most firms are faced with the challenge of maximizing survivability and performance in light of tight budgets and troubled economies. As a result, network continuity planning becomes a practice of prioritizing; assigning dollars to those portions of the network that are most mission critical and that can directly affect service delivery. If done indiscriminately, network continuity planning can waste time and money, leading to ineffective solutions that produce a false sense of security.

1.3 Network Continuity Versus Disaster Recovery

Although the terms disaster recovery, business continuity, and network continuity are used interchangeably, they mean different things. Disaster recovery focuses on immediate restoration of infrastructure and operations following a disaster, based heavily on a *backup and restore* model. Business continuity, on the other hand, extends this model to the entire enterprise, with emphasis on the operations and functions of critical business units. Disasters are traditionally associated with adverse conditions of severe magnitude, such as weather or fires. Such adverse events require recovery and restoration to operating levels equal or functionally equivalent to those prior to the event. In the context of IT, the definition of a disaster has broadened from that of a facility emergency to one that includes man-made business disruptions, such as security breaches and configuration errors.

A disaster recovery plan is a set of procedures and actions in response to an adverse event. These plans often evolve through a development process that is generally longer than the duration of the disaster itself. Revenue loss and restoration costs accumulate during the course of an outage. Immediate problems arising from a disaster can linger for months, adding to further financial and functional loss. For this reason, successful disaster recovery hinges on the immediacy and expediency of corrective actions. The inability to promptly execute an effective disaster recovery plan directly affects a firm's bottom line. As the likelihood of successful execution can be highly uncertain, reliance on disaster recovery as the sole mechanism for network continuity is impractical.

For one thing, disaster recovery plans are based on procedures that convey emergency actions, places, people, processes, and resources to restore normal operations in response to predetermined or known events. The problem with this logic is that the most severe outages arise from adverse events that are unexpected and unknown. In the last couple of years, the world has seen first-hand how events that were once only imaginable can become reality. With respect to networking, it is impractical to devise recovery plans for every single event because it is simply impossible to predict them all. This is why IT disaster recovery is a subset of a much larger, high-level strategic activity called network continuity planning.

Network continuity is the ability of a network to continue operations in light of a disruption, regardless of the origin, while resources affected by the disruption are restored. In contrast to disaster recovery, network continuity stresses an avoidance approach that proactively implements measures to protect infrastructure and systems from unplanned events. It is not intended to make disaster recovery obsolete. In fact, disaster recovery is an integral part of network continuity. Network continuity helps avoid activating disaster-recovery actions in the first place, or at least buys some time to carry out disaster recovery while processing continues elsewhere. The best disaster-recovery mechanisms are those manifested through network design.

Network continuity planning means preparing ahead for unexpected disruptions and identifying architectures to ensure that mission-critical network resources remain up and running. Using techniques in distributed redundancy, replication, and network management, a *self-healing* environment is created rivaling even the most thorough disaster-recovery plans. Although it is often easier to build avoidance into a brand new network implementation, the reality is that most avoidance must be added incrementally to an existing environment.

Network continuity addresses more than outages that result in discontinuities in service. A sluggish, slow performing network is of the same value as an outage. To the end user, the effect is the same as having no service. In fact, accumulated slow time can be just as costly, if not more costly, than downtime. Network continuity focuses on removing or minimizing these effects. It promotes the concept of an *envelope of performance*, which is comprised of those conditions that, if exceeded or violated, constitute the equivalent of an outage.

1.4   The Case for Mission-Critical Planning

Since the era of year 2000 (Y2K) remediation, network contingency planning has become of great interest to industries. This interest was heightened in ensuing years from the onslaught of the telecommunication and high-tech industry insolvencies, numerous network security breaches, terrorist attacks, corporate accounting fraud scandals, and increasing numbers of mergers and acquisitions. But in spite of all of this, many firms still do not devote ample time and resources towards adequate network continuity planning, due largely in part to lack of funds and business priority [1].

There have been a plethora of studies conveying statistics to the effect. But what is of more interest are the findings regarding the causes of inadequate planning, even for those organizations that do have some form of planning in place:

- Although most businesses are dependent on their IT networks, IT comprises only a small percentage of a typical organization's budget—usually no more than 5%. Furthermore, only a small percentage of an IT budget is spent on continuity as a whole, typically no more than 10% [2, 3]. For these reasons, continuity may not necessarily gain the required attention from an enterprise. Yet, studies have shown that 29% of businesses that experience a disaster fold within two years and 43% never resume business, due largely in part to lack of financial reserves for business continuity [4, 5].
- Although three out of every four businesses in the United States experience a service disruption in their lifetime, a major disruption will likely occur only once in a 20-year period. For this reason, many companies will take their chances by not investing much in continuity planning. Furthermore, smaller companies with lesser funds will likely forego any kind of continuity planning in lieu of other priorities. Yet, most small businesses never recover from a major outage [6].
- Companies that do have adequate funding often do not fully understand the financial impact of a possible IT outage. Many companies are not well prepared to recover from a major disruption to their top earnings drivers. Those that do still lack true network continuity and broader business-recovery plans. Many plans focus on network data recovery, overlooking the need to protect and recover key applications and servers, which are even more prone to disruption.

Network continuity planning is a cost-effective, yet underutilized, practice. Ultimately, the main benefits of network-continuity planning can be measured in dollars. It can significantly reduce the overall cost of ownership of an infrastructure. Those organizations that are thoroughly prepared for outages ultimately exhibit significantly lower expected loss of revenues and services, and encounter less frequent variances in budget. They are less subject to penalties arising from infractions in legal and regulatory requirements such as those imposed by the Internal Revenue

Service, U.S. Patent Office, and Securities and Exchange Commission (SEC). They are also more likely to meet their contractual commitments with customers, partners, and suppliers.

While traditional disaster recovery methods focus on data protection and major outages, a firm's profitability is driven more so by its ability to handle less pronounced, yet more common disruptions. Over time, these can have greater cumulative impact on a firm than a single major disruption. Because they are less newsworthy, they often go unnoticed and unreported. Network continuity planning entails how to make the most of limited resources to protect against such events. To do so, an evaluation of current operations is necessary to identify where an organization is most at risk and those critical resources that require protection. This book discusses in further detail the types of safeguards that can be used and their pros and cons.

1.5   Trends Affecting Continuity Planning

The true case for mission-critical network planning rests on some underlying phenomena that persist in today's business and operational IT environments:

- Most industries rely heavily on computing and networking capabilities, to the point where they have become inseparable from business operations. Customer-relationship management (CRM), enterprise resource planning (ERP), financial, e-commerce, and even e-mail are a few examples of applications whose failure can result in loss of productivity, revenue, customers, credibility, and decline in stock values. Because such applications and their associated infrastructure have become business critical, network continuity has become a business issue.

- Business need is always outpacing technology. Major initiatives such as Y2K, Euro-currency conversion, globalization, ERP, and security, to name a few, have resulted in a flood of new technology solutions. But such solutions often exceed the users' ability to fully understand and utilize them. They evolve faster than an organization's ability to assimilate their necessary changes to put them in service. This gap can manifest into implementation and operational mishaps that can lead to outages.

- The need for 24 × 7 network operation has become more prevalent with e-commerce. While high-availability network architectures for years focused on the internal back end of the enterprise, they are now finding their way into the front end via the Internet. Outages and downtime have become more visible. But many venture-funded e-businesses focus on revenue generation versus operational stability. Those that are successful become victims of their own success when their e-commerce sites are overwhelmed with traffic and freeze up, with little or no corrective mechanisms.

- Consequently, user service expectations continue to grow. Now, 24 × 7 service availability has become a standard requirement for many users and organizations. Many who did not own computers several years ago are now on-line shoppers and users of network communications and Web applications. Those who experience a failed purchase at a particular Web site shop elsewhere.

While expectations for data service in the residential market might remain low when compared to voice, availability standards for both data and voice remain high in the commercial arena.

- The trend in faster and cheaper systems and components has led many organizations to distribute their infrastructure across business units in different locations. Mission-critical applications, data, and systems no longer reside in the sanctity of a corporate headquarters. Instead, they are spread among many locations, exposing them to many threats and vulnerabilities. IT has become network centric, integrating diverse, isolated systems using networking to enable computing not only inside the enterprise, but outside as well. This greatly complicates continuity and recovery planning, particularly for those firms fixated on centralized IT management and control.

- Growing churn in organizations stemming from corporate mergers, reorganizations, downsizing, and turnover in personnel leads to gaps. Like technology changes, sweeping business shifts can outpace the managerial, administrative, and technology changes required to realize them. Gaps and oversights in planning, consolidation, and rearrangement of IT infrastructure and operations are likely to ensue, creating the propensity for errors and disruption.

In light of the recent flurry of corporate accounting malpractice scandals, organizations will be under greater scrutiny regarding tracing expenditures to the organization's bottom line. This makes the practice of network continuity ever the more challenging, as now all remedial actions will be tested for their added value. It is hoped that this book will assist IT planners in choosing cost-effective solutions to combat IT outage and improve performance.

1.6 Mission Goals and Objectives

Before we begin, know your mission goals and objectives. Remedial measures must be aligned with mission goals to avoid placebo effects and undesirable outcomes. For instance, firms will institute database replication to instill a sense of high availability, only to find out that it may not necessarily protect against transaction overload. Continuity strategies and tactics should be devised in light of well-defined mission goals. The following are examples of some goals objectives and how they can impact continuity [7]:

- *Maximize network performance subject to cost:* This objective requires satisfying different service performance requirements at the lowest cost. If not carefully implemented, it could result in multiple service-specific networks deployed within the same enterprise.

- *Maximize application performance:* Applications critical to business success will require an optimized network that satisfies their bandwidth and quality of service (QoS) requirements. Service applications will have to be prioritized based on their importance.

- *Minimize life cycle costs:* Choosing continuity solutions solely on the basis of their life cycle costs can lead to less robust or ineffective solutions, or those that cannot live up to their expectations for effectiveness.

- *Maximize time to value:* Selecting solutions on the basis of their ability to deliver return on investment sooner can lead to a series of short-lived quick fixes that can evolve into an unwieldy network environment.
- *Minimize downtime:* Unless further qualified, this objective can lead to over-spending and overprotection. It is better to first identify the most critical areas of the organization where downtime is least tolerated.

IT organizations will standardize on combinations and variations of these items. Careful thought should be given when setting objectives to understand their implications on network continuity planning to avoid superfluous or unwanted outcomes.

1.7 Organization of the Book

Survivability and performance should be addressed at all levels of an enterprise's computing and communication environment. For this reason, the book is organized according to IT functional areas that are typical of most enterprises:

- *Network topologies and protocols (Chapter 4):* This chapter discusses some basic network topology and protocol concepts as they relate to continuity to provide background for subsequent discussions on networking.
- *Networking technologies (Chapter 5):* This chapter focuses on how to leverage different network technologies for continuity, related to local area networks (LANs) and wide area networks (WANs).
- *Processing, load control, and internetworking (Chapter 6):* This chapter discusses various other networking and control technologies and how they can be used to enhance continuity.
- *Network access (Chapter 7):* This chapter presents technologies and techniques that can be used to fortify voice and data network access. It also presents some discussion on wireless options.
- *Platforms (Chapter 8):* This chapter focuses on hardware platforms and associated operating systems for various computing and communication components that are utilized in an enterprise. It talks about platform features pertinent to continuity.
- *Software applications (Chapter 9):* This chapter reviews those features of application software relevant to service continuity.
- *Storage (Chapter 10):* This chapter discusses the types of storage platforms, media, operations, and networking that can aid data protection and recovery.
- *Facilities (Chapter 11):* This chapter discusses implications of geographically locating facilities and focuses on the physical and environmental attributes necessary to support continuous operation of IT infrastructure. This includes discussions on power plant, environmental, and cable plant strategies.
- *Network management (Chapter 12):* This chapter reviews key aspects of network monitoring, traffic management, and service level management as they relate to performance and survivability.
- *Recovery sites (Chapter 13):* This chapter discusses options for selecting and using recovery sites, along with their merits and their caveats.

• *Testing (Chapter 14):* This chapter reviews the types of tests that systems and applications should undergo to ensure stable performance in a mission-critical environment.

To set the stage for these chapters, they are preceded by two chapters that focus on some of the underlying principles of continuity. Chapter 2 reviews some fundamental tenets that provide the foundation for understanding many of the practices discussed in this book. They are further amplified in Chapter 3, which presents formulas of key network measures that can be used to characterize continuity and performance.

References

[1]   Chabrow, E., and M. J. Garvey, "Regeneration: Playing for Keeps," *Information Week*, November 26, 2001, pp. 39–50.

[2]   Silverthon, A., "Disaster Recovery: Not Just Backup," *Business Solutions*, February 2002, pp. 39–44.

[3]   Chandler, D., and C. Hoffman, "High Availability: Oracle Addresses the Demand for Application Availability," IDC White Paper, International Data Corporation, 1999, pp. 2–5.

[4]   Simmons, J., "Information Availability: When Failure Is Not an Option," Sunguard Availability Services White Paper, April 2002, pp. 1–3.

[5]   Hinton, W., and R. Clements, "Are You Managing the Risks of Downtime?" *Disaster Recovery Journal,* Summer 2002, pp. 64–67.

[6]   Butler, J., "In Depth: How to Decrease Downtime and Increase Profits," *Tech Republic,* August 21, 2001, www.techrepublic.com.

[7]   Rybczynski, T., "Net Value—The New Economics of Networking," *Computer Telephony Integration*, April 1999, pp. 52–56.

CHAPTER 2

Principles of Continuity

This chapter introduces some basic principles behind network continuity planning. Although many of these concepts have been around for quite some time, they have only found their way into the IT environment in the last several years. Many arose from developments in telecommunications and the space and defense programs. These concepts are constantly replayed throughout this book, as they underpin many of the network continuity guidelines presented herein. Acquiring an understanding of basic principles helps any practitioner maintain a rational perspective that can easily become blurred by today's dense haze of networking technologies.

2.1   Fault Mechanics

Nature has often demonstrated the laws of entropy that foster tendency toward eventual breakdown. Those of us in the IT business witness these in action daily in our computing and communication environments. These laws are displayed in various ways: hardware component failures, natural disasters (e.g., fire and floods), service outages, failure to pay InterNIC fees (resulting in unexpected disconnects), computer viruses, malicious damage and more [1].

Regardless of the adverse event, disruptions in a network operation occur mainly due to unpreparedness. Inadequacy in planning, infrastructure, enterprise management tools, processes, systems, staff, and resources typically drives disruption. Poor training or lack of expertise is often the cause of human errors. Process errors result from poorly defined or documented processes. System errors such as hardware faults, operating system (OS) errors, and application failures are inevitable, as well as power outages or environmental disasters. Thus, network continuity planning involves creating the ability for networks to withstand such mishaps through properly designed infrastructure, systems, and processes so that operation disruption is minimized.

Because almost any adverse event or condition can happen, the question then remains as to what things must happen in order to identify and respond to a network disruption. The answer lies somewhere within the mechanics of responding to a network fault or disruption (Figure 2.1). These mechanics apply at almost any level in a network operation—from integrated hardware components to managerial procedures. The following sections consider some of these mechanisms.

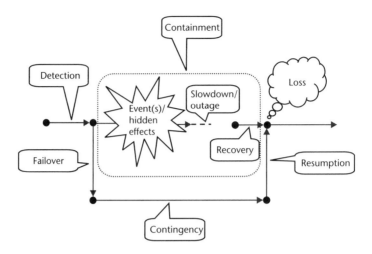

Figure 2.1 Fault mechanics.

2.1.1 Disruptions

Before a fault can be resolved, it must be discovered through a *detection* mechanism. Quite often, a problem can go undetected for a period of time, intensifying the potential for damage. *Early detection* of conditions that can lead to a disruption is fundamental to network continuity. In fact, some network management systems are designed to anticipate and predict problems before they happen. Some of these capabilities are addressed in a subsequent chapter on network management.

Disruptions produce downtime or slowtime, periods of unproductive time resulting in undue loss. The conditions that can lead to disruption can comprise one or many unplanned circumstances that can ultimately degrade a system to the point where its performance is no longer acceptable. Criteria must be collected to establish a *performance envelope* that, once violated in some way, qualifies an operation as disrupted. A good example of this at a global level is the performance envelope the Federal Communications Commission (FCC) places on telephone carriers. It requires carriers to report network outages when more than 90,000 calls are blocked for a period of 30 minutes or more [2, 3].

Sources of unplanned downtime include application faults, operation errors, hardware failures, OS faults, power outages, and natural disasters. In recent years, several nationwide network outages resulted from system instability following upgrades. Many server outages result from disk or power problems. Software application problems usually arise out of improper integration and testing. Security intrusions and viruses are on the rise and are expected to skyrocket in ensuing years.

Slow time is often largely unnoticed and for this reason comprises most productivity loss versus hard downtime. Slow response and inefficiency caused by degraded performance levels in servers, networking, and applications can equate to denial of service, producing costs that even exceed downtime.

A mission-critical network operation must be designed to address both unplanned and planned disruptions. Planned disruptions must occur for the purposes of maintenance. Planned system downtime is necessary when a system needs to be shut down or taken off-line for repair or upgrade, for data backup, or for moves, adds, or changes. Consequently, a mission-critical network must be designed so that operations can be maintained at acceptable levels in light of planned shutdowns. This is a driving factor behind the principles of *redundancy* discussed later in this chapter. Components, systems, applications, or processes will require upgrade and repair at some point.

Uptime is the converse of downtime—it is the time when an operation is fully productive. The meaning of uptime, downtime, and slowtime are a function of the context of the service being provided. Many e-businesses view uptime from the end user's perspective. Heavy process businesses may view it in terms of production time, while financial firms view it in terms of transaction processing. Although an important service metric, uptime is not the sole metric for scrutinizing network continuity. Regardless of the context, developing an envelope of performance should be one of the first orders of business in planning network continuity. This initiative should constitute collecting objectives that convey successful operation.

2.1.2   Containment

When a disruption occurs, several activities must take place. The process of *containment* controls the effects of the disruption so that other systems are unaffected. But more importantly, it should also minimize customer impact so that the disruption is virtually invisible to them. Many organizations assume that no news is good news—meaning that things are fine as long as there are no customer complaints. However, studies have shown that only a small percentage of customers affected by an outage actually report it [4]. Using a customer base as a detection mechanism can prove futile. Early detection will aid containment. The longer a disruption goes unanswered, the greater difficulty there will be in containment and recovery.

2.1.3   Errors

Errors are adverse conditions that individually or collectively lead to delayed or immediate disruption. *Self-healing errors* are those that are immediately correctable and may not require intervention to repair. Although they may have no observable impact on a system or operation, it is still important to log such errors because hiding them could conceal a more serious problem. *Intermittent errors* are chronic errors that will often require some repair action. They are often corrected on a retry operation, beyond which no further corrective action is needed. Persistent intermittent errors usually signal the need for some type of upgrade or repair.

Errors that occur in single isolated locations, component, connectivity, or some other piece of infrastructure are referred to as *simplex* errors. Multiple independent simplex errors not necessarily caused by the actions of one another can and will occur simultaneously. Although one might consider such instances to be rare, they almost invariably arise from external events such natural disasters, storms, and environmental factors.

If unanswered, simplex errors can often create a chain reaction of additional errors. An error that cascades into errors in other elements is referred to as a *rolling error*. Rolling errors are often characterized by unsynchronized, out-of-sequence events and can be the most damaging. Recovery from rolling errors is usually the most challenging, as it requires correction of all of the simplex errors that have

occurred. The *rolling window*, the length of time between the onset of the first simplex error and the resulting disruption, is often an indicator of the magnitude of the
recovery effort required.

A network system, no matter how well constructed and maintained, will eventually encounter an error. In large complex networks with numerous interoperable
systems, errors that plague one system are likely to affect others. A disruption in one
network can also create problems in other networks, creating a domino effect. This
has been demonstrated time and again during service provider network outages.

2.1.4 Failover

Failover is the process of switching to a backup component, element, or operation
while recovery from a disruption is undertaken. Failover procedures determine the
continuity of a network operation. Failover mechanisms can be devised so that they
take place immediately or shortly after a disruption occurs. Many systems use automatic failover and data replication for instant recovery. *Preemptive* failover can also
be used if an imminent disruption is detected.

Failover requires the availability of a backup system to eventually take over
service. The type of failover model required dictates the backup state of readiness
(Figure 2.2). There are three basic types of failover model. Each has implications on
the amount of information that must be available to the backup system at the time of
failover:

- *Hot or immediate failover* requires a running duplicate of the production system as a backup to provide immediate recovery. Consequently, it is the more
  complex end expensive to implement. The backup system, referred to as a *hot
  standby*, must constantly be updated with current state information about the
  activity of the primary system, so that it is ready to take over operation quickly
  when needed. This is why this type of failover is sometimes referred to as a
  stateful failover. Applications residing on the backup system must be designed
  to use this state information when activated. For these reasons, hot standby
  systems are often identical to the primary system. They are sometimes
  designed to *load share* with the primary system, processing a portion of the
  live traffic.
- *Cold failover*, on the other hand, is the least complex to implement but likely
  results in some disruption until the backup is able to initiate service. A *cold
  standby* backup element will maintain no information about the state of the
  primary system and must begin processing as if it were a new system. The
  backup must be initialized upon failover, consuming additional time. For these
  reasons, a cold failover model is usually the least expensive to implement.
- *Warm failover* uses a backup system that is not provided with state information on the primary system until a failover takes place. Although the backup
  may already be initialized, configuration of the backup with the information
  may be required, adding time to the failover process. In some variants of this
  model, the standby can perform other types of tasks until it is required to take
  over the primary system's responsibilities. This model is less expensive than
  the hot standby model because it reduces standby costs and may not necessarily require a backup system identical to the primary system.

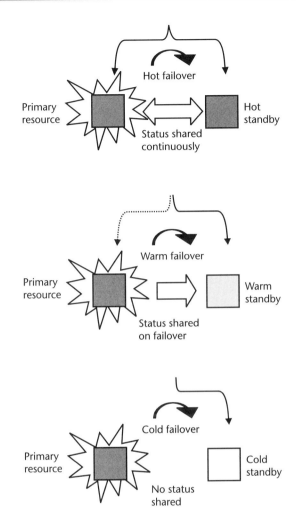

Figure 2.2   Types of failover.

2.1.5   Recovery

Recovery is the activity of repairing a troubled component or system. Recovery activity may not necessarily imply that the element has been returned to back to its operational state. At the system level, recovery activities can include anything from automatic diagnostic or restart of a component or system, data restoration, or even manual repair. Disaster recovery (DR) activities usually focus on reparations following a disaster—they may not necessarily include contingency and resumption activities (discussed later), although many interpret DR to include these functions.

2.1.6   Contingency

Contingency is the activity of activating a backup element or process upon failover from the primary failed element. Depending on the context, a contingency can take on various forms. It can represent a backup component, system, network link or path, service provider, or even a recovery data center or hot site. A contingency serves one main purpose—a fallback that buys time for recovery. Greater availability implies instantaneous failover to a contingency, reducing service interruption. It also implies higher system and infrastructure costs to implement backup capabilities.

2.1.7 Resumption

Once a recovery activity is completed, the process of resumption returns the repaired element back into operational service. It is the process of transferring operations over to the restored element, either gradually or instantaneously. A repaired element, although operational, should not be thrown immediately back into productive service until there is confidence that will function properly. This strategy should be followed at almost every level—from component to data center. Transfer to live service can almost assure the occurrence of other problems if not done correctly.

Flash-cut to full load can often heighten the emergence of another glitch. A wise approach is to transfer live load gradually and gracefully from the contingency element to back a restored element. Yet another alternative is to simply maintain the currently active element as the primary and keep the restored element as a backup. This approach is less disruptive and safeguards against situations where a restored element is still problematic after recovery. It assumes that both the primary and backup elements have identical, or at least equivalent, operating capacity.

2.2 Principles of Redundancy

Redundancy is a network architectural feature whereby multiple elements are used so that if one cannot provide service, the other will. Redundancy can be realized at many levels of a network in order to achieve continuity. Network access, service distribution, alternate routing, system platforms, recovery data centers, and storage all can be instilled with some form of redundancy. It can also include the use of service providers, suppliers, staff, and even processes. Redundancy should be applied to critical resources required to run critical network operations. It should be introduced at multiple levels, including the network, device, system, and application levels. Redundancy in networking can be achieved in a variety of ways, many of which are discussed further throughout this book.

Because redundancy can inflate network capital and operating costs, it should be effectual beyond the purposes of continuity. It is most cost effective when it is intrinsic to the design of a network, supporting availability and performance needs on an ordinary operational basis, versus an outage-only basis. Management can better cost-justify a redundant solution when it provides operational value in addition to risk reduction, with minimal impact on current infrastructure and operations. Redundant network systems and connectivity intended solely for recovery could be used for other purposes during normal operation, rather than sitting idle. They should be used to help offload traffic from primary elements during busy periods or while they undergo maintenance.

2.2.1 Single Points of Failure

A single point of failure is any single isolated network element that, upon failure, can disrupt a network's productive service. Each single point of failure represents a "weakest link" in an IT operation. The greater the importance or responsibility of the element, the greater the impact of its failure or degradation. Single points of failure should be secured with redundancy in order to deter a disruption before it occurs.

Single points of failure are best characterized as a *serial path* of multiple elements, processes, or tasks where the failure or degradation of any single one can cause a complete system disruption (Figure 2.3). Serial paths appear in operational procedures as well as in logical and physical system implementations. An element that cannot be recovered while in productive service will likely require a redundant element. Application software, for example, can be restarted but cannot be repaired while processing. Redundant processing is often required as a contingency measure.

Minimizing single points of failure through redundancy is a fundamental tenet of mission-critical network design. Redundancy must be applied, as appropriate, to those aspects of the network infrastructure that are considered mission critical.

A general misconception is that redundancy is sufficient for continuity. *False redundancy* should be avoided by applying the following criteria:

- *The redundancy eliminates single points of failure* and does not inherently have its own single points of failure. A solution involving a server fitted with two network interface cards (NICs) feeding two separate switches for redundancy still retains the server as a single point of failure, resulting in only a partial redundancy. Another example is redundant applications that might share a single database image that is required for processing.

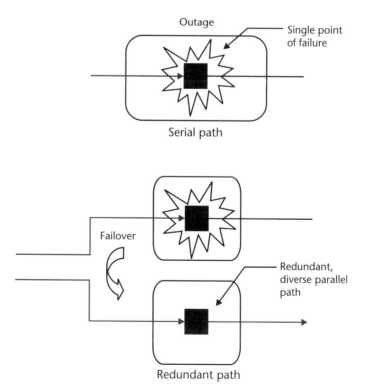

Figure 2.3   Removal of serial path through redundancy.

- *An adequate failover process to the redundant element is required* because it makes no sense to have redundancy unless it can be readily accessed when needed. For example, a server should be able to failover to a standby server under acceptable criteria of time and transaction loss.
- *The redundant element must provide an equivalent level of service* or one that meets continuity criteria. Elimination of a serial path will divert processing to a surviving path with other elements, temporarily increasing their workload. A properly designed network will limit the occasion for diversions and assure that the surviving elements can withstand the extra workloads placed upon them, within accepted operating criteria.
- *The redundancy should be diverse to the extent possible.* Replicated resources should not be collocated or share a common infrastructure or resource. Commonality introduces other single points of failure, subjecting them to the same potential outage. For example, a backup wide area network (WAN) access link connecting into a facility should not share the same network connectivity as the primary link, including the same cabling and pathway.

2.2.2 Types of Redundancy

Fundamentally, redundancy involves employing multiple elements to perform the same function. Because redundancy comes with a price, several types of redundancy can be used depending on the cost and level of need. If N is the number of resources needed to provide acceptable performance, then the following levels of redundancy can be defined.

2.2.2.1 *kN* Redundancy

This type of redundancy replicates N resources k times (i.e., *2N, 3N, 5N,* and so on), resulting in a 1 to 1 (1:1) redundancy. *kN* redundancy can be employed at a component, system, or network level, using k identical sets of N resources (Figure 2.4). For continuity, standby resources are updated with the activities of their primary

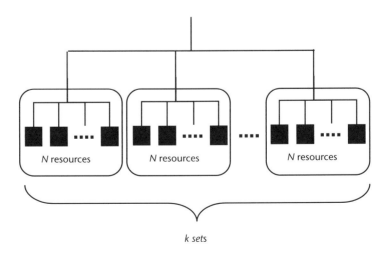

Figure 2.4 *kN* redundancy.

resource so that they can take over when failure or degradation occurs. Load sharing can take place among k sets of components if needed.

N is the minimum number of resources required for service—if a resource in the set fails or degrades, then it is assumed that the set cannot adequately provide service and must rely on another set N resources. Depending on operating requirements and how the resources are interconnected, the same number of resources can be partitioned in different ways.

For example, 2N redundancy duplicates a resource for each that is required for service. The 2N network switches can be deployed in isolated zones comprised of switch pairs ($N=1$) so that if one fails, the other can continue to switch traffic. On the other hand, a network of N switches can be backed up by a second network of N switches ($k=2$), so that if one network fails, the other can continue to switch traffic. As another example, if $k=2$ and $N=3$, there are two sets of three resources for a total of six resources. The same number of resources can be deployed in a configuration with $k=3$ and $N=2$ (Figure 2.5).

kN redundancy typically involves less fault and failover management by the individual resources. Because there is a one-to-one correspondence between component sets, one set can failover to the other in its entirety. A more global fault management can be used to convey status information to a standby set of resources and to decide when failover should occur. Reliance on global fault management requires less managerial work by the individual resources.

Although kN redundancy can provide a high level of protection, it does so at greater expense. Replication of every resource and undue complexity, especially in cases where a large number of connections are involved, can add to the cost.

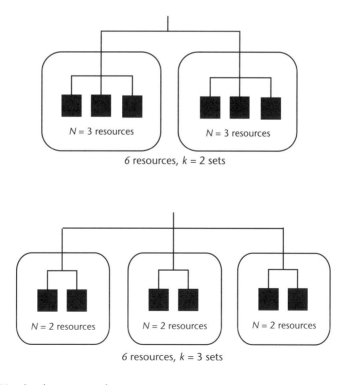

Figure 2.5   *kN* redundancy example.

2.2.2.2   N + K Redundancy

In situations where kN redundancy is not economical, resources can be spared in an $N + K$ arrangement. $N + K$ redundancy involves having K spare resources for a set of N resources. The K spare resources can load share traffic or operate on a hot, warm, or cold standby basis. If one of the N resources removes, fails, or degrades, one of the K spares takes over (Figure 2.6). For example, if an application requires four servers ($N=4$), then a system should be configured with five servers ($K=1$), so that losing one server does not affect service. The simplest and most cost-effective form of redundancy is $N + 1$ redundancy, also referred to as 1 to N or 1:N redundancy.

As in the case of kN redundancy, $N + K$ arrangements can be applied at all levels of an operation across many types of systems and devices. Data centers, servers, clusters, and networking gear can all be deployed in $N + K$ arrangements. Disk arrays, central processor unit (CPU) boards, fans, and power supplies can be deployed similarly as well.

$N + K$ redundancy allows a resource to be swapped out during operation, often referred to as *hot swapping*. Although maintenance could be performed on a resource without service interruption in an $N + 1$ arrangement, a greater risk is incurred during the maintenance period, as failure of one of the remaining N resources can indeed disrupt service. For this reason, having an additional redundant component ($K > 1$) in case one fails, such as an $N + 2$ arrangement, enables failover during maintenance (Figure 2.7).

$N + K$ arrangements are more likely to involve complex fault management. A more complicated fault management process cycle is necessary, requiring more effort for managing faults at a network or system level as well as the individual resource level. Hot failover requires a standby resource to know the states of many other resources, requiring more managerial work and extra capacity from the resource.

Boundaries must define the level of granularity that permits isolation of one of the N resources. A virtual boundary should be placed around a resource that masks its complexity and encapsulates faults, so that its failure will be benign to the rest of the system. In many systems and operations, there is often a hierarchy of resource dependencies. A resource failure can affect other resources that depend on it. $N + K$ fault management must identify the critical path error states so that recovery can be implemented.

2.2.2.3   N + K with kN Redundancy

$N + K$ arrangements can be applied within kN arrangements, offering different granularities of protection. Figure 2.8 illustrates several approaches. A kN arrangement can have K sets of resources, each having an $N + K$ arrangement to ensure a

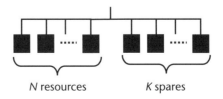

Figure 2.6 $N + K$ redundancy.

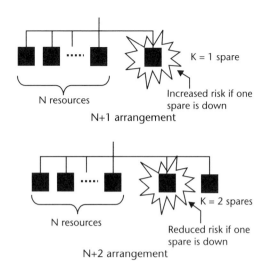

Figure 2.7 $N + 1$ versus $N + 2$ arrangement.

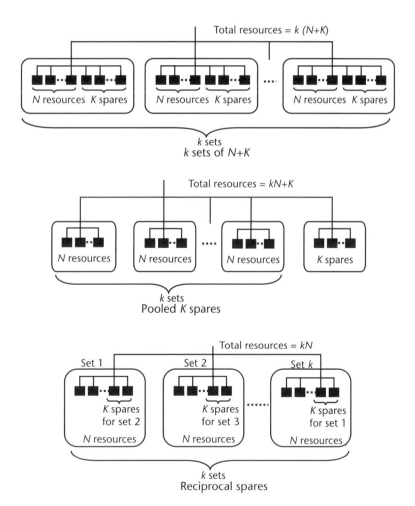

Figure 2.8 $N + K$ with kN redundancy.

higher degree of reliability within a set. But yet again, this configuration suffers from the coarse failover granularity associated with kN redundancy. More importantly, having additional K resources within a set can significantly increase the costs associated with this arrangement.

A more economical approach to redundancy is to pool K spares into a single set that can be used by the k sets of N resources. Yet another more economical but more complex approach is to assign K resources from one set to back up N resources in another set, in a reciprocating spare arrangement. The $N + K$ resources can assume an active/passive or load-sharing arrangement. This requires additional management complexity among resources. For hot failover, standby resources must be designated in advance and must be kept aware of resource states in other sets, in addition to their own. This adds to the management complexity inherent to $N + K$ arrangements, requiring additional work capacity on the part of the resource.

The choice of which strategy to use is dependent on a variety of criteria. They include the level at which the strategy will be applied (i.e., component, system, or network); the complexity of systems or components; cost; the desired levels of tolerance; and ultimately availability. These criteria are discussed in the following sections.

2.3   Principles of Tolerance

Tolerance is a concept that is revisited throughout this book. Simply put, it is the ability of an operation to withstand problems, whether at the component, system, application, network, or management level. The choice of tolerance level is often a combination of philosophical, economic, or technical decisions. Because tolerance comes with a price, economic justification is often a deciding factor. Philosophy will usually dictate how technology should be used to realize the desired levels of tolerance. Some organizations might rely more heavily on centralized management and integration of network resources, while others feel more comfortable with more intelligence built into individual systems.

The concepts of fault tolerance, fault resilience, and high availability are discussed in the following sections. Their definitions can be arbitrary at times and can include both objective and subjective requirements depending on organizational philosophy and context of use. Although they are often used to characterize computing and communication platforms, they are applicable at almost any level of network operation—from a system component to an entire network. They are discussed further in the chapter on platforms in the systems context.

Tolerance is often conveyed as *availability*, or the percentage of time that a system or operation provides productive service. For example, a system with 99.9% (three 9) availability will have no more than 8.8 hours of downtime per year. A system with 99.99% (four 9) availability will have no more than 53 minutes of downtime a year. A system with 99.999% (five 9) availability will have about 5 minutes of downtime a year. Availability is discussed in greater detail in the chapter on metrics.

The relationship between tolerance and availability is illustrated in Figure 2.9. The ability to tolerate faults without any perceivable interruption implies continuous availability. Continuous availability often entails avoiding transaction loss and reconnection of users in the event of a failure. Minimal or no visible impact on the

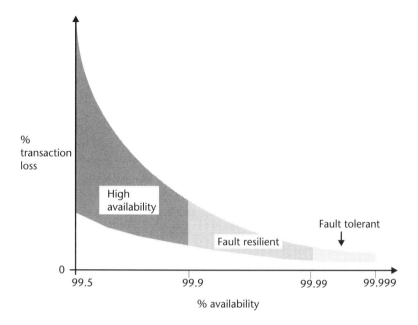

Figure 2.9   Tolerance versus availability.

end user is required [5]. A system of operation must transparently maintain user transactions in the original, prefailure state.

2.3.1 Fault Tolerance

Fault tolerance (FT) is a network's ability to automatically recover from problems. For this reason, FT is usually associated with availability in the range of four to five nines (or 99.99% to 99.999%). FT must be designed into a network through infrastructure and operations. To do this, an organization must understand which faults are tolerable. This requires determining which computing and communications processes are critical to an operation. Furthermore, it requires an understanding of how a network should behave during adverse events, so that it can be designed to behave in predictable ways.

FT systems are designed so that a single fault will not cause a system failure, allowing applications to continue processing without impacting the user, services, network, or OS. In general, a fault tolerant system or operation should satisfy the following criteria [6]:

- It must be able to quickly identify errors or failures.
- It must be able to provide service should problems persist. This means isolating problems so that they do not affect operation of the remaining system. This could involve temporarily removing problematic components from service.
- It must be able to repair problems or be repaired and recover while continuing operation.
- It must be able to preserve the state of work and transactions during failover.
- It must be able to return to the original level of operation upon resumption.

FT can be achieved in a variety of ways. Component redundancy, continuous error checking, and automatic recovery mechanisms are typically the most popular tactics. They frequently include things like dual power supplies, redundant CPUs, backup disks, and automatic failover software, all designed with the intent of eliminating service interruption. Notification of self-healing faults or failures is mandatory, as their masking could conceal a chronic problem. Furthermore, if a backup component or process is initiated, recovery of the problematic element, although automatic, still requires discovery and understanding of a root cause. The reason is simple—if a problem caused failure of a primary element, it can just as easily cause failure of a backup element. A failure of a backup element while a primary element is inactive can disrupt service.

2.3.2 Fault Resilience

Fault resilience (FR) is a term sometimes used synonymously with FT, but it differs with regard to the preservation of transactions and work during failover. FR is used to characterize systems or operations that can provide nearly the same level of availability as FT, yet not necessarily guarantee no transaction loss. Failover to a contingent system of operation requiring warm initialization is usually involved. For this reason, FR is typically associated with availability of approximately four nines (or 99.99%). Because the backup must learn the state of the primary element at the time of failover, transactions held by the primary system can be lost. FR is further discussed in the chapter on platforms.

2.3.3 High Availability

High availability (HA) is typically associated with availability in the range of 99.5% to 99.9%, although it sometimes is attributed to systems with 99.99% (or four 9) availability. HA services are usually designed under the assumption that transactional loss will occur during failover, usually to a cold-standby system or operation. The intent is not to prevent disruption, but to minimize the disruptive impact. Systems or operations characterized as HA cost less to implement than those under FT or FR criteria.

HA is usually applied at a localized level versus an entire operational or platform level. HA systems are usually located in the same operational environment with other redundant HA systems. Failover is thus confined to a single system or group while others pick up the gap in processing. HA systems will typically rely on higher level resources to ensure availability at a more global level. HA is discussed further in the chapter on platforms.

2.4 Principles of Design

User, throughput, and bandwidth requirements are becoming more challenging to predict and characterize for the purpose of planning and designing network infrastructure. To compensate, overprovisioning is often used to build in extra protective capacity as an insurance policy against the unforeseen. Some might say that this is justified in light of the declining unit costs of technology. But, on the other hand, designing a network with significantly more capacity that doesn't see efficient use can be ineffective use of money.

Capacity can be most effective when it is intelligently situated in the right places. Throwing extra capacity at a problem can often result in short-lived solutions or even be counterproductive. For instance, impedance mismatches can result if capacity is inequitably placed among systems; increasing the throughput of a network link is futile unless an end system can process data at the same transmission rate. If link capacity exceeds a system's processing capacity, performance degradation, congestion, and data loss are likely.

The following sections discuss some basic concepts of network design. They are intended as general guiding principles to design network resource capacity, review network architectures, and spot potential problems at the component, system, or network level.

2.4.1 Partitioning

A *partitioned* design is usually the best approach to realize a mission-critical network, system, or operation. Partitioned architectures involve separating a network or system into zones or modules. A *module* is the lowest acceptable level of independent operation. It can be manifested from a single component board to a network segment. Although modular design is key to cost-effective replacement, it does not necessarily imply rapid recovery. A management process must still be employed to induce failover and recovery of a downed module.

Partitioned designs that are *compartmental* further reduce risk by confining the impact of a problem to a module or group of modules, sometimes referred to as a *failure group*. The familiar "water-tight" doors on naval vessels are created so that an adverse event in one group does not affect operation of the entire network or system. Figure 2.10 illustrates these concepts.

2.4.2 Balance

The best network designs are those that *balance* distribution of resources with cost. This means providing an architecture that is neither centralized nor decentralized, but has adequate redundancy for survivability and performance [7]. The following sections describe the merits of each approach.

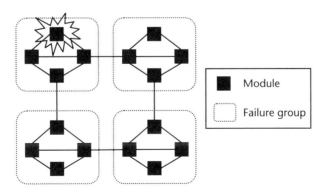

Figure 2.10 Partitioned network design example.

2.4.2.1 Centralization

Centralization is an approach that consolidates network resources into single locations, be it a node, platform, or data center, to reduce cost and complexity (Figure 2.11). It offers more efficient use of resources and reduces operational costs. But with consolidation comes less flexibility and higher risk, as loss of a consolidated resource can result in service loss. Consequently, customers of a centralized design require higher service guarantees. Loss of centralized or highly interconnected network nodes can effectively wipe out or degrade performance of the entire network. Recent studies of the Internet revealed that knocking out only 1% of the most highly connected nodes could cut average Internet performance in half [8].

Although a centralized architecture is often perceived as cost effective, hidden costs do prevail. For example, a consolidated node that is $2N$ redundant must be paired with an identical twin or equivalent, which can effectively double costs. Furthermore, consolidating resources into a single node may only replace connectivity costs with intranodal capacity costs in the form of CPU and memory needed to support processing of aggregated traffic. But, on the other hand, centralized architectures can require lower maintenance and operational costs and provide more stable performance [9].

2.4.2.2 Decentralization

In recent years, the use of centralized mainframe architectures in networking and computing has declined, giving rise to more decentralized architectures involving distributed systems (Figure 2.12) [10]. Where once single mainframes handled the processing of multiple functions, now many function-specific servers are used. Despite the trend towards distributed computing, firms are physically centralizing site resources, such as the case of server farms with large numbers of servers housed within a single site. In either case, distributed architectures entail more network interfaces and connectivity, require the integration of many disparate systems, and consequently lead to greater operational costs. On the other hand, distributed computing offers greater flexibility with the ability to use systems from multiple vendors featuring different OSs, applications, and hardware.

At first glance, a decentralized architecture might imply greater survivability and performance, as an outage in one location can be compensated by systems in

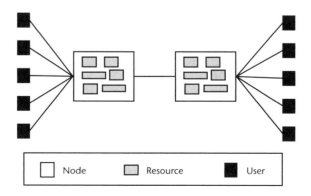

Figure 2.11   Centralized network architecture.

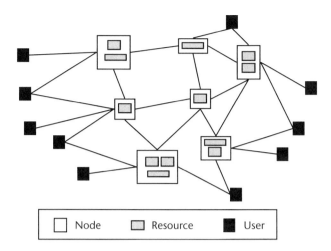

Figure 2.12   Decentralized network architecture.

other locations. But unless redundancy, failover, and recovery measures are implemented with other systems in the correct fashion, a false sense of security will exist. In fact, distributed architectures can increase the number of points of failure on a network [11]. The greater the number of nodes, the greater the likelihood of failures or faults across the overall network [12]. Decentralized networks require compartmental designs to limit the effect of nodal failures.

2.4.3   Network Scale

A network architecture that is scalable has the ability to cost-effectively change to accommodate increased demand. It should exhibit economies of scale, or in other words have the ability to expand gracefully to support capacity while achieving declining costs per unit of traffic growth. Scalability implies sustaining acceptable performance under changing traffic patterns and conditions. But a network's scale has implications on outage vulnerability. *Scale* is a measure of the number of end users or user nodes served by a network. Network *scope* is the number and location of nodes that provide services (i.e., applications) to the end users. A *scale factor* (σ) can be estimated as scale per unit scope, or the average percent of users served per service node:

$$\sigma = 1/N \tag{2.1}$$

σ can be applied in almost any context. It can measure the service or application density per node, or the density of connectivity. Networks having a small σ, sometimes referred to as *scale-free* networks, have most service nodes supporting a handful of users. σ is to be used as a gross number—it is transparent to the distribution of users across service nodes. A network with a uniform distribution of users where each service node handles the same number of users is sometimes referred to as an *exponential network*. Figure 2.13 illustrates these concepts.

Networks having high values of σ are said to be *limited in scale* and are difficult to cost-effectively grow to accommodate demand. The phenomenon of *net effect*,

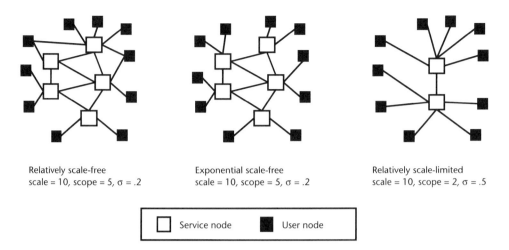

Relatively scale-free
scale = 10, scope = 5, σ = .2

Exponential scale-free
scale = 10, scope = 5, σ = .2

Relatively scale-limited
scale = 10, scope = 2, σ = .5

☐ Service node   ■ User node

Figure 2.13 Network scale factor.

expressed as *Metcalfe's Law*, states that the potential value of a network is proportional to the number of active user nodes. More precisely, the usefulness of a network to new and existing users equals the square of the number of user nodes. Although networks with higher scale factors provide greater potential value per node, they are less economical to grow [13].

σ has implications on network risk. Large, scale-free networks imply distributed architectures, which, we see from prior discussion, have a greater likelihood of outage. But because there are fewer users served per node, the expected loss from a nodal outage is less. In a network of limited scale, where users are more concentrated at each node, greater damage can happen.

These points can be illustrated by estimating the minimum expected loss in a network of N nodes, each with an outage potential p, equal to the probability of a failure. The probability of f failures occurring out of N possible nodes is statistically characterized by the well-known Binomial distribution [14]:

$$P(f) = N!\, p^{f} (1-p)^{N-f} \,/\, f!(N-f)! \qquad (2.2)$$

where $P(f)$ is the probability of f failures occurring (this formula will be revisited in the next chapter with respect to the discussion on network reliability). If $P(0)$ indicates the percentage of the time no failures occur ($f = 0$), then $1 - P(0)$ is the percentage of time that one or more failures occur. If we next assume that σ is the average nodal loss per outage, measured in terms of percentage of users, then on a broad network basis, the minimum risk (or percent expected minimum loss) ρ for a network is given by:

$$\rho = \sigma[1 - P(0)] \qquad (2.3)$$

Figure 2.14 graphically shows how ρ varies with network scale at different nodal outage potentials p. It shows that investing to reduce nodal outage potentials, regardless of scale, can ultimately still leave about one percent of the users at risk. Expanding the size of the network to reduce risk is more effective when the scale is

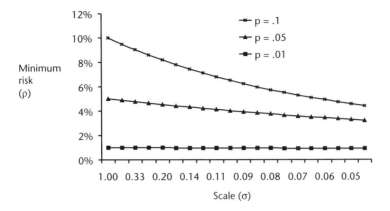

Figure 2.14 Network risk example.

limited or, in other words, users are more concentrated. Concentration will dominate risk up to a certain point, beyond which size can greater influence network risk.

This analysis assumes uniform, random outage potential at each node. Of course, this assumption may not hold true in networks with a nonuniform user distribution and nodal outage potential. An intentional versus random outage at a concentrated node, such as one resulting from a directed security attack, can inflict great damage. Overall, scale-free networks with less concentrated nodal functionality are less vulnerable to directed, nonrandom outages than networks of limited scale.

2.4.4 Complexity

Complexity in a network is characterized by variety. Use of many diverse network technologies, circuits, OSs, vendors, interfaces, devices, management systems, service providers, and suppliers is quite common and often done for valid reasons. Many networks are grown incrementally over time and result in *à la carte* designs. Such designs usually do not perform as well as a single integrated design. Even if a complex design has strong tolerance against errors, extra complexity can impede recovery activities and require more costly contingency.

The problems with complexity are attributed to not only system variety. They are also attributed to whether components are chosen correctly for their intended purpose and how well they are *matched*. A matched system is one in which components fit well with each other and do not encroach upon the limits of one another. Turnkey systems delivered by a single vendor or integrator are usually delivered with intentions of behaving as matched systems. On the other hand, unmatched systems are usually characterized by poorly integrated piecemeal designs involving different vendor systems that are plugged together. The cost of maintaining the same levels of service increases exponentially as more complexity is used to deliver the service.

Consolidation of multiple components, interfaces, and functions into a single system or node can reduce the number of resources needed and eliminate potential points of failure and the interoperability issues associated with matching. But from previous discussion, we know that consolidation poses greater risk because the consolidated resource can result in a single point of failure with high damage potential, unless a redundant architecture is used.

2.5   Summary and Conclusions

This chapter reviewed some fundamental concepts of continuity that form the basis for much of the remaining discussion in this book. Many of these concepts can be applied to nearly all levels of networking. A basic comprehension of these principles provides the foundation for devising and applying continuity strategies that use many of the remedial techniques and technologies discussed further in this book.

Adverse events are defined as those that violate a well-defined envelope of operational performance criteria. Service disruptions in a network arise out of lack of preparedness rather than the adverse events themselves. Preparedness requires having the appropriate capabilities in place to address such events. It includes having an early detection mechanism to recognize them even before they happen; the ability to contain the effects of disruption to other systems; a failover process that can transfer service processing to other unaffected working components; a recovery mechanism to restore any failed components; and the means to resume to normal operation following recovery.

Redundancy is a tactic that utilizes multiple resources so that if one resource is unable to provide service, another can. In networks, redundancy can add to capital and operating costs and should therefore be carefully designed into an operation. At a minimum, it should eliminate single points of failure—particularly those that support mission-critical services. There are various ways to implement redundancy. However, if not done correctly, it can be ineffective and provide a false sense of security.

Tolerance describes the ability to withstand disruptions and is usually expressed in terms of availability. A greater level of tolerance in a network or system implies lower transaction loss and higher cost. FT, FR, and HA are tolerance categories that are widely used to classify systems and services, with FT representing the highest level of tolerance. FR and HA solutions can be cost-effective alternatives to minimizing service disruption, but they may not guarantee transaction preservation during failover to the same degree as FT.

There are several key principles of network and system design to ensure continuity. Capacity should be put in the right places—indiscriminate placement of capacity can produce bottlenecks, which can lead to other service disruptions. Networks should be designed in compartments that each represent their own failure groups, so that a disruption in one compartment does not affect another. The network architecture should be balanced so that loss of a highly interconnected node does not disrupt the entire network.

Finally, the adage "the simpler, the better" prevails. Complexity should be discouraged at all levels. The variety of technologies, devices, systems, vendors, and services should be minimized. They should also be well matched. This means that each should be optimally qualified for its intended job and work well with other components.

References

[1]   Henderson, T., "Guidelines for a Fault-Tolerant Network," *Network Magazine*, November 1998, pp. 38–43.

[2]   Slepicka, M., "Masters of Disasters—Beating the Backhoe," *Network Reliability—Supplement to America's Network*, June 2000.

[3]   Barrett, R., "Fiber Cuts Still Plague ISPs," *Interactive Week*, May 31, 1999, p. 36.

[4]   Campbell, R., "Cable TV: Then and Now," *TV Technology*, September 20, 2000, p. 34.

[5]   Brumfield, R., "What It Takes to Join the Carrier Class," *Internet Telephony*, May 1999, pp. 80–83.

[6]   Glorioso, R., "Recovery or Tolerance?" *Enterprise Systems Journal*, July 1999, pp. 35–37.

[7]   Klein, D. S., "Addressing Disaster Tolerance in an E-World," *Disaster Recovery Journal*, Spring 2001, pp. 36–40.

[8]   Whipple, D., "For Net & Web, Security Worries Mount," *Interactive Week*, October 9, 2000, pp. 1–8.

[9]   Oleson, T. D., "Consolidation: How It Will Change Data Centers," *Computerworld—Special Advertising Supplement*, 1999, pp. 4–19.

[10] Nolle, T., "Balancing Risk," *Network Magazine*, December 2001, p. 96.

[11] Sanborn, S., "Spreading Out the Safety Net," *Infoworld*, April 1, 2002, pp. 38–41.

[12] Porter, D., "Nothing Is Unsinkable," *Enterprise Systems Journal*, June 1998, pp. 20–26.

[13] Rybczynski, T., "Net-Value—The New Economics of Networking," *Computer Telephony Integration*, April 1999, pp. 52–56.

[14] Bryant, E. C., *Statistical Analysis*, New York: McGraw-Hill Book Company, Inc., 1960, pp. 20–24.

CHAPTER 3

Continuity Metrics

Metrics are quantitative measures of system or network behavior. We use metrics to characterize system behavior so that decisions can be made regarding how to manage and operate them efficiently. Good metrics are those that are easily understood in terms of what they measure and how they convey system or network behavior. There is no single metric that can convey the adequacy of a mission-critical network's operation. Using measures that describe the behavior of a single platform or portion of a network is insufficient. One must measure many aspects of a network to arrive at a clear picture of what is happening.

There is often no true mathematical way of combining metrics for a network. Unlike the stock market, use of a computed index to convey overall network status is often flawed. For one thing, many indices are the result of combining measures obtained from ordinal and cardinal scales, which is mathematically incorrect. Some measures are obtained through combination using empirically derived models. This is can also be flawed because a metric is only valid within the ranges of data from which it was computed. The best way of combining measures is through human judgment. A network operator or manager must be trained to observe different metrics and use them to make decisions. Like a pilot, operators must interpret information from various gauges to decide the next maneuver.

Good useful metrics provide a balance between data granularity and the effort required for computation. Many statistical approaches, such as experimental design, are aimed at providing the maximum amount of information with the least amount of sampling. The cost and the ability to obtain input data have improved over the years. Progress in computing and software has made it possible to conduct calculations using vast amounts of data in minimal time, impossible 20 or 30 years ago. The amount of time, number of samples, complexity, and cost all should be considered when designing metrics.

Metrics should be tailored to the item being measured. No single metric is applicable to everything in a network. Furthermore, a metric should be tied to a service objective. It should be used to express the extent to which an objective is being achieved. A metric should be tied to each objective in order to convey the degree to which it is satisfied.

Finally, computing a metric should be consistent when repeated over time; otherwise, comparing relative changes in the values would be meaningless. Repeated calculations should be based on the same type of data, the same data range, and the same sampling approach. More often than not, systems or network services are compared based on measures provided by a vendor or service provider. Comparing different vendors or providers using the measures they each supply is often difficult and sometimes fruitless, as each develops their metrics based on their own methodologies [1].

3.1 Recovery Metrics

Recovery is all of the activities that must occur from the time of an outage to the time service is restored. These will vary among organizations and, depending on the context of use, within a mission-critical network environment. Activities involved to recover a component are somewhat different than those to recover an entire data center. But in either case, the general meaning remains the same. General recovery activities include declaration that an adverse event has occurred (or is about to occur); initialization of a failover process; system restoration or repair activities; and system restart, cutover, and resumption of service. Two key recovery metrics are described in the following sections.

3.1.1 Recovery Time Objective

The recovery time objective (RTO) is a target measure of the elapsed time interval between the occurrence of an adverse event and the restoration of service. RTO should be measured from the point when the disruption occurred until operation is resumed. In mission-critical environments, this means that operation is essentially in the same functional state as it was prior to the event. Some IT organizations may alter this definition, by relaxing some of the operational state requirements after resumption and accepting partial operation as a resumed state. Likewise, some will define RTO based on the time of recognizing and declaring that an adverse event has occurred. This can be misleading because it does not take into account monitoring and detection time.

RTO is an objective, specified in hours and minutes—a target value that is determined by an organization's management that represents an acceptable recovery time. What value an organization assigns to "acceptable" is influenced by a variety of factors, including the importance of the service and consequential revenue loss, the nature of the service, and the organization's internal capabilities. In systems, it may even be specified in milliseconds. Some will also specify RTO in transactions or a comparable measure that conveys unit throughput of an entity. This approach is only valid if that entity's throughput is constant over time.

RTOs can be applied to any network component—from an individual system to an entire data center. Organizations will define different RTOs for different aspects of their business. To define an RTO, an organization's managers must determine how much service interruption their business can tolerate. They must determine how long a functional entity, such as a business process, can be unavailable. One may often see RTOs in the range of 24 to 48 hours for large systems, but these numbers do not reflect any industry standard. Virtual storefronts are unlikely to tolerate high RTOs without significant loss of revenue. Some vertical markets, such as banking, must adhere to financial industry requirements for disruption of transactions [2].

Cost ultimately drives the determination of an RTO. A high cost is required to achieve a low RTO for a particular process or operation. To achieve RTOs close to zero requires expensive automated recovery and redundancy [3]. As the target RTO increases, the cost to achieve the RTO decreases. An RTO of long duration invites less expensive redundancy and more manual recovery operation. However, concurrent with this is business loss. As shown in Figure 3.1, loss is directly related to RTO—the longer the RTO, the greater the loss. During recovery, business loss can be realized in many ways, including loss productivity or transactions. This topic is discussed further in this chapter. At some point, there is an RTO whose costs can completely offset the losses during the recovery [4, 5].

It becomes evident that defining an RTO as a sole measure is meaningless without some idea of what level of service the recovery provides. Furthermore, different systems will have their own RTO curves. Critical systems will often have a much smaller RTO than less critical ones. They can also have comparable RTOs but with more stringent tolerance for loss. A tiered-assignment approach can be used. This involves defining levels of system criticality and then assigning an RTO value to each. So, for example, a three-level RTO target might look like this:

- Level 1—restore to same service level;
- Level 2—restore to 75% service level;
- Level 3—restore to 50% service level.

A time interval can be associated to each level as well as a descriptor of the level of service provided. For example, a system assigned a level 2 RTO of 1 hour must complete recovery within that time frame and disrupt no more than 25% of service. A system can be assigned a level 1 RTO of 1 hour as well, but must restore to the same level of service. Level 1 may require failover procedures or recovery to a secondary system.

Assuming that the service level is linearly proportional to time, RTOs across different levels can be equated on the same time scale. A time-equivalent RTO, RTO_E, can thus be computed as:

$$RTO_E = RTO / (1 - \alpha) \tag{3.1}$$

where α is the maximum percentage of service that cannot be disrupted. In our example, an RTO (level 2) requires an α of 25%, which equates to an RTO_E of 80 minutes.

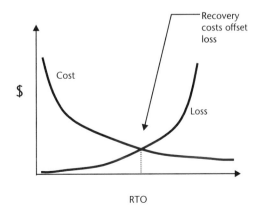

Figure 3.1   RTO versus loss and cost.

3.1.1.1   Recovery Time Components

The RTO interval must incorporate all of the activities to restore a network or component back to service. A flaw in any of the component activities could lead to significant violation of the RTO. To this end, each component activity can be assigned an RTO as well. The addition of each of these component RTOs may not necessarily equal the overall RTO because activities can be conducted in parallel. Some of these component RTOs can include time to detection and declaration of an adverse event, time to failover (sometimes referred to as a failover time objective), time to diagnose, and time to repair.

The last two items are typically a function of the network or system complexity and typically pose the greatest risk. In complex networks, one can expect that the likelihood of achieving an RTO for the time to diagnose and repair is small. Failover to a redundant system is usually the most appropriate countermeasure, as it can buy time for diagnostics and repair. A system or network operating in a failed state is somewhat like a twin-engine airplane flying on one engine. Its level of reliability is greatly reduced until diagnostics and repairs are made.

Figure 3.2 illustrates the continuum of areas activity relative to a mission-critical network. Of course, these may vary but are applicable to most situations. The areas include the following:

- *Network recovery.* This is the time to restore voice or data communication following an adverse event. Network recovery will likely influence many other activities as well. For instance, recovery of backup data over a network could be affected until the network is restored.

- *Data recovery.* This is time to retrieve backup data out of storage and deliver to a recovery site, either physically or electronically. It also includes the time to load media (e.g., tape or disk) and install or reboot database applications. This is also referred to as the time to data (TTD) and is discussed further in the chapter on storage.

- *Application recovery.* This is the time to correct a malfunctioning application.

- *Platform recovery.* This is the time to restore a problematic platform to service operation.

- *Service recovery.* This represents recovery in the broadest sense. It represents the cumulative time to restore service from an end user's perspective. It is, in essence, the result of an amalgamation of all of the preceding recovery times.

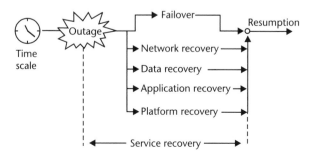

Figure 3.2   Continuum of recovery activities.

All of these areas are discussed at greater length in the subsequent chapters of this book.

3.1.2   Recovery Point Objective

The recovery point objective (RPO) is used as target metric for data recovery. It is also measured in terms of time, but it refers to the age or freshness of data required to restore operation following an adverse event. Data, in this context, might also include information regarding transactions not recorded or captured. Like RTO, the smaller the RPO, the higher the expected data recovery cost. Reloading a daily backup tape can satisfy a tolerance for no more than 24 hours' worth of data. However, a tolerance for only one minute's worth of data or transaction loss might require more costly data transfer methods, such as mirroring, which is discussed in the chapter on storage.

Some view the RPO as the elapsed time of data recovery in relation to the adverse event. This is actually the aforementioned TTD. RPO is the point in time to which the data must be recovered—sometimes referred to as the *freshness window*. It is the maximum tolerable elapsed time between the last safe backup and the point of recovery. An organization that can tolerate no data loss (i.e., RPO = 0) implies that data would have to be restored instantaneously following an adverse event and would have to employ a continuous backup system.

Figure 3.3 illustrates the relationship between TTD and RPO using a timeline. If we denote the time between the last data snapshot and an adverse event as a random variable $\mathring{a}$, then it follows that the TTD + ε must meet the RPO objective:

$$TTD + \varepsilon \leq RTO \tag{3.2}$$

A target RPO should be chosen that does not exceed the snapshot interval (SI), and at best equals the SI. If data is not restored prior to the next scheduled snapshot, then the snapshot should be postponed or risk further data corruption:

$$TTD + \varepsilon \leq RTO \leq SI \tag{3.3}$$

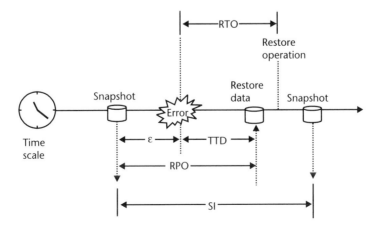

Figure 3.3   Relationship of RPO, RTO, and TTD.

The variable ε represents a margin of error in specifying an RPO, assuming that TTD is usually a fixed quantity—which may not necessarily be true in all cases.

There are some caveats to using an RPO. RPO assumes that the integrity of the data is preserved at recovery. The value does not necessarily convey the quality and quantity of data that is to be restored [6]. It is conceivable that an operation can provide service with partial transaction data, until transactions are reconstructed at a later time. If data is restored to a level such that full operation can take place, then in the strictest sense the RPO has been satisfied.

RPO also assumes a uniform transaction-recording rate over time, which may not necessarily be true. In other words, an RPO of one hour implicitly assumes that no more than one hour's worth of transaction loss can be tolerated. In fact, if an adverse event took place during off hours, the likelihood is that hardly any transactions would be lost within an hour's time. For this reason, different levels of RPO may need to be specified depending on the time of day.

3.1.3 RTO Versus RPO

RTO and RPO are not necessarily tied to each other, but they can be interrelated [7]. Figure 3.3 also illustrated the relationship between RTO and RPO. Specifying an RTO that is short in duration does not necessarily imply a short RPO. For example, although a system can be restored with working data within an RTO of an hour following an adverse event, it is acceptable for that data to be four hours old—the RPO. RTO specifies the maximum time duration to recover a network, system, or component. RPO defines how much working data, in terms of time, can be lost in the process. A system or network working under an RTO and RPO both equivalent to zero requires instantaneous recovery and essentially no data loss. Specification of both the RTO and RPO is driven mainly by economics.

3.2 Reliability Metrics

Reliability is defined as the probability (or likelihood) that a network (or component) will perform satisfactorily during a specified period of time. It is measured by how long it takes for a network or system to fail (i.e., how long it continues to function until it ceases due to failure). Reliability and *availability* are often used interchangeably, but there is a subtle difference between them. Availability (discussed in Section 3.3) is the probability that a network is in service and available to users at any given instant in time.

The difference between reliability and availability is best explained through an analogy. A car, for example, may break down and require maintenance 5% of the time. It is therefore 95% reliable. However, suppose the same car is equally shared between two family members. To each, the car is available only 47.5% ($50\% \times 95\%$) of the time, even though the car is very reliable. Even if the car was 100% reliable, the availability to each is still only 50%. To improve availability, the family members can purchase an additional car, so that each now has 100% autoavailability.

3.2.1   Mean Time to Failure

Mean time to failure (MTTF) is a metric that is often used to characterize the operating life of a system. It is the amount of time from the placement of a system or component in service until it permanently fails. Passive components are not often considered in MTTF estimations. They can have lifetimes on the order of 20 years or so. Network cables are known to have lifetimes from three to 100 years or so, depending on where and how they are used. Active components, on the other hand, may likely have shorter MTTFs.

Ideally, accurately calculating MTTF requires the system to be monitored for its expected useful lifetime, which can be quite long. Whether estimation of the MTTF involves monitoring a system for several years or a hundred years, accurate MTTF computation is often impractical to obtain. Some standard prediction methods to estimate MTTF are found in Military Standards (MIL-217) and Naval Surface Warfare Center (NSWC) specifications, and telephony standards such as Telcordia Specifications (TR-332 Version 6) and French Telecommunications (RDF 2000). If a system is in operation until it is no longer of use, then one can say that the *mission time* of the device is assumed to be the same as the MTTF. In many military network installations, a system's mission time may complete much sooner than the MTTF.

MTTF is viewed as an elapsed time. If a network element or system is not used all of the time, but at a periodic rate (e.g., every day during business hours), then the percentage of time it is in operational use is referred to as the *duty cycle*. The duty cycle is defined as:

$$\delta = \text{OT} / \text{MTTF} \tag{3.4}$$

where δ is the duty cycle and OT is the total operating time of the element. For a network circuit, for example, it is the fraction of time the circuit is transmitting. For a system or component such as a disk drive, it is the percentage of time the drive spends actively reading and writing. If, for example, the drive has a MTTF of 250,000 hours and is in use 5% of the time ($\delta = .05$), the same drive would have a MTTF of 125,000 hours if it were used twice as much ($\delta = .10$). In other words, the more a system or device is in use, the shorter the life expectancy.

3.2.2   Failure Rate

Systems will fail or be placed out of service for many reasons. A system upgrade, maintenance, or system fault may require placing a system out of service. A failure rate, *F,* can be defined to express failure frequency in terms of failures per unit time, say percentage of failures per 1,000 hours. System vendors often use statistical sampling methods to estimate average failure rates over large populations of components. These populations can be on the order of tens of thousands of components. Once components are embedded in a complex platform or network, their significance in the overall reliability becomes ambiguous. The more complex a system or network grows, the greater likelihood of failure, even though the individual subsystems are highly reliable.

A failure may not necessarily mean that a system has stopped operating. A failure can also be associated with those circumstances in which a system is producing service at an unsatisfactory performance level and hence is of little or no use. The failure rate, F, of a system can be estimated as follows:

$$F = f / \text{System's useful life} \qquad (3.5)$$

where f is the number of failures experienced during a system's useful life or mission time (i.e., the total time a system is performing service operations.) Many failure estimation techniques assume an exponential distribution where the failure rate is constant with time.

3.2.3 Mean Time to Recovery

Mean time to recovery (MTTR) is sometimes referred to as mean time to repair or restore. In either case, it means the same thing. It is the time required to restore operation in a component that has stopped operating or that is not operating to a satisfactory performance level. It includes the total time it takes to restore the component to full operation. It could include things like diagnosing, repairing, replacement, reboot, and restart. MTTR is expressed in units of time. The time to diagnose can typically present the most uncertainty in estimating MTTR and can thus have a profound effect on MTTR and, ultimately, system availability.

MTTR can be estimated from observed data in several ways. The most common method is to simply obtain the sum total of all observed restoration times and divide by the number of reported outages or trouble tickets. MTTR can be used to estimate the restoration rate, μ (sometimes referred to as the recovery rate) of a system as follows:

$$\mu = 1 / \text{MTTR} \qquad (3.6)$$

where μ is used to convey the recoverability of a system. Systems that minimize MTTR or that have a high recoverability μ should be favored. MTTR is also a primary measure of availability. The availability of systems with numerous components will be bound by those having the longest MTTR.

3.2.4 Mean Time Between Failure

The mean time between failure (MTBF) is a metric that conveys the mean or average life of a system based on the frequency of system outages or failures. For this reason, it is different than MTTF, although the two are quite often used interchangeably. Also, MTBF is sometimes referred to as the mean time between system outages (MTBSO) [8], depending on the context of use. For our purposes, we will use MTBF because it is the more recognizable metric.

MTBF is a measure that system vendors often use to compare their product to another [9]. System vendors will quote an MTBF without any basis or justification. Many system vendors may quote an MTTF for a product, which may actually be the computed MTBF. Because of the complexity of today's systems, computation of a true MTBF for a platform can be daunting. Another issue is that mission-critical network systems do not, by definition, function as isolated items. An MTBF usually conveys stand-alone operation. If the MTBF is reached, considerable operational risk is incurred.

A system with a low MTBF will require more servicing and consequently additional staffing, monitoring, and spare components. This typically implies higher maintenance costs but lower capital costs. A high MTBF, on the other hand, indicates that a system will run longer between failures and is of higher quality. This may imply a higher capital cost but lower maintenance cost. Some systems will try to integrate high-quality components having the highest MTBFs, but their level of integration is such that the MTBF of the overall system is still low.

MTBF is measured based on the number of failures during the service life of a system, or simply the inverse of the failure rate, F:

$$\text{MTBF} = 1 / F \qquad (3.7)$$

For example, if a system has a MTTF of 100 years and experiences three failures in that time ($f = 3$), then the MTBF is approximately 33.3 years. Many use MTBF to convey the reliability of a system in terms of time. The higher the MTBF, the more reliable a system is.

The MTBF for a system can be estimated in various ways. If MTBF is estimated as an arithmetic mean of observed MTBF values across N systems, one could assume that MTBF represents the point in time that approximately half of the systems have had a failure, assuming that F is uniform over time. In general, the percentage, p, of devices that could fail in a given year is then estimated as:

$$p = .5 / \text{MTBF} \qquad (3.8)$$

So, for example, in a large network with an estimated MTBF of 20 years, one would expect on average about 2.5% of the devices to fail in a given year. It is important to recognize that some components might fail before reaching the MTBF, while others might outperform it without problem. It is best to use MTBF with the most critical components of a system or network, particularly those that are potentially single points of failure.

To plan recovery and network-management operations and resources, it is often valuable to have a feel for how many simultaneous outages or failures can occur in a network, consisting of N nodes. If there were only one node ($N = 1$) in a network, then p is the probability of that node failing in a day. However, as the number of nodes N in a network increases, so will the likelihood of having more failures in a given day. In general, the probability of f discrete events occurring out of N possible outcomes, each with a probability of occurrence p, is statistically characterized by the well-known Binomial distribution [10]:

$$P(f) = N! \, p^f \left(1 - p\right)^{N-f} / f!(N - f)! \qquad (3.9)$$

where $P(f)$ is the probability of f events occurring. If we assume that N is the maximum number of possible node failures in a given day and p is the probability of an individual node failing, then $P(f)$ is the probability of f failures in a given time frame. If we substitute the expression for p obtained in (3.8) into (3.9), then the probability of having f failures in a network (or what percentage of time f failures are likely to occur) is [11]:

$$P(f) = N! \left(2 \, \text{MTBF} - 1\right)^{N-f} / f!(N - f)!(2 \, \text{MTBF}) \qquad (3.10)$$

This expression assumes that all nodes have an equal probability of failing, which obviously may not always be true. However, it can be used as an approximation for large networks. It also assumes that all failures are independent of each other, which is also another simplifying assumption that may not necessarily be true. In fact, many times a failure will lead to other failures, creating a rolling failure. In the expression, it is assumed the total number of failure outcomes N is the same as if all nodes were to fail simultaneously.

This expression can be used to gain insight into large networks. If $P(0)$ indicates the percentage of the time no failures will occur, then $1 - P(0)$ is the percentage of time that one or more failures will occur. Figure 3.4 shows how this probability varies with the number of network nodes for different values of nodal MTBF. An important concept is evident from the figure. The marginal gain of improving the nodal MTBF is more significant with the size of the network; however, the gains diminish as the improvements get better.

Variants of MTBF will be used in different contexts. For example, mean time to data loss (MTDL) or mean time to data availability (MTDA) have often been used, but convey similar meaning. Highly redundant systems sometimes use the mean time between service interruptions (MTBI).

3.2.5 Reliability

Reliability is the probability that a system will work for some time period t without failure [12]. This is given by:

$$R(t) = \exp(-t / \text{MTBF}) \tag{3.11}$$

where $R(t)$ is the reliability of a system. This function assumes that the probability that a system will fail by a time t follows an exponential distribution [13]. Although this assumption is commonly used in many system applications, there are a number of other well-known probability distributions that have been used to characterize system failures.

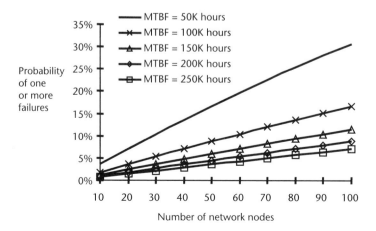

Figure 3.4 Probability of failure versus network size.

Reliability deals with frequency of failure, while availability is concerned with duration of service. A highly reliable system that may infrequently fail can still have reduced availability if it is out of service for long periods of time. Keeping the duration of outages as short as possible will improve the availability. Mathematically, availability is the probability of having access to a service and that the service operates reliably. For a mission-critical network, availability thus means making sure users have access to service and that service is reliable when accessed. If a component or portion of a network is unreliable, introducing redundancy can thus improve service availability [14]. (Availability is discussed in Section 3.3.)

Figure 3.5 illustrates the reliability function over time for two MTBF values. Reliability at any point in time t is essentially the probability percentage that a system with a particular MTBF will operate without a failure or the percentage of the systems that will still be operational at that point in time.

When different network components or platforms are connected in series, the overall system reliability is reduced because it is the product of component system reliabilities. Failure of any one component could bring down the entire system. In large networks or platforms with many systems and components, a high level of reliability may be difficult to achieve. Improving the reliability of a single component will marginally improve the overall system reliability. However adding a redundant component will improve the overall reliability.

A reliability block diagrams (RBD) is a tool used for a first-pass computation of reliability. Figure 3.6 illustrates two RBD examples. If a system is operational only if all components are operating, the relationship is conveyed as a *serial* relationship. If the system is operational if either component is operating, then a *parallel* relationship is made. Both arrangements can be generalized to greater numbers of N components, or to systems with components having a mix of parallel or serial relationships. The following formulas are used to convey those relationships:

$$\text{Serial Relationship: } R(t) = \prod_{i=1}^{N} R_i(t) \tag{3.12}$$

$$\text{Parallel Relationship: } R(t) = 1 - \prod_{i=1}^{N} \left[1 - R_i(t) \right] \tag{3.13}$$

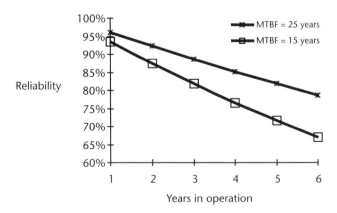

Figure 3.5   Reliability function.

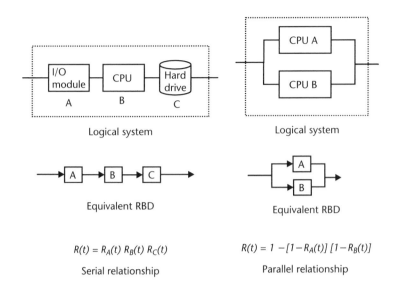

Figure 3.6 Reliability block diagrams.

RBDs can be used to construct a generalized abstract model of a system as an aid to understanding the reliability of a system. But they become impractical to model large complex systems with numerous detailed interactions.

3.3 Availability Metrics

Availability is the proportion of time that a system or network will provide service. It is the percentage of required mission time that a system actually provides service. Reliability is the likelihood that a system will continue to provide service without failure. Availability is the likelihood that a system will provide service over the course of its lifetime. The availability, A, of a system or component can be calculated by the following [15]:

$$A = \text{MTBF} / (\text{MTBF} + \text{MTTR}) \tag{3.14}$$

The unavailability of a system or component is simply $1 - A$. This ends up being numerically equivalent to amortizing the MTTR over the MTBF [16]. For example, if a critical system in a network has an MTBF of 10,000 hours and an MTTR of 2 hours, it is available 99.98% of the time and unavailable .02% of the time. This may seem like a highly available system, but one must consider the absolute service and outage times that are implied. Assuming this system must provide service all the time (i.e., $7 \times 24 \times 365$), implying a total of 8,760[1] hours per year of service, then it is unavailable (or down) 1.75 hours per year. This could be significant for a mission-critical network. The relationship between availability and MTBF is shown in Figure 3.7.

---

1.   Use 8,766 hours per year to account for leap years.

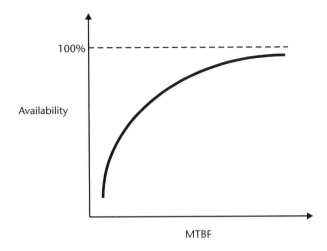

Figure 3.7   Availability versus MTBF.

Redundancy can have a significant effect on availability. In the example mentioned earlier, if a redundant system were to be placed in parallel operation in the same network, then the percentage of time the network is available is now equivalent to the percentage of time either or both systems are operating. Figure 3.8 shows an example of the use of availability block diagrams (ABDs). ABDs can be used in the same fashion as RBDs. In fact, the mathematical relationship between systems working in parallel or series is the same as the RBD.

The relationship of availability among systems can be generalized in the same way as an RBD. For a network having N components or systems, the following formulas can be used:

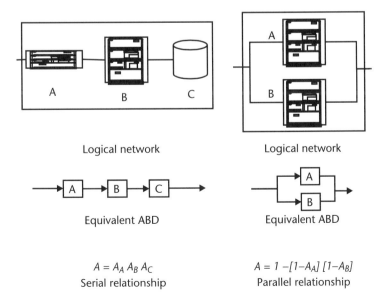

Figure 3.8   Availability block diagram.

$$\text{Serial Relationship: } A = \prod_{i=1}^{N} A_i \tag{3.15}$$

$$\text{Parallel Relationship: } A = 1 - \prod_{i=1}^{N}\left(1 - A_i\right) \tag{3.16}$$

Figure 3.9 shows a hypothetical example that illustrates the impact of redundancy on availability. The example shows a Web site network configured with and without parallel redundancy. By adding a dual server cluster, downtime is improved to less than a day. Adding parallel redundancy to systems with low availability rates has greater impact than adding redundancy to systems that already have high availability. Additionally, improving the availability of individual parallel systems will only marginally improve overall network availability.

Systems in mission-critical networks, be they computing or networking platforms, require rapid MTTRs. Availability is affected by system or network recovery procedures. Figure 3.10 shows the relationship between MTTR and availability. Decreasing the MTTR can have a profound effect on improving availability. Thus, any tactic that can be used to reduce the MTTR can help improve overall availability. Systems needing a high MTTR may thus require a back up system.

Over the years, the IT industry has used several known levels of availability [17]. These are listed in Table 3.1. If a system is carrier class, for example, it is considered 99.999% available (colloquially known as five nines). This level is the standard for public switched telephone network (PSTN) systems [18]. This means that there could be one failure during a year that lasts just over five minutes or there can be five failures that each last one minute.

Organizations will typically define several levels of availability, according to their severity and impact on operations. For example, the following levels could be defined, each with an associated availability or downtime [19]:

- Level 1: Users and service are interrupted, data is corrupted;
- Level 2: Users and service are interrupted, data remains intact;

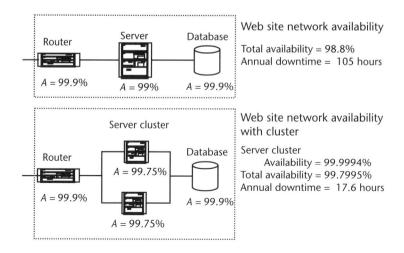

Figure 3.9   Availability example.

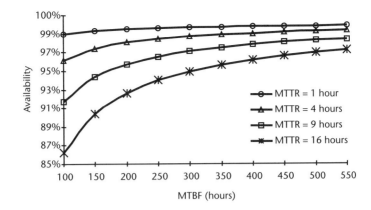

Figure 3.10 Effect of MTTR on availability.

Table 3.1 Classes of Availability

Availability (%)	Annual Downtime	Description
98	175.2 hours	Failures too frequent
99	87.6 hours	Failures rare
99.5	43.8 hours	Considered high availability
99.9	8.8 hours	Three nines (often used for storage systems)
99.99	52.6 minutes	Considered fault resilient
99.999	5.3 minutes	Fault tolerant (also called carrier class for PSTN infrastructure)
99.99966	1.8 minutes	Six sigma (often used in manufacturing) [20]
99.9999	31.5 seconds	Six nines
100	0	Continuous availability

- Level 3: Users interrupted, service remains intact;
- Level 4: No interruptions, but performance degradation;
- Level 5: No interruptions, failover is implemented.

One of the fundamental laws of availability is the law of diminishing returns. The higher the level of availability, the greater the incremental cost of achieving a small improvement. A general rule is that each additional nine after the second nine in an availability value will cost twice as much. As one approaches 100% availability, return on investment diminishes. The relationship between cost and availability is illustrated in Figure 3.11. In the end, to achieve absolute (100%) availability is cost prohibitive—there is only so much availability that can be built into an infrastructure. This view does not consider the potential savings resulting from improved availability.

In a network where systems operate in series or in parallel, the availability of the network is dependent on the ability to continue service even if a system or network

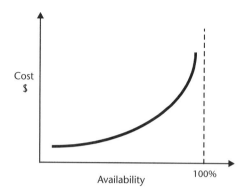

Figure 3.11   Cost versus availability.

link becomes inoperable. This requires a mix of redundancy, reliable systems, and good network management. We have seen that the overall impact of adding redundant systems in parallel is significant, while adding components in series actually reduces overall network availability. Because all systems connected in series must operate for overall service, there is greater dependency on the availability of each component.

The availability calculations presented here can be used to estimate the contribution of each system to the overall network availability. Although the resulting availability generally yields a high-side estimate, it can flag those portions of a network that can be problematic. More complex statistical models are required to obtain more precise estimates. Markov chain models, which enumerate all possible operating states of each component, can be used. Transitions between states (e.g., between an operational or failure state) are assumed to be probabilistic instead of deterministic, assuming some probability distribution. Availability is then measured by the fraction of time a system is operational.

The problem with this approach, however, is that it is limited to small numbers of components and states. As both of these grow in number, the problem cannot be solved easily. In this case, other approaches, such as network simulation, can be used. This requires building a computerized model of a network, which often reflects the network's logical topology. Calibration of the model is required in order to baseline the model's accuracy.

Availability for a network can be estimated from field data in many ways. A general formula for estimating observed availability, A_o, from a large network is the following:

$$A_o = (N \times MT - OT) / (N \times MT) \tag{3.17}$$

where N is the total number of nodes, MT is the overall mission or service time of the network, and OT is the total observed outage times across all nodes.

When using availability as a metric to characterize service behavior, the following are some assumptions that must be kept in mind:

- Availability is a relative measure—it is how one entity perceives the operation of another. End users ultimately determine availability. It should be computed from the end users' perspective, conveying their perception of a network's service.

- Availability does not necessarily capture network management and operational procedures—things that can profoundly affect network availability.
- Availability represents an ideal value. No single network element can ensure high availability. It is achieved through a blending of systems, connectivity, and operations. Regardless of the value obtained, networks are still prone to single points of failure.
- Availability can mask outage frequency. A system with five nines availability is subject to about 5 minutes of downtime per year. This can equate to one outage of 5 minutes, five 1-minute outages, or even 10 half-minute outages.
- In conjunction with the previous point, availability also masks outage impact. Even though 5 minutes of downtime per year might seem impressive, there is no indication of the potential severity and damage caused by the outage. An outage that takes place during off hours may have minimal impact when compared to one that occurs during peak hours.
- Improvements made to drive up availability and reliability can be quite costly. As was discussed earlier, marginal improvements in availability gradually decrease with the invested cost.

3.4 Exposure Metrics

Exposure is a metric that is used to convey a network's vulnerability to a particular threat [21]. As there is no formal definition, its meaning is typically left up to the context at hand. It conveys a network's ability to reduce the likelihood of a particular threat affecting service operation. For example, adding security capabilities to a network can reduce the exposure to security threats, such as intrusions or viruses. Another example could be adding redundant network access links to a facility in order to reduce exposure to cable cuts. In each case, incorporating capabilities that minimize the possibility of known threats reduces exposure.

The number of single points of failure in a network has been used to convey network exposure to outages. Some IT organizations contrive their own vulnerability indices that suit their individual needs. Exposure is best measured using a relative comparison of two metrics. For example, one can compare an observed MTTR with a target RTO to determine exposure to unexpected outages. Similarly, comparing A_o with an estimated A (i.e., A_o / A) can gauge a network's current exposure to service disruption (sometimes referred to as percent *lost availability*).

3.5 Risk/Loss Metrics

Risk is the potential monetary loss resulting from a particular threat. The classic formula for estimating risk, ρ, is as follows [22]:

$$\rho = T \times V \times C \tag{3.18}$$

where T is the threat, or frequency, of a problem expressed as a percentage, V is vulnerability or probability that a threat will impact a network's operation, and C is the expected cost or damage resulting from the threat. The result, ρ, is a monetary value.

The value T is the probability that a threat will occur. A threat can represent a broad range of adverse events that can produce damage. C represents lost assets, productivity, or missed earnings or sales that result as a direct consequence of the threat. V is a measure of a network's ability to withstand the threat (i.e., the probability of the threat inflicting damage on a network). It is, in essence, the ineffectiveness of a network to withstand the threat. As seen in (3.18), both T and V modify the outage cost.

This formula has been used quite often to compare individual risks and gain better understanding of known threats by comparing dollars. ρ is often rightly referred to as an expected loss, as it is a quantification of loss based on the likelihood of a threat. The higher ρ is, the more important it is to mitigate a threat. If a threat has a small chance of occurring, the result is little damage. This approach can be used to filter out those threats that pose little risk versus the "killer risks." It is also used to weigh the cost of proposed preventive measures by comparing them based on their expected reduction in risk.

These three elements must be present in order to view a threat as a considerable risk. An event with little or no chance of occurrance should not be viewed as a risk. If there is a strong possibility that an event will occur, but a network is invulnerable to it, the event should not pose much risk. Likewise, if a threat poses no negative financial impact, it too should not pose much risk [23].

For example, suppose that 75% of network outages are due to intrusions. Suppose also that half of the time, intrusions are successfully blocked. The threat T of an intrusion is therefore 75% and the network's vulnerability to intrusion outage, V, is 50%. If downtime costs are estimated at $750,000 per intrusion, then the risk posed intrusions is estimated to be:

$$\rho = .75 \times .50 \times \$750,000 = \$281,250 \qquad (3.19)$$

This is essentially the expected monetary loss of intrusion-based outages. It represents possible monetary impacts, weighted by the probability that intrusion-based outages will inflict damage on a network. Over the course of a large number of intrusions, we would expect average losses to equal to ρ per intrusion. Estimating the cost C is the topic of the next section.

3.6 Cost Metrics

The difference between risk/loss, as just discussed, and cost is expectation. In other words, risk encompasses uncertainty, whereas cost is deterministic. Sudden changes in circumstances produce the uncertainty. Unexpected change in a network's operation produces the unfortunate opportunity to incur cost. Cost is the calculated payout for a single service disruption. The cost of an outage (also referred to as downtime costs) can be far reaching. It is presented and normalized in various ways. Many quantify it in terms of hourly cost of service unavailability. Others specify it as unavailability cost per user or per company site. E-businesses have stated it in terms of costs per hour per application or cost per transaction. The actual units of measure depend on the context of use and the situation at hand.

Although monetary units may seem to be the most universally understood measure, other units can be used as well. For example, units that measure production, such as transactions, calls, widgets, and megabytes are often used by operational organizations for performance or expense planning. This often removes many of the ambiguities associated with costs and conveys cost in more precise terms.

From discussion in the previous chapter, we know that the costs of degraded service or slow time should not be overlooked. The occurrence of slow time, which is the degradation of service below accepted levels, is often more frequent than service disruption. Although service is degraded, it is not halted. However, system productivity and throughput are adversely affected. Slow time is often more costly than downtime: it is often unreported, yet in transaction-based environments, for example, it can be debilitating. A conservative approach to assigning values to cost metrics should involve some specification of a performance envelope—those levels that constitute thresholds of acceptable service availability and service interruption.

Downtime costs vary among organizations. Many studies have been done to quantify hourly downtime costs across different industries. Hourly costs can range from thousands to millions of dollars, depending on the type and size of the firm. For example, financial and brokerage downtime costs are directly related to sheer disasters. E-businesses, on the other hand, focus on transactional disruption and Web site availability. In addition to lost transacting costs, they are very sensitive to corporate image, legal costs of resulting lawsuits, stock devaluation, brand erosion, and customer dissatisfaction. In the end, each industry will place emphasis on particular downtime cost components.

Some of the key findings of downtime cost studies are not the actual downtime cost quantities, but identification of what factors have a pronounced effect on downtime costs for different types of organizations. Some of these factors include [24]:

- *Dependency on technology*. Firms that are technology driven, such as telecommunications, energy, IT-dependent manufacturing, and financial industries often exhibit high downtime costs.
- *Labor intensity*. Firms that are less dependent on IT infrastructure and more labor intensive, such as health care and service industries, are less prone to downtime costs.
- *Capital intensity*. Capital-intensive firms may exhibit lower direct downtime costs, but higher indirect costs that reflect protective capacity can burden profitability.
- *Margin intensity*. Low-margin, high-volume firms are prone to serious downtime costs. These often represent commodity-oriented businesses, such as retail. In such businesses, there is little room for incentives to draw back lost customers following an outage. Low-margin and capital-intensive businesses typically exhibit high indirect downtime costs.
- *Captive market*. Firms that have an exclusive customer base through either a highly specialized service or product or a monopoly can typically withstand the possibility of business loss due to service disruptions. Although experience

has shown captive customers are not likely to defect following an outage, these firms are often subject to more regulatory and legal scrutiny.

There is really no standard regarding categorizing downtime costs. Different industries, states, countries, company policies, and accounting practices place different requirements on how costs are stated. There are typically three types of costs associated with downtime: direct, indirect, and intangible costs. Direct costs are those that immediately materialize as a result of a disruption. Indirect costs are those that represent secondary effects and can often be quite significant. Direct and indirect costs that are quantifiable are considered *tangible costs*, while *intangible costs* are those that are more qualitative in nature or not readily quantifiable. *Avoided costs* are those that are not incurred due to the outage. The true downtime cost, C, is then sum of these costs:

$$C = C(\text{direct}) + C(\text{indirect}) + C(\text{intangible}) - C(\text{avoided}) \qquad (3.20)$$

The following lists some of the component costs that comprise downtime costs [25]. The definition of these categories and the inclusion and exclusion of certain costs in the overall cost of downtime has been the topic of much debate.

1. Direct costs:
- *Recovery costs*. These are the time and out-of-pocket expenses associated with restoring normal network service. Recovery costs are usually directly related to the degree and extent of a service disruption. The longer the disruption, the less likely a recovery and the higher the recovery cost.
- *Revenue loss*. This is the value of missed sales that would have otherwise occurred without a disruption. Stating a monetary value of lost transactions, or some other unit of production, often assumes a constant hourly sales rate that is applied to the time of the outage. Because this loss is not viewed as a "booked" expense, its inclusion as a direct cost is debatable.
- *Asset damage*. This is the cost to recover assets that are destroyed, damaged, or wasted as a result of an outage. Costs associated with corrupted or lost data, software, systems, equipment, materials, and other related infrastructure are included.
- *Settlement costs*. These are refunds, credits, penalties, and fees associated with breach of service commitments to customers or other entities as a result of an outage.
- *Regulatory exposure*. These are fines, penalties, and fees as a result of violations of laws or regulations from government or regulatory agencies.

2. Indirect costs:
- *Protective capacity costs*. This is the capital and operational expenses associated with protective measures to avoid service disruption. It can include the extra staff, systems, spares, and infrastructure capacity. These costs are typically realized on an annual basis versus the exact time of an outage.
- *Productivity loss*. This represents the cost of paying for assets that would otherwise produce service if not for the outage. It is usually broken down into two categories:

- Costs representing the time for employees who cannot produce useful work during the interruption (i.e., unused people time) as well as unproductive management time. This cost should also include the extra time for employees to recover work that could have been performed during the disruption.
- Costs representing unused systems, software, and equipment that would otherwise provide revenue-producing service.

- *Legal exposure.* These are fees and costs for settling disputes with customers, suppliers, shareholders, or other entities surrounding the service disruption. This can also be viewed as a direct cost depending on the circumstance.
- *Liability exposure.* This is the increase in insurance premiums and costs resulting from the service disruption.
- *Market capitalization.* This includes stock devaluation and loss of investors as a result of the service disruption.

3. Intangible costs:
- *Market perception.* A service disruption, particularly a major one, can be an indication that there is something generally wrong with an operation. Others in the industry will question the integrity management and staff and even hesitate to conduct business. Publicity, lost goodwill, reputation, public image, customer confidence, and brand identity can all be adversely affected.
- *Customer attrition.* Customer dissatisfaction as a result of a service disruption creates customer loss. Existing customers that represent the current sales, as well as potential customers that represent future sales, are motivated to take their business elsewhere.
- *Customer retention.* The costs to retain customers can be manifested in several ways. These include costs such as those to address increased service challenges from customers in terms of complaints, service calls, costs to provide consolation services, and incentives for customers to stay.
- *Lost financing.* In addition to hurting a firm's bottom line, many of the aforementioned factors can also affect a firm's credit rating and the ability to seek financing.
- *Employee morale.* Organizational culture can be affected if many of the aforementioned factors impact a firm's bottom line and cause subsequent loss of personnel. Poor morale affects service quality and throughput.
- *Market position.* Loss of market share and position relative to competitors could affect overall competitiveness.
- *Asset utilization.* Nonuse of productive assets during a service disruption ultimately affects the return on assets in a balance sheet. Systems, applications, network and technology services, and infrastructure are viewed as assets.

Surprisingly, there are monetary gains that can arise out of a service disruption, in certain cases. These usually take the form of avoided costs of doing business, much in the same way that canceled flights do not burn fuel or an idle factory does not consume raw materials. Likewise, IT operations that consume usage-based network services need not pay for those services during an outage, depending on the service agreement. Rebates or deferred payments can also apply, depending on the terms of the agreement. Other forms of monetary gains include filing losses and reduced taxes from lost income. Interestingly, some firms have leveraged their recovery from an outage as a marketing ploy to distinguish themselves from competitors.

3.7   Capacity Metrics

Measures of capacity differ by the context of their use. In networking, it is usually defined as how fast a network element is per unit time or how much work per unit time can be achieved. The following sections discuss some of the more commonly used capacity metrics in networking.

3.7.1   Utilization

Capacity utilization is a measure of congestion, workload, or usage of a network element [26]. Percentage of CPU, memory, and bandwidth usage are all examples of capacity utilization. Capacity utilization is the ratio of the observed capacity value relative to the theoretical capacity. For any given unit of measure q, the utilization, U, is computed as:

$$U = \left[q(\text{observed}) / q(\text{theoretical}) \right] \times 100\% \tag{3.21}$$

The denominator here is sometimes referred to the as the *burst capacity* or *throughput* of a system. Throughput is the theoretical carrying capacity corrected for overhead (discussed in a later section). For data traffic, obtaining actual values for q is not easy. Because of the variability of data traffic and transmission overhead that is required, a computation of U may not be adequate to characterize a network's utilization. Controlled rate applications, such as voice, on the other hand will exhibit less variability in the traffic pattern. A metric used to characterize this variability is the *peak-to-average ratio*, P:

$$P = q(\text{peak}) / q(\text{average}) \tag{3.22}$$

where q (peak) is the observed peak or burst traffic value. In the case of bandwidth, for example, peak bandwidth is the largest amount of data transmitted at any given time. q (average) is the observed average. The average can be taken over time or another scale depending on the context of use. Figure 3.12 illustrates these concepts. P is typically expressed as a ratio rather than a percentage. For example, P for voice traffic is typically in the range of 3:1 [27], while P for data traffic is in the range of 10:1 to 15:1.

The variability in data traffic, particularly Internet protocol (IP) traffic, has made network engineering quite challenging. Studies have shown that IP traffic exhibits the same distribution regardless of the measurement period. This is referred to as a *self-similar* trait, characterized by high variation. Whereas voice connection times can be modeled by an exponential distribution, the sampling and histograms for IP transmission times have yet to be accurately characterized.

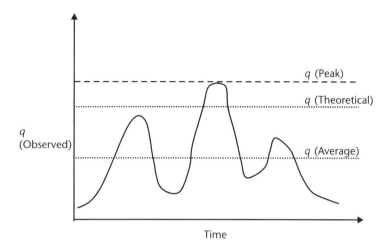

Figure 3.12   Illustration of capacity utilization.

3.7.2   Bandwidth

Depending on the area of networking, bandwidth can take on a couple of different meanings. The concept of bandwidth originated in wireless applications, such television broadcast, microwave, and radio. It was often used to convey a spectrum of frequencies, often expressed in units of hertz, available for use by a transmitter or receiver. This spectrum is often referred to as a *channel*. This concept was later extended to the voice and data networking worlds as well. In these worlds, bandwidth is often used to convey data speed. This is actually not exactly accurate. Because it indicates how much data, in bits or bytes, flows between two points in a given time, it is actually a transfer rate.

The capacity of a data link is representative of the total amount of information that can be transmitted at any one time. If a link operates under *half duplex*, the capacity is the maximum amount of data that can be transferred in one direction at any given time. If a link operates under full duplex, the capacity is the maximum amount of data that can be transferred in both directions at any given time. For example, a half duplex link operating at 10 Mbps has capacity of 10 Mbps. However, a full-duplex link operating at 10 Mbps has a capacity of 20 Mbps, as information can be sent and received at the same time.

To compute the required bandwidth q (peak) for a network traffic flow, the following formula would apply:

$$q(\text{peak}) = \text{number of requested peak flows} \times \text{flow bandwidth} \times d \qquad (3.23)$$

where d is a duplex factor. A flow could equate to a channel if the traffic is connection oriented, as in the case of voice. The duplex factor is simply 1/2 if the flows are half duplex (i.e., both ends of the flows take turns transmitting). This assumption would hold in the case of voice traffic because in a typical conversation each end would take turn speaking. So, for example, assuming 1,000 peak voice calls, each requiring 20 Kbps of IP packets, the total bandwidth required for the voice over IP (VoIP) traffic on a network would be 10 Mbps [28].

3.7.3 Overhead

As stated earlier, throughput is a measure of bandwidth corrected for overhead and other factors. It is the actual quantity of information content transferred in a unit of time. Overhead includes conditions that must be factored in to convey the actual payload capacity of a network. For connectionless networks, such as Ethernet, this can be quite challenging. It is more applicable to connection-oriented networks, such as circuit-switched voice, channel-based TDM, and packet-based virtual circuit networks such as frame relay or ATM. The following are some of the variables that need to be considered in overhead:

- *Duplexing.* Duplexing can sometimes compound overhead effects. For example, if data transfer in one direction of a half-duplex 10-Mbps link consumes 2 Mbps of overhead, the link throughput is reduced down to 6 Mbps, a 40% reduction. Because an overhead operation usually requires both sending and receiving overhead information to validate a data transfer, a total of 4 Mbps of overhead is required. In a full-duplex operation where links collectively provide 20 Mbps of bandwidth, the transfer of data in each direction includes the overhead information, so the throughput is reduced by only 20% down to 16 Mbps. This phenomenon depends on the protocols in use. Some protocols include features to correct for this. Data transfers involving unidirectional protocols, which send information in one direction, are known to be less reliable.

- *Collisions.* Network communication over wireline typically involves modulating a carrier signal. A carrier signal inherently lets both ends of a conversation know when to transmit. When multiple conversations take place over the same wire, as in the case of Ethernet, performance of half-duplex devices degrades because of the increase of packet loss resulting from packet collisions. A collision occurs when both ends of a conversation try to transmit data at the same time over the same wire. Because the opportunity for collisions grows as the number of users grows, so does the reduction in link throughput. Collisions are further discussed in Chapter 5, which focuses on networking technologies.

- *Duty cycle.* Loss in actual bandwidth can be incurred due to a circuit or network element not operating all the time. In applications such as wireless, which sometimes use pulsed radio frequency (RF), or polling applications, where circuits are used intermittently, the full bandwidth utilization may not be realized.

- *Impedance mismatch.* Although a circuit can be rated at a particular line capacity, the devices at each end might process information at a far less rate. Thus the full line capacity will never be realized. For example, if a T1 line connects two hosts, but each can only process information at only half the speed of a T1, then the effective throughput is only half that of a T1.

- *Protocol overhead.* When a sending device transmits a message to a receiving device, protocols will insert additional information with the message to govern the transmission. This extra overhead information will consume line capacity in addition to the message and affect service throughput.

3.8   Performance Metrics

Although sometimes viewed as a performance metric, availability alone does not characterize the performance of a network. While the focus of availability is the degree in which a network is operable, it is a function of the network's performance, conveyed through several key metrics that are discussed in this section. Numerous factors affect overall network performance. The metrics discussed in this section are used to describe in summative fashion the behavior of a network. They should be used in conjunction with many of the aforementioned metrics.

Consistency in performance is a true indicator of a mission-critical network. It means providing the same performance over time and having predictability in response to various situations. However, performance measurements, regardless of the metric, may not always reflect consistency. Measurements should be taken at constant load levels, spanning at least several minutes and long enough to remove any transient effects. Measurements taken over a range of load levels should span at least several hours. These factors should be taken into account in relative comparisons of the same metric.

3.8.1   Latency

Delay is the duration of time from when a signal is transmitted to when it is received. It is used as a strict measure of a network element's ability to process a signal or request. *Latency* is the time it takes for a partuiclar signal, often in the form of a data packet or voice signal, to get from a network origin point to a destination point, measured in units of time (usually milliseconds). In essence, latency is the cumulative delay posed by all of the network elements in a transmission path. In an IP network, for example, it is the time it takes for data packets to go from a client to an intended server.

Zero latency is a coined term to describe an ideal goal for processing information through a network infrastructure. This term is most commonly used to describe individual network element characteristics, such as a nonblocking Ethernet switch. It represents instantaneous real-time response to transaction requests. Unfortunately, this ideal is virtually impossible to achieve. In a mission-critical network, the connectivity between elements need not be the quickest path. Instead, it should be the most reliable path based on the network's state. This is why routing algorithms often do not use latency as a prime consideration for choosing a path. For one thing, there is no true standard for measuring latency. Latency, as measured or stated by different network elements, may only reflect capacity levels and not necessarily a range of levels.

Latency is the cumulative time to process a given packet or signal across stages along a transmission path. This includes network link, switches, hub, routers, host platforms, and other devices or components that are required to process the transmission. Excessive latencies are often attributed to problems from a single element or location. In these situations, it is often a good exercise to list out these elements and their respective delay times to determine which one could potentially contribute the most. Contributors to latency include the following:

- *Encryption* and *decryption* of packets, particularly at ingress and egress routers, which requires additional processing, can add to latency.

- *Propagation* latency includes the signal transmission time delay, plus any processing delay by transmission devices. Propagation across network links can contribute to latency, but, surprisingly, is not usually a prime factor. Some applications, such as data mirroring, can be sensitive to link or network latency.
- *Network hops,* which are connection points between networks, can increase latency as they grow in number. The greater number of networks that must be traversed, the greater opportunity for latency.
- *Peering points or POP sites,* where multiple service providers connect, as well as hops can also be possible points of congestion. Congestion can cause packets to drop. Although latency is not a true measure of packet loss, it can still convey how well congestion-management mechanisms are working in a network.
- *Devices such as routers and firewalls,* found at hop locations, must often inspect every packet passing through them, adding to latency. Domain name servers (DNSs) are often placed behind firewalls and routers, adding latency to the domain name resolution process.
- *Impedance mismatches* between devices and network links, for reasons discussed earlier, will also contribute to latency. Performance imbalances across different networks in a transmission path can constrain throughput. A single host processing an application whose throughput is not matched with other elements in a path can pose a bottleneck.

Jitter refers to variation in latency or delay. Arrival times of network data or packets will vary, which can pose a problem for real-time streamed services such as voice or video transmissions. Because these applications require nearly constant arrival times, they require minimum latency variation. Sensitivity to jitter can be high; even slight variations can negatively impact service performance.

Because latency will vary in time and across elements, using a deterministic value to convey latency may not be apropos. Instead, a latency distribution or histogram can better describe a network service at its best and worst levels. Using an approach similar to (3.22), the variability or latency, L, for a service can be conveyed as:

$$L = l(\text{peak}) / l(\text{average}) \qquad (3.24)$$

where l (peak) is the peak observed latency and l (average) is the average over the observed samples. For example, the L for a wide area network (WAN) consisting of three permanent virtual circuits (PVCs) with latency values of 40 ms, 60 ms, and 180 ms is about 2:1. Services such as VoIP can be greatly affected by latency. For voice calls, latency above 250 ms can be problematic. For this reason, conventional wisdom leans towards handling this traffic apart from other IP traffic, until accurate traffic engineering and management techniques can be developed.

Many service providers might quote gross latency values for their networks to advertise their capabilities. These values, however, often misstate the carrier's actual performance. First, the value is usually specific only to the carrier's network. In reality, many networks, including the origin network, several carrier networks, and the network at which the destination host resides, can be traversed in a transaction.

Second, they usually represent values averaged over periods, as long as a month, which are not reflective of true real-time latency. This is why many service providers are unlikely to guarantee latency in their service agreements.

3.8.2 Response Time

Depending on the context, latency is sometimes used interchangeably with *response time*, which is the time from when a request is sent to when a response is received. Response time can be thought of as round-trip latency from the perspective of the user, or the sending device. For this reason, the same caveats that applied to latency also apply to response. The PING program is often used to measure network response time. This is an Internet control message protocol (ICMP) message that sends packets a specific host at an IP address and times the response. Although this program can be indicative of network-based processing such as connection setup, routing, and transmission delay, it may not be truly reflective of overall response from a service perspective. Server processing, application processing, and data transfer from storage should also be included.

3.8.3 Loss

Data or packet *loss* can arise out of network congestion or transmission problems. Loss would occur in the case of a buffer overflow in a device where overflowed data is discarded altogether, versus fragmenting it prior to retransmission. A data transmission that exceeds a committed burst size and excess burst size is likely to experience data loss.

Loss can also occur when two or more devices attempt to simultaneously seize a network interface, trunk, or link. The data intended to be sent by the losing device could be discarded. This phenomenon is manifested in many types of networks, even voice networks. In Ethernet networks, they are called collisions. In general, collisions are a result of too many devices trying to access the same resource at the same time. Many devices use collision-detection routines to sense when collisions are taking place and keep track of lost packets or data. When collisions occur at inordinate rates, devices will intentionally slow transmitting and receiving rates and retransmit data in response. This results in ineffective use of bandwidth.

Some services may exhibit high sensitivity to packet loss. A small degree in loss could cause significant deterioration in performance. Again, streamed services such as voice or video can exhibit this kind of sensitivity. Not only do transmissions become choppy, but noise can be generated in the received signal as well.

3.8.4 Error

Error is a general term that refers to data that has been altered in some fashion during transmission such that when it is received at a destination, it is different from what was originally sent. *Error* rates, particularly bit error rates (BERs), are typically stated as a percentage of the transmitted traffic. Errors usually arise out of some problem along the transmission path and are usually characterized by corrupted data or out-of-sequence packets.

Data corruption, although more rare now than in previous decades, has the same effects as packet loss (discussed in the next section). It results in packets that are simply of no use. Packets or data that are out of sequence is characteristic of situations where a device along a transmission path was unable to accommodate the amount of received data. Some devices might overflow their buffers and have to further segment the data they received prior to retransmission. The segments might not all arrive at the destination in the right sequence, particularly if the segments were routed on different paths. The receiving device may either try to reassemble the data or discard it altogether if reassembly is not possible.

Errors usually result in retransmission of the data. A high error rate is indicative of ineffective use of resources and lower transmission rates. When error rates are unusually high, some real corrective measures are to alter the packet routing or lower transmission rates.

3.8.5 Throughput

Estimating the throughput of a network service could provide insight as to where potential bottlenecks might occur. It requires understanding each network component, protocol, and transmission technology. Components may become CPU, memory, or I/O bound at the disk and network interface levels. Knowing the operating limits of a device can help match its impedance with that of the connecting network links.

As discussed in the section on utilization, throughput is the net carrying capacity of an element corrected for overhead. Throughput is a theoretical value, calculated based on the operating characteristics of a particular network. It represents the effective capacity of a connection or service once all things are considered.

Figure 3.13 illustrates this concept and how it relates to estimating the throughput of a device or network link [29]. For a network link, the following formula can be used:

$$q_L = \left[(Q / K) - \sum_{i=1}^{L} \theta_i \right] d \tag{3.25}$$

where q_L is the realized channel throughput at protocol layer L, Q is the gross data rate based on the transmission technology, K is the number of channels or traffic flows, θ_i is the channel protocol overhead at layer i, and d is the duplex factor that was discussed earlier. θ_i is the accumulated protocol loss over layer L and subtending layers. So, for example, assuming 11 Kbps accumulated protocol loss up to the IP layer (layer 3), the actual per-channel throughput of voice over a T1 link is about 53 Kbps (simplex) or 107 Kbps (duplex).

Throughput does not reflect the actual transfer rate when a channel is in use. This is sometimes referred to as *goodput*, q_A [30]. Goodput represents the amount of payload data net any bits that had to be retransmitted or reordered due to errors, loss, latency, jitter, or other prevailing transmission factors. It is often prudent to assume that perfect transmission rates are rarely realized, and tweaking the calculated throughput for such factors can provide a conservative view. Goodput can come into play when designing specialized applications that rely on sustained rates of incoming or outgoing data streams.

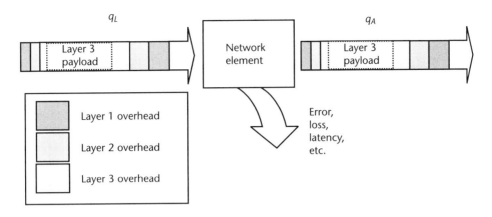

Figure 3.13   Throughput illustration.

3.9   Summary and Conclusions

Metrics are quantitative measures used to convey system or network behavior. The better metrics are those that strike a balance between data granularity and computation effort. Metrics should be associated with a service objective—usually several are required to express how well an objective is being met. This chapter discussed several classes of metrics that are used to assess network continuity.

Recovery metrics include the RTO and RPO. The RTO is a target measure of the time interval between an adverse event and service restoration. The interval should encompass all relevant restoration activities. Lower RTOs imply greater recovery costs. The RPO, on the other hand, is used specifically for data recovery. Measured in units of time, the RPO conveys the desired age of data required to restore operation following a service interruption. RPO does not necessarily express the integrity or quality of data required for restoration.

There are several metrics used to convey network and system reliability. The MTTF is used to characterize the operating life of a system. The failure rate F expresses the frequency of failures or outages. The MTTR is an observed measure of the time required to restore operation. The MTBF measures the average life of a network component. A low MTBF implies greater maintenance costs. Improving the MTBF of each node in a network can provide greater value as the network grows in size.

Availability is the proportion of time service is provided. Availability as a measure can conceal outage frequency, so it should be used carefully. Availability is increased through redundancy and improved restoration practices. Many organizations define several levels of availability for their IT environments. Although it is used interchangeably with reliability, the two differ. Reliability is suggestive of failure frequency, while availability is indicative of service duration.

Risk is a metric used to convey the expected loss resulting from a particular threat. It is a product of the frequency of the threat, an organization's vulnerability to the threat, and the potential loss or damage arising from the threat. The last item is comprised of direct, indirect, intangible, and avoided costs resulting from a service disruption or outage.

Capacity and throughput metrics can vary according to their context of use, but generally any such metric should reflect "useful" versus raw capacity. This means that raw capacity should be corrected for overhead and other factors that are necessary for operation, but consume raw capacity as well. In networks, this includes such things as protocol overhead, impedance mismatches, and collisions.

References

[1]   Smith, J., "A Brand New Role for IT," *Optimize*, April 2002, pp. 71–74.

[2]   Bakman, A., "The Weakest Link in Disaster Recovery," *Disaster Recovery Journal*, Spring 2002, pp. 26–30.

[3]   Douglas, W. J., "Systematic Approach to Continuous Operations," *Disaster Recovery Journal*, Summer 1998.

[4]   La Pedis, R., "Disaster Recovery: No Longer Enough," *Disaster Recovery Journal*, Summer 2001, pp. 14–18.

[5]   Toigo, J., *Disaster Recovery Planning*, New York: John Wiley & Sons, Inc., 1996, pp. 99–122.

[6]   Beeler, D., "The Internet Changes Everything! Techniques and Technology for Continuous Offsite Data Protection," *Disaster Recovery Journal*, Fall 2000.

[7]   Oracle, Inc., "Database Disaster Recovery Planning," *Oracle Technical White Paper*, January 2000, pp. 1–9.

[8]   Rehbehn, K., "Defining Availability in Packet Networks," *Network World*, April 12, 1999, p. 35.

[9]   Thorson, E., "Why the Industry Needs Open Service Availability API Specifications," *Integrated Communications Design*, April 15, 2002, pp. 20–21.

[10] Bryant, E. C., *Statistical Analysis*, New York: McGraw-Hill Book Company, Inc., 1960, pp. 20–24.

[11] Dooley, K., *Designing Large-Scale LANs*, Sebastopol, CA: O'Reilly & Associates, Inc., 2002, pp. 14–44.

[12] Blanchard, B. S., *Logistics Engineering and Management*, Englewood Cliffs, NJ: Prentice Hall, 1981, pp. 24–58.

[13] Paulsen, K., "Understanding Reliability in Media Server Systems," *TV Technology*, October 9, 2002, pp. 29–33.

[14] Wiley, J., "Strategies for Increasing System Availability," *Cable Installation & Maintenance*, September 2001, pp. 19–23.

[15] Oggerino, C., *High Availability Network Fundamentals*, Indianapolis, IA: Cisco Press, 2001.

[16] Hill, C., "High Availability Systems Made Easy," *Communications Systems Design*, November 2000, pp. 47–55.

[17] DeVoney, C., "Power & Pain: Dollars & Sense," *Smart Partner*, September 18, 2000, pp. 56–57.

[18] Nawabi, W., "Avoiding the Devastation of Downtime," *America's Network*, February 2001, pp. 65–70.

[19] Young, D., "Ante Uptime," *CIO—Section 1*, May 1, 1998, pp. 50–58.

[20] Paul, L. G., "Practice Makes Perfect," *CIO Enterprise—Section 2*, January 15, 1999, pp. 24–30.

[21] Hoard, B., "In an Internet World…You Can't Hide Downtime," *CIO Magazine—Comdisco Special Advertising Supplement*, pp. 2–11.

[22] Liebmann, L., "Quantifying Risk," *Network Magazine*, February 2001, p. 122.

[23]   Kuver, P. B., "Managing Risk in Information Systems," *Tech Republic,* June 26, 2001, www.techrepublic.com.

[24]   Bell, D. R., "Calculating and Avoiding the Hidden Cost of Downtime," *Control Solutions*, January 2002, pp. 22–27.

[25]   Hinton, W., and R. Clements, "Are You Managing the Risk of Downtime," *Disaster Recovery Journal*, Summer 2002, pp. 64–67.

[26]   Schwartz, M., *Computer-Communication Network Design and Analysis*, Englewood Cliffs, NJ: Prentice Hall, 1977, pp. 286–318.

[27]   Mina, R., *Introduction to Teletraffic Engineering*, Chicago, IL: Telephony Publishing Corporation, 1974, pp. 67–74.

[28]   Hills, M. T., "Traffic Engineering Voice for Voice over IP," *Business Communications Review*, September 2002, pp. 54–57.

[29]   Held, G., *LAN Performance Issues and Answers*, New York: John Wiley & Sons, Inc., 1996, pp. 110–128.

[30]   Newman, D., "Measure for Measure: Making Metrics Matter," *Network World*, January 14, 2002, p. 31.

CHAPTER 4

Network Topology and Protocol Considerations for Continuity

Designing mission-critical networks is often more art than science, involving integration of different networking technologies. Networking technologies typically specify how network elements communicate with each other. This communication often involves an organizational framework of information, routing, and communication protocols. This framework also influences how network elements must be interconnected (i.e., topology) and in what quantity (i.e., capacity) to achieve optimal communication and has implications on how traffic should flow through the network. In this chapter, we discuss some basic concepts in network topology and protocols that are fundamental to mission-critical network planning. The chapter reviews concepts that provide the foundation for understanding many of the networking technologies discussed further in this book.

4.1 Network Topology

A topology defines the ways in which nodes or elements of a network are associated or interconnected. The interconnections between nodes are graphically represented using a line that represents a communications link. Because communication networks use layered protocol architectures, they are often comprised of different layers of networks. A layer can consist of a logical or physical topology. In a *logical topology*, links specify the way in which nodes communicate with each other. In a *physical topology*, links specify how nodes physically connect to one another.

A path or *route* through a network is comprised of a series of links. It is a sequence or series of paths that are traversed. When designing a network for mission-critical purposes, the challenge is to develop redundant paths between two nodes in a network, at both the logical and physical layers. When protecting a link, a redundant link or path is usually defined. The purpose of redundant link design is to provide at least one alternate link between the nodes so that information can be transported between the two in the event the primary link fails. The protection can be achieved by duplicating the link over the same lower layer, creating a secondary link at a lower layer, or by creating completely different links using different physical or logical layers (see Figure 4.1).

Protecting a path is more complex than protecting a link. When a link fails, the end nodes are simply connected using an alternate link. Protecting a path, however, implies more recovery intelligence in the network components. If a protection path is required, a key question is whether to define that route adaptively, immediately

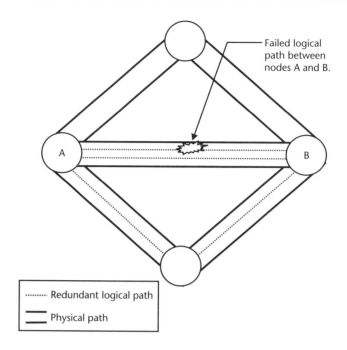

Figure 4.1   Logical and physical path redundancy.

after the failure, or to use predefined or static routes. When routes are defined dynamically, it implies that the network technology in use is focused primarily on maintaining connectivity in the event of a failure. An example of this is use of spanning tree protocols, which are used in switches to compute multiple paths though a local area network (LAN), or open shortest path forwarding (OSPF), which is used to compute multiple paths through a complex IP network. Although it can define paths that are free of loops, the spanning tree protocol can often take an inordinate amount of time to recompute a spanning tree following a change in LAN topology.

By using preconfigured or static routes, a certain level of performance can be guaranteed, in addition to maintaining connectivity, as these routes can be traffic-engineered beforehand. Using preplanned routes eliminates the need to compute routes in real time, enabling more time for other slower, but more thorough, recovery mechanisms.

4.1.1   Fundamental Topologies

There are several types of fundamental network topologies, shown in Figure 4.2, that are building blocks for larger topologies. They include:

- *Point-to-point.* Two nodes communicate with each other directly or are directly connected by a link.
- *Bus.* This is a collection of nodes that share a common link at the same logical layer. When one node transmits, all of the other nodes receive the transmission. A bus network is typically viewed as the most cost-effective way to connect more than two nodes together.

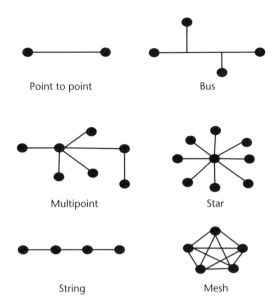

Figure 4.2   Fundamental network topologies.

- *Multipoint.* This is a collection of point-to-point topologies that share common nodes.
- *Star.* This is a special case of a multipoint topology, where all point-to-point links share one common node. In practice, many times this node controls communications, as in the case of a switch. For this reason, star networks are highly vulnerable, as loss of the central node will cause network failure. This configuration is also frequently called *hub and spoke*.
- *String and ring.* This is a series of point-to-point links, where one node connects to another node, which connects to another node, and so on. A string topology that is a subset of larger network can define a route. A ring is essentially a string topology, whereby the end nodes are one and the same.

4.1.2   Mesh Topologies

In a *full mesh* topology, each node is connected to every other node, eliminating single points of failure as well as bottlenecks. If a network is comprised of *n* nodes, then the number of links *l* that can be built to achieve a full mesh is computed using the general formula $l = n\,(n–1)/2$. A mesh topology allows more efficient use of network resources because multiple routes can be defined over a common set of nodes and links. This allows flexibility in providing different service levels between pairs of nodes. Because multiple routes can be created through a mesh network, it is better suited to for packet-switched traffic, versus circuit-switched traffic, where a constant connection required as in the case of traditional voice telephone networks [1].

Mesh topologies can be built over time because the entire network does not have to be planned all at the same time. Links can be added as capacity is needed, sometimes on demand. Because protection routes are shared and multiple routes can be defined, uniformity in link capacity is not an issue. If capacity is added, only links that serve the more heavily used routes need to be upgraded, without having to upgrade the entire network.

Although mesh topologies offer the best protection characteristics, they are typically the most expensive to implement [2]. Because they require the maximum amount of resources manifested in the nodes and links, they do not scale very efficiently [3]. They also require greater network intelligence to route traffic. Furthermore, protection routes can end up being significantly longer than primary routes, adding latency. For this reason, variations or subsets in mesh architectures have been defined.

Different mesh topologies are shown in Figure 4.3. The degree of "meshiness" of a network is usually measured in terms of the distribution of connected nodes. In a *flat mesh*, the distribution is uniform. In a *partial mesh*, the distribution is less uniform. Because failure of a highly connected node can be catastrophic, a mesh network allows many candidate protection routes to be defined, and thus protection capacity costs can be shared among the many routes.

In a *constrained* or *bounded mesh*, there is a consolidation in either links or nodes, which can reduce implementation costs [4]. After a failure, traffic is rerouted on alternate routes that use the consolidated links or node. Constrained meshes are more typical in practice and offer a balance between cost and protection. Many times, one will find a constrained logical mesh overlaid on a fuller physical mesh. Constrained meshes are typical in backbone and long-haul carrier networks.

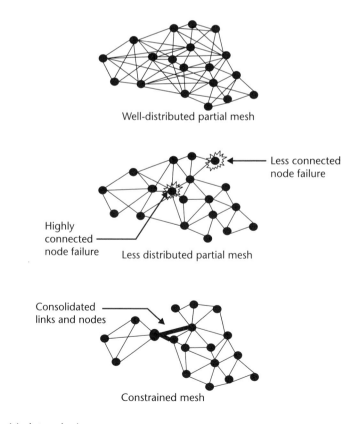

Figure 4.3   Mesh topologies.

4.1.3   Ring Topologies

A ring is essentially a string topology whereby the end nodes are one and the same. Communication between nodes in ring network typically *loops* around the ring. For reliability, dual loops are created where transmission in the redundant loop is in an opposite direction to that of the primary loop [5]. In synchronous optical networks (SONETs), for example, this second loop is often referred to as a *counterrotating ring*. If one node in the ring fails, traffic can loop back around the ring to the desired destination. This concept is illustrated in Figure 4.4.

Ring topologies evolved from voice-based network technology and were conceived with voice traffic in mind. Voice traffic patterns are typically local and hierarchical in nature. Ring networks are often found in the transport portion of most large networks to improve vulnerability against large failure groups.

Although rings provide protection against node and link failures, they do have an inherent vulnerability: if one link on the ring is congested, the entire ring can be affected. To circumvent this, a second ring, which is overlaid or "stacked" on the existing ring, is created. These stacked rings are often interconnected at the nodes. SONET networks use cross-connect devices to interconnect different rings.

Ring topologies can be expensive to build as well. The extra expense is necessary to safeguard against congestion and node or link failures by adding extra protection capacity to accommodate rerouted traffic. In SONET rings, which will be discussed later, this can result in the doubling of ring capacity, which is often

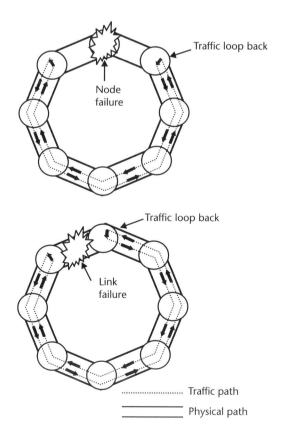

Figure 4.4   Counterrotating ring topology.

underutilized. This is often referred to as *stranded capacity*. In contrast, the amount of protection capacity in a mesh topology is far less than that required for rings. Furthermore, ring capacity upgrades must be made to all of the node and links in ring.

4.1.4 Tiered Topologies

A single-tier network is a multipoint network where selected nodes are connected in a string topology. Each of these nodes is then connected to several end nodes. In practice, the end nodes are typically users and the nodes comprising the string segment are switches. In a two-tier topology, for example, the highest tier switches are charged with interconnecting the lower tier switches. Multiple tiers are usually defined to improve network manageability and reduce switch hops. A sound approach to implementing a network survivability program is to optimize each tier. Because traffic is aggregated the higher the tier, switches grow larger in capacity and more efficient in port utilization.

Tiered networks evolved out of the toll hierarchy used in telephone networks. Mesh networks can be used in conjunction with tiered topologies to reduce vulnerability and improve performance. Each switch in one tier can be homed to all or some of the higher tier switches. Tier networks are often found in large switched LANs. Figure 4.5 shows a three-tier network with a middle tier consisting of four nodes and a top tier of two nodes.

As a general rule, links in a network should be engineered not only to accommodate their working traffic, they should have excess capacity to handle additional load displaced by an outage in the network. In general, the working traffic load of any failed link between two nodes should be spread in some manner across the links comprising the alternate paths between those nodes. In the example of Figure 4.5, there are two alternate paths consisting of two links and two alternate paths consisting of three links. If the top tier's working traffic load is d, then each subtending link in the lower tier should be engineered for additional capacity, assuming the working load is spread uniformly across the alternate paths.

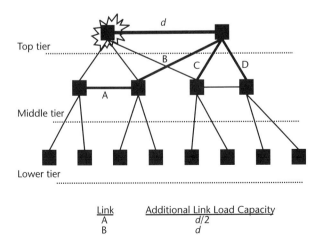

Link	Additional Link Load Capacity
A	$d/2$
B	d
C	$d/2$
D	$d/2$

Figure 4.5 Tier topology example.

4.1.5   Edge Topologies

The edge network is the access portion to a network. The edge network topology is that portion of a network that connects remote end points, typically users or other networks, to a main network. It is in the edge network where survivability is the most problematic.

The edge network is the most vulnerable portion of any network. Any effort to improve the reliability of a network can be useless if the edge network is isolated in the event of a failure. Edge networks typically have lower capacity and nonredundant connections to a core network, making them a barrier to improving network performance. An edge node, especially one that aggregates traffic from the edge network onto a core or backbone network, can be a most worrisome single point of failure. If the edge network is home to a large number of users who connect to an edge node, often a switch or router, failure of that device or link can be catastrophic. Figure 4.6 illustrates these concepts.

This issue is further compounded by the effects such a failure can have on the core network. Switches or routers that connect to the edge node must somehow notify other network elements (or network management) of the loss. All traffic in the core network destined to the edge network must then be discouraged.

In an Internet protocol (IP) network, for example, a failed router's neighbors would report that the affected destinations via the failed router are no longer available. If the edge router recovers, this process must be repeated. In traditional telephone networks, calls to a failed end office are often blocked or throttled until the problem is resolved. In either case, throttling traffic to the affected location can keep the remaining network stable until the problem is resolved. A common way around this is to simply establish redundancy in how the edge network connects to the core network. Redundant connections and/or edge nodes can achieve this.

4.1.6   Peer-to-Peer Topologies

As of this writing, there is growing renewed interest in peer-to-peer networks. Peer-to-peer networking is logical topology that is layered over another topology. In peer networks, nodes can behave as clients or servers or both [6]. The intent is to make the most use of available computing power in a network (especially at the network edge). There are no rules as to what services can be provided by which nodes.

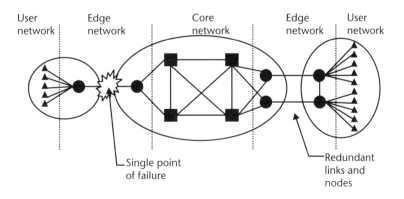

Figure 4.6   Edge network example.

Examples of such services include registration, searching, storage, network management, and many other types of activity found in networking. Peer schemes can blend together many of the topologies that were previously discussed. For example, a peer topology can take on a tiered, hierarchical look or even a mesh look.

4.2 Network Protocol Considerations

The best strategy for working with network protocols is simplicity. Simplicity in network design is often the most efficient, cost-effective, and reliable way to design networks. Reducing the number of different protocols used can further assure interoperability and reduce management headaches. As one proceeds up the protocol stack, vendor products tend to become more specific to the protocols they support, particularly above layer 2. One may find interoperability issues in using different vendor products. As the popularity of appliance-based networking grows, there will be a tendency for one vendor's product to dominate a network. This strategy is very sound, but it creates the inherent vulnerability related to sole sourcing a product or service to a single vendor.

Fundamental to implementing today's Internet architecture protocol model is how well the different layers of protocol can freely communicate and interact with each other. Given the mission-critical nature and fluidness of today's networking, new technologies and features are always being developed to enhance and leverage this interaction, particularly in network switching equipment. These features can also make network implementation and management easier and more cost effective.

Different protocol layers have inherent reliability features, and different protocols at the same layer will also have protection and reliability features. The question then arises as to how to assure the right protection is being provided at the right layers while avoiding over protection or conflicting mechanisms. Extensive management coordination between the network layers can introduce unwanted costs or resource consumption, as well as more network overhead. For each network service, a general rule is if a lower layer cannot provide needed protection, then apply protection at the next highest layer.

The following are some general strategies to follow to coordinate the survivability mechanisms among different layers of protocol [7]. They are illustrated in Figure 4.7.

- *Selective strategy:* Apply the recovery or protection mechanism on one layer at a time.
- *Sequential strategy:* If a protection or recovery mechanism at a particular layer fails, apply a mechanism in another layer. This would be the next higher layer if the previous layer were unable to recover.
- *Parallel strategy:* Allow every layer to apply the protection or recovery mechanism. This can consume extra resources in all layers. Overprotecting may cause oscillations in the provided service and an unnecessary throttling of traffic.
- *Interlayer coordination strategy:* Exchange alarm and state information between layers in order to know how and where to activate the survivability

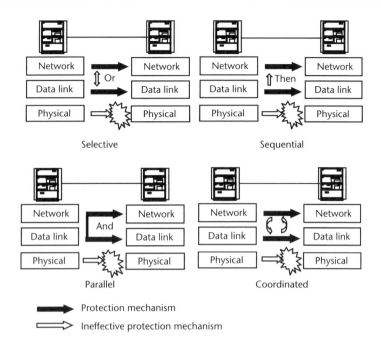

Figure 4.7   Network protocol protection strategies.

mechanism. Although this seems like the best strategy, it can be quite complex to implement, particularly because it may be quite difficult to exchange information between different network vendor products—even those that use the same protocol.

Network layer protocols should come into play if a transmission link has high a bit error rate (BER) or if a link or node fails. Conflicting survivability mechanisms between layers should be avoided. An example is the case of using IP over a SONET network. A fiber cut will cause a SONET ring to invoke automatic protection switching (APS). If IP links are affected by the cut, this can cause routing changes to be broadcast and then rebroadcast once the APS is completed, causing a flapping condition. APS usually can take up to 50 ms, which was once considered sufficient to avoid switching contention because the IP layers, which switch at slower speeds, would be unaware that the APS has taken place. However, as IP switching times continue to decrease, it may become difficult to ensure that lower layer protection will be able to serve all higher layer schemes.

4.3   Summary and Conclusions

Network topology defines how individual nodes and elements within a network interconnect with each other using links. Routes are comprised of a sequence of links and require greater failure recovery intelligence than an individual link. Mesh topologies are the most robust in eliminating single points of failure. Because every node is connected to every other node to some degree, many alternate traffic routes can be defined. However, mesh networks are typically the most expensive to build.

In a ring topology, traffic loops around to each node on the ring. For survivability, multiple loops of the same traffic are used, typically traveling in opposite directions. For this reason, physical ring topologies are popular in fiber-optic networks. However, the use of multiple loops can result in stranded capacity, which unfavorably impacts the economics of the ring solution.

Network topologies are often layered in tiers to improve manageability. Multiple tiers can reduce backbone switch hops as well as aid survivability. For effective survivability, links should be engineered to accommodate excess capacity to handle load displaced from a failed node in the event of an outage. This particularly holds true for edge networks, which are traditionally the most critical (and vulnerable) portion of a network topology. Establishing redundant access links coupled with the ability to divert traffic away from a failed link are two classic remedial measures for edge survivability.

Protocols are fundamental to network operation—yet they can add to network management complexity. Minimizing the number of different protocols in use can reduce complexity and aid interoperability. However, for survivability, they should be carefully chosen so that each provides the right protection for the protocol layer and does not overprotect or conflict with protection mechanisms at other layers. Several possible scenarios were discussed to this effect.

References

[1]   Saleh, A., and J. Simmons, "All-Optical Mesh Backbone Networks Are Foundation of the Next-Generation Internet," *Lightwave*, June 2000, pp. 116–120.

[2]   Sweeney, D., "Viable and Reliable," *America's Network*, October 1, 2001, p. 22.

[3]   Whipple, D., "For Net & Web, Security Worries Mount," *Interactive Week*, October 9, 2000, pp. 1–8.

[4]   Richards, K., "Choosing the Best Path to Optical Network Profits," *Fiber Exchange*, July 2000, pp. 11–16.

[5]   Woods, D., "Going Toward the Light," *Network Computing*, January 22, 2001, pp. 97–99.

[6]   Schwartz, M., "Peer Pressure," *CIO Insight*, March 2002, pp. 55–59.

[7]   Fontalba, A., "Assessing the Impact of Optical Protection with Synchronous Resilience," *Lightwave*, May 2000, pp. 71–78.

CHAPTER 5

Networking Technologies for Continuity

In this chapter, we discuss a variety of networking technologies in terms of their mission-critical characteristics. We explore elements and techniques of redundancy, routing, and transport that can be leveraged for use in mission-critical networks and their relative merits and pitfalls. It is assumed that the reader already has some familiarity with these technologies. While this chapter does not provide a comprehensive review of these technologies, we present sufficient overviews to establish a basis for subsequent discussion.

Numerous networking technologies are available, each with their own merits and caveats. It was stated earlier that simplification through a minimal mix of protocols is one of the best approaches to network survivability and performance. On the other hand, overreliance on a single protocol or technology is unwise. In planning and designing mission-critical networks, the challenge is to find that happy medium where the minimal mix of multiple technologies provides the best protection for the least cost. In this section, we will review the capabilities and techniques involving the more popular networking technologies with respect to performance and survivability.

5.1   Local Area Networks

Local area networks (LANs) are gradually becoming cluttered with a growing mix of hosts, peripherals, and networking appliances. Dedicated application servers, load-balancers, hubs, and switches each have their impact on data traffic in the LAN. Redundant, fail-over devices are used in many cases, adding to the number of nodes using the LAN. The growth in the diversity and quantity of LAN devices has a pronounced effect on the quantity and predictability of LAN traffic. LAN traffic estimates place the average annual growth in excess of 40%.

As LAN technologies improve, adding bandwidth to the LAN becomes less expensive but may not necessarily resolve traffic issues. Use of Web-based applications, centralization of applications, and the introduction of new services such as voice over Internet protocol (VoIP) and video have shifted the percentage of intra-LAN traffic to well below the traditional 80%. Prioritizing these different services such that bandwidth utilization and performance are optimized becomes the real challenge. For example, layer 3 switches, routers, and firewalls, which must process the interLAN traffic, can become bottlenecks regardless of the amount of available LAN bandwidth.

For the purposes of this book, we focus discussion on Ethernet, as it is the most widely used LAN technology. Other technologies, such as fiber distributed data interface (FDDI) and token ring are still in use, but not to the same magnitude as Ethernet.

5.1.1   Ethernet

Developed in the 1970s, Ethernet is by far the most popular layer 2 LAN technology in use today and is gradually finding its way in wide area network (WAN) use as well. Ethernet operates on a *best-effort* principle of data transmission. In a best-effort environment, reliable delivery of data is not guaranteed. Its use in LANs is popular much for this reason, as LAN environments in the past have been internal to organizations and thus were not subject to the high data delivery requirements demanded by external clients. Its *plug and play* ease of operation made it affordable and easy for firms to implement computer networks and manage them easily. However, things have changed in recent years.

Ethernet transports data in frames containing header and trailer information and payload of up to 1,500 bytes. As each Ethernet frame is transmitted on to the physical medium, all Ethernet network adapters on the network receive the first bits of the frame and look at the destination address in the header information. They then compare the destination address with their own address. The adapter having the same address as the destination address will read the entire frame and present it to the host's networking software. Otherwise, it discards the frame entirely.

It is possible for more than one adapter to start transmitting their frames simultaneously. Ethernet employs rules to allow hosts accessing the physical media to decide when to transmit a frame over the media. These media access control (MAC) rules are typically embedded within the network adapters and are based on a protocol called carrier sense multiple access with collision detection (CSMA/CD). CSMA/CD allows only one network adapter to talk at a time on a shared media. The adapter first senses a carrier on the media, if the media is in use. If it is, it must wait until 9.6 ms of silence have passed before transmitting. This is sometimes referred to as an *interframe gap*. After the interframe gap, if two network adapters start transmitting at the same time, they detect each other's presence and stop transmitting. Each device employs a *backoff algorithm* that causes it to wait a random amount of time before trying to send the frame again. This keeps the network adapters from constantly colliding during retransmission.

In a busy network, many network adapters use an *expanding backoff process*, also known as the *truncated binary exponential backoff*, which enables the adapter to adjust for network traffic conditions. The adapter will discard the Ethernet frame after 16 consecutive collisions for a given transmission attempt, which can happen if the network is overloaded for a long period of time or if a failure of a link or node has taken place.

Hubs are devices used to connect multiple hosts to a segment of physical media. Because all hosts share the same physical media, they also share the same bandwidth as well as the same opportunity for collisions to take place, sometimes referred to as a *collision* or *broadcast domain*. In a heavily loaded network, an Ethernet switch should be used in place of a shared media hub because a switch splits up the media into different segments, reducing the opportunity for collisions.

When using Ethernet for mission-critical implementations, there are many caveats that must be kept in mind:

- Ethernet, as a protocol, cannot on its own provide redundant connections.
  Ethernet assumes that the physical media is unreliable and relies on higher layers of the network protocol to deliver data correctly and recover from errors.
  Thus, if a physical link fails, Ethernet cannot provide an immediate work
  around on its own and must depend on layer 3 routing protocols to get
  around the failure. In the end, to have working redundant routes in an Ethernet network, you must employ routers in addition to switches.

- Ethernet was not designed to carry connection-oriented traffic, such as that seen
  in voice or video. Capabilities in higher protocol layers must be used to encapsulate such traffic and ensure that packets are streamed in the correct fashion.

- A good policy to follow is to be consistent with the types of network adapters
  used wherever possible. Many adapter manufacturers advertise smaller interframe gap cycles than their competitors. Inequity among interframe gap cycles
  could foster unwanted collisions.

- Collisions and multiple collisions are expected for a given transmission
  attempt, even in a lightly loaded network. As network traffic load increases,
  collisions become more frequent. Once network traffic reaches overload, the
  addition of a few more nodes can cause the network to cease functioning. This
  phenomenon is the Achilles' heel of Ethernet. Although, 10BaseT might have
  an advertised bandwidth of 10 Mbps, this congestion phenomenon is known
  to reduce Ethernet's effective capacity to about 60% of the advertised capacity. In a network where links operate at half duplex, the effect can be even
  more pronounced.

Although many companies are moving to fast Ethernet (100BaseT) to improve
LAN performance, bottlenecks at aggregation points such as server connections or
switches can still result. While Gigabit Ethernet (1000BaseT) can further improve
the effective bandwidth over an existing copper infrastructure, it too can be subject
to the same types of bottlenecks that can be created due to impedance mismatches in
hosts and networking equipment.

Problems in Ethernet networks can typically fall into three categories: hardware
problems, which typically affect frame formation; transmission problems, which
typically lead to corrupted data; and network design deficiencies, which usually
involve cascading more than four cascaded repeaters—an inherent limitation in
Ethernet. Ethernet employs a *cyclic redundancy check* (CRC) procedure to verify
the integrity of a frame when it is transmitted. A transmitting device calculates a
frame check sequence (FCS) number based on the frame's contents and is transmitted in the Ethernet frame. The receiving device does the same calculation and com-
pares the FCS value with that received. A discrepancy in the values is an indication
that the frame was corrupted during transmission. With Ethernet, some of the types
of problems that can arise include the following:

- *Out-of-window or late collisions* can occur when a station receives a collision
  signal while still transmitting beyond the maximum Ethernet propagation

delay. This can occur if the physical length of the link exceeds 100m or if a device is late in transmitting.

- *Giants* are frames that exceed the maximum Ethernet frame size. They usually occur due to faulty adapters sending erroneous transmissions or corrupted packets. On the other hand, *runts* are frames that are less than the minimum required Ethernet frame size. Runts can occur from collisions, improper network design, or faulty hardware.

- *Misaligned* frames contain bytes having inordinate numbers of bits. This occurs from data corruption, usually stemming from faulty equipment or cabling.

5.1.2 Switching Versus Segmenting

Moving servers and users to switched connections, versus segmenting through the addition of hubs, enables each user to have more bandwidth through dedicated physical media. Hubs are still a good, cost-effective way of linking different hosts. However, in large heavily loaded networks, moving to a switched environment can reduce the effects of collisions and avoid some of the transmission latency associated with hubs. Figure 5.1 illustrates the differences between a LAN using a hub versus a switch [1].

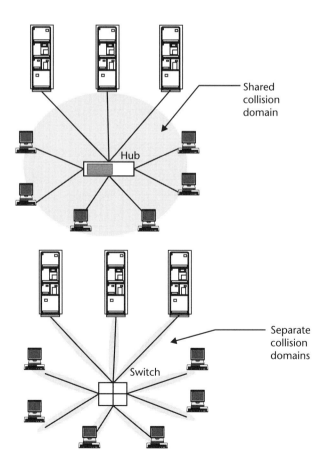

Figure 5.1 Shared versus switched LANs.

Layer 2 switching can cause added complexity to network troubleshooting and fault isolation. Protocol analyzers and tools typically can only view traffic on a single physical media, such as a switch port. Many Ethernet switches have monitoring capabilities built into each port, which makes it possible to view utilization levels, errors, and multicast properties of the traffic. Some products can capture full-duplex traffic at line speeds. *Port mirroring* is a technique where the traffic on one port can be duplicated on an unused port to which a network-monitoring device is connected. Port mirroring can affect switch performance and quite often will not enable physical-layer problems to be reproduced at a mirrored port. Furthermore, full-duplex Ethernet often cannot be mirrored successfully. There are variants of port mirroring that mirror only the traffic between an ingress port and an egress port or that can mirror multiple ports to a single monitoring port.

5.1.3   Backbone Switching

As was stated earlier, the 80% to 20% ratio of internal-to-external traffic in a LAN is rapidly shifting in the reverse direction, affecting network backbone traffic. As in our discussion of tiered networks, backbones consist of a set of core switches tied together with single or multiple higher speed connections. Inefficient traffic patterns over a backbone can often lead to surprise surges in bandwidth utilization. Much care should be given to constructing backbones and assigning traffic streams to backbone transport. Gigabit Ethernet links between switches should stay under 15% utilization and not exceed 25%. Higher utilization levels increase the potential for collisions.

Layer 3 switches should be used in locations where there is a concentration of traffic, such as in front of server farms, or in place of routers where uplinks to a WAN or the Internet are required. Routers have a higher per-port cost than switches and must perform route calculations within software, which can consume central processing unit (CPU) and memory resources. They can often present bottlenecks for large complex networks. Many LAN topologies use layer 2 switches in the lowest network tier and use layer 3 switches in the remaining upper tiers. Although layer 2 switches could be used in the next tier up from the lowest, layer 3 switches can provide better utilization and load sharing over parallel links. Figure 5.2 illustrates these concepts.

As shown in Figure 5.2, links stemming from the middle tier to top layer 3 tier would be switched at layer 2. However, the spanning tree algorithm prevents using parallel paths from each layer 2 switch to redundant layer 3 switches. As layer 2 uses the spanning tree protocol to discourage traffic to redundant links in order to avoid looping of frames, the redundant devices may end up being underutilized. Layer 3 or multilayer switches should be considered in the middle tier to reroute traffic versus using redundant layer 2 links.

Asynchronous transfer mode (ATM) and Gigabit Ethernet are popular backbone layer 2 technologies. Although ATM has inherent quality of service (QoS) capabilities, ATM has been known to have more management complexity and does not offer the plug-and-play characteristics of Ethernet. Furthermore, Gigabit Ethernet can interwork naturally with an existing Ethernet LAN versus ATM.

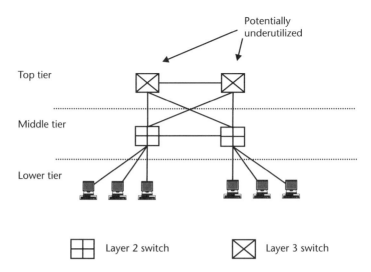

Figure 5.2   Layer 2 and layer 3 networks.

5.1.4   Link Redundancy

Multiple links between switches devices can ensure redundancy in the event a switch link fails. If possible, the primary and backup links should be used simultaneously through load sharing to avoid having an idle link. Load sharing is not typically found in traditional layer 2 switches, but newer devices are beginning to incorporate this capability. Nevertheless, a hardware-based restoration should switch immediately from a failed link to a good link, without loss of the session. A software-based solution, such as that found in server switches, could be used not only to load share traffic, but can also restore the failed links [2].

5.1.5   Multilayer LAN Switching

Multilayer switches consist of a switch with a layer 3 routing functionality installed. When a layer 2 frame is received, it is forwarded to a route processor. The route processor determines where to forward the frame, based on the Internet protocol (IP) address. The router's MAC address is inserted in the frame as the source address and the frame is sent to its destination. All future frames are then forwarded accordingly, without having to query the route processor again.

Multilayer switching was designed to overcome some of the problems associated with two-tier network design. For one thing, the routing lookup is conducted only once by the route processor. Routing decisions are made using application-specific integrated circuits (ASICs) instead of software, providing significant performance improvement gains. Furthermore, multilayer switches offer a lower cost per-port than routers.

5.1.6   Virtual LANs

Virtual LANs (VLANs) arose out of the IEEE 802.1Q and 802.1p standards [3]. VLANs were intended as a way to simplify MAC address management by associating hosts in different subnets into virtual groups, enabling these devices to be deployed anywhere in a network. VLANs have evolved into a means of controlling network traffic by segmenting it. Traffic that is bursty, chatty, or streamed can be assigned to separate VLANs so that quality on other parts of the network is unaffected. VLAN membership can be identified within an Ethernet frame. The IEEE 802.1p standard allows traffic to be prioritized.

If VLANs are dispersed over too many devices, it could create undesired increases in backbone traffic, as in the case of a tiered network, and create complexity with respect to subnet configurations (see Figure 5.3) [4]. The security of data between different VLANs is not necessary ensured—data has been known to leak between different VLANs. With VLANs, the best policy is to try to keep VLANs on the same physical switch. Setting up a VLAN inside of a switch usually requires defining the VLAN on a port-by-port basis. This approach best works in a fixed environment where hosts always reside on the same port. Consequently, VLANs should be used mainly with static, versus dynamically assigned, IP addresses. In a dynamic environment, users are unlikely to retain the same IP address, making it difficult to define IP addressing rules.

5.1.7 Transceivers

Transceivers operate at the physical layer of the OSI model and are used in Ethernet networks to connect a device to a fiber or copper cable. *Redundant or fault-tolerant* transceivers can be used to create backup links between critical devices [5]. Redundant transceivers typically have three ports. One port links to the device node and the other two ports connect to the primary and secondary links across the network. If the primary link fails, the secondary port is automatically activated. Failover is typically within nanoseconds for Gigabit Ethernet and milliseconds for fast Ethernet. Upon restoration, the primary link is restored to operation.

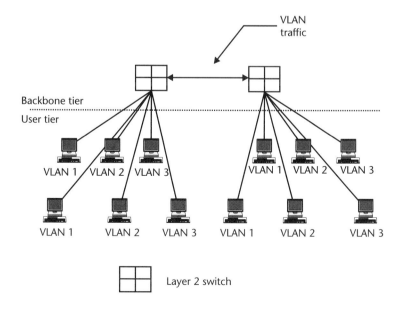

Figure 5.3 VLAN backbone traffic.

Using redundant transceivers can be a cost-effective option for establishing a redundant link versus doubling the number of network adapter cards. Not only do they not require configuration or additional software, but their installation involves minimal network disruption. If multiple ports require redundancy, multiple transceivers can be used (see Figure 5.4). Additional redundant paths can be created if the transceivers are configured back to back.

5.1.8   Media Translators

Media translators (or converters) are devices that are used to integrate fiber optics with Ethernet networks [6]. These devices are typically connected to a network interface card (NIC) in a server using either copper of fiber media. Copper-to-fiber media converters are often used to increase the distance of a copper link. They convert a signal from copper cable to a format suitable for fiber-optic cable. Translators can also be used to convert multimode fiber to single-mode fiber. They translate signals without regenerating or retiming them in any significant way. There is no IEEE standard for media translators, and for this reason they are often viewed as a crude means of extending the reach of a copper network.

However, media translators can be used to create redundant links between devices in Ethernet networks [7]. In fast Ethernet (100BaseTX) and Gigabit Ethernet (1000Base TX) networks, copper-to-fiber media translators can be used to establish redundant paths between core backbone switches (see Figure 5.5). Some devices even duplicate layer 2 for extra reliability. These devices monitor the primary link and upon failure automatically redirect traffic to the secondary link, virtually instantaneously. However, such translators have been known to improperly default to half-duplex operation when full duplex is required in links using autonegotiation.

5.1.9   Network Adapter Techniques

Network adapter cards, also referred to as NICs, can be used to connect a host device directly to another device, such as a hub or switch. Because NICs are also known to fail, a second NIC can serve as a backup. As in the case of the transceiver, a redundant link is established, but this time redundancy is established at layer 2.

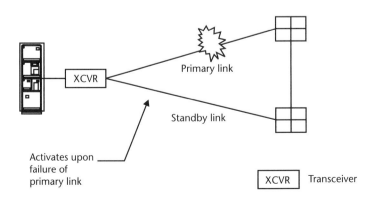

Figure 5.4   Use of transceiver for link redundancy.

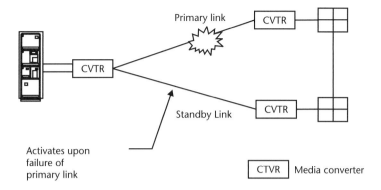

Figure 5.5   Use of media converter for redundancy.

Use of a redundant or multiport NIC is often accompanied by special software that allows the backup NIC or NIC port to take effect if the primary link fails. In clusters, multiple NICs are grouped into sets so that if one fails, another one in the group takes over. This reduces the need for application failover in the event of minor network-related problems, reducing cluster disruption. Figure 5.6 illustrates NIC failover. Failover software typically incorporates several features, such as binding a single network address to multiple NICs, load balancing across multiple NICs, and using only the active connections to a switch for reliability and better performance [8]. Failover times can be slow, in the order of one to six seconds.

Because the use of an additional NIC can consume a slot on the server, multiport NICs can be used in these situations to conserve slot usage. Multiport NICs, in general, can mask multiple MAC addresses into one, avoiding the need to recalculate routes. They can also increase throughput through *link aggregation*, which involves aggregating multiple ports from a single adapter, resulting in greater throughput while conserving bus slots and decreasing bus usage [9]. Furthermore, network links can be created without using additional bus slots by connecting multiple NIC ports on the same NIC to different network locations.

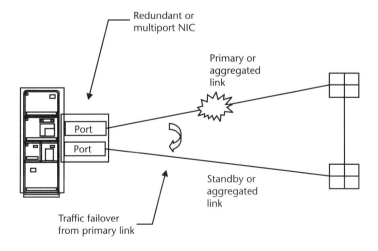

Figure 5.6   Use of NICs for redundancy.

NICs are considered single points of failure and can create bottlenecks when many users are accessing a host. This is especially true in the case of a 10/100 NIC connected to a Gigabit Ethernet network. Using faster NICs, such as gigabit NICs, can improve performance. Using autonegotiating 10/100/1,000-gigabit NICs can provide the additional advantage of deploying 1000BaseT incrementally in a network. As network speeds grow, the more susceptible the network becomes to cabling and connection problems. This has placed tighter operational tolerances on NICs. NICs have been known to bring networks down, sometimes broadcasting erroneous frames throughout a network. Server-class NICs are available that have greater reliability features with on-board memory and processing that offloads transmission control protocol/IP (TCP/IP) functions from the host server.

Duplex mismatch between NIC devices is a frequent problem in Ethernet networks [10], one commonly overlooked in troubleshooting. Collisions can occur on links that operate at half duplex, as only one end of the link can transmit at a time. Because full-duplex links allow each end to transmit at the same time, collisions are avoided and full link utilization can be achieved. If NIC duplex settings at each end do not match, or if they are both set for autodetection, incorrect duplex could be assumed by the devices. This could result in using the link at only half duplex and losing over half of the link's available capacity, versus using it at full duplex. In the end, inspection of all Ethernet links for consistency in duplex and speed detection is almost mandatory for any network installation.

5.1.10 Dynamic Hierarchical Configuration Protocol

Dynamic hierarchical configuration protocol (DHCP) is a layer 3 protocol that dynamically assigns (or leases) an IP address with the MAC address of a network device. Most devices on LANs today use IP addresses to communicate. Unless fixed IP addressing is used, loss of the DHCP process or the server that provides that service to a network could be disastrous. Because an IP address can be leased for up to 12 days, the DHCP protocol verifies host connection with the DHCP server every 6 days. If a DHCP server fails, there is at least 6 days to restore the DHCP service. Otherwise, IP addresses for new users will be affected.

Deploying a second DHCP server can resolve this problem [11]. The redundant server should assign addresses that do not overlap with those leased by the primary server, so that any new users to the network would still receive IP addresses while the primary server is down. Although dedicating the redundant DHCP server solely for this task could seem wasteful, there is no way to figure out which stations have leases from what servers at any given time.

Request For Comment (RFC) 2131 is a draft standard that allows multiple DHCP servers to assign leases for the same IP address space (sometimes referred to as a *scope*). In the case of two servers, each server can lease addresses while sharing lease information with each other. Each server has its own pool of IP addresses, but the scope information must be provisioned manually with each server. Changes to the lease scopes must be synchronized manually between the two machines, which could be an arduous task in large networks. Each server also monitors each other's heartbeat. If the primary server fails, it is important to be sure that the packets of new users joining the network are forwarded to the secondary server. Because RFC 2131 is a draft, it has yet to be standardized as of this writing. In the meantime, vendor-specific implementations are available that could be used to improve DHCP survivability.

5.2   Wide Area Networks

WANs transport data traffic between LANs and other adjoining WANs. WANs typically constitute the highest backbone tier in a tiered topology. Because WANs have no geographical boundaries, they can transcend many countries and connect large organizations or groups of organizations. Poor WAN performance or outages can have far-reaching effects. Duplicating WAN links between sites not only adds reliability but also improves performance if the links are load shared.

5.2.1   WAN Technologies

There are a variety of WAN technologies in use today. The following sections review some of the most popular WAN technologies with respect to some of their survivability features.

5.2.1.1   Frame Relay

Frame relay is a layer 2 connection-oriented protocol designed for transmitting intermittent data traffic between LANs or adjoining WANs. *Connection-oriented* services typically establish logical links between hosts. *Connectionless-oriented* services, on the other hand, depend on best-effort delivery of individual packets or messages between hosts. Frame relay transmits data in variable-size frames. Error correction and retransmission of data is left up to the end points, enabling faster transmission. Customers see a continuous, dedicated connection called a permanent virtual circuit (PVC). Customers do not have to pay for a dedicated leased line—the service provider determines how each frame travels to its destination and charges the client only for the access circuit and a committed level of bandwidth.

Redundancy can be introduced using backup PVCs, switched virtual circuits (SVCs), integrated services digital network (ISDN) failover, or any equivalent, temporary connection [12]. ISDN can provide near-equivalent switched service to what most companies buy in terms of frame relay bandwidth. Using an alternative carrier service or even V.90 modems operating multilink with equivalent bandwidth are reasonable alternative solutions to WAN outages.

Traffic can be rerouted to a backup PVC if congestion or a failure occurs. PVCs and SVCs can also be configured in mesh topologies, creating greater redundancy. Installing backup PVCs on two predefined paths between two locations on a network can avoid outage in the event a network path fails (see Figure 5.7) [13]. Key to successfully implementing this redundancy is the ability to detect failover and automatically remap to correct data link connection identifiers (DLCIs) for rerouting over the other path. Traffic en route over the failed link or at the failed node could be lost during the failover process [14].

There is much debate over the use of backup PVCs. Many firms regard backup PVCs as poor value and only want to pay for them when they are used. Because most WAN providers already embed network redundancy in their networks, many

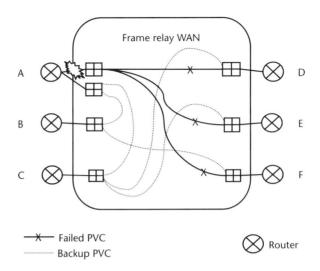

Figure 5.7 Use of backup frame relay links.

companies feel that their single PVCs can carry traffic quite reliably, as long as their traffic stays within the bounds of the contracted rate. Furthermore, complete frame relay redundancy cannot be achieved using the services of only one carrier. If a carrier's backbone fails, redundant paths are of no value. In the end, the most critical links that do require redundancy will be the access links that connect a customer premise to the carrier's network.

When planning a frame relay WAN, historical data from service providers and their equipment vendors should be reviewed for outage history. The following are some known problems to occur in frame relay networks:

- A traffic surge in a frame relay network can cause a congested state. When this happens, frame relay's link integrity verification protocol local management interface (LMI) frames are issued, known as forward explicit congestion notification (FECN) and backward explicit congestion notification (BECN). These notify users of the congestion and that their traffic will be discarded unless flow control is applied.

- Physical connectivity problems can cause the LMI to fail due to improper handshake between the frame relay switch and the customer premise equipment (CPE), causing all connections using the interface to also fail.

- Frame relay links can fail due to problems within the network. These conditions are conveyed through messages sent by the frame relay switch to the CPE indicating the status of each DLCI that uses the link as either active or inactive. An inactive DLCI can be caused by either a remote CPE or network problem. The messages typically do not indicate the cause of the problem.

Carriers will also have difficulty tracing errors when there is a mix of different customer WAN and public network traffic across their network. It is imperative that when planning the use of frame relay WAN services, information should be obtained regarding the carrier's network management practices, the mix of services over their network, ways that levels of service are managed, and ways that traffic is measured. Sampling traffic over long time intervals and averaging them over 15-min intervals can often mask traffic peakedness and make the traffic look smoother than it really is.

5.2.1.2 ATM

ATM is a layer 2 connection-oriented protocol, but unlike frame relay it provides higher bandwidth and better tracking of connection availability. ATM transmits data in 53-byte units called *cells*. ATM uses rerouting for survivability. Before rerouting takes place, the source node is notified of the network failure. ATM can use two reroute mechanisms. A centralized reroute mechanism called *global repair* sends the failure notification throughout the network. This mechanism can cause undue loss of cells due to the time required to propagate a failure notification through the network. On the other hand, a decentralized mechanism called *local repair* attempts to fix the problem within a very short distance of the failure, reducing cell loss. Figure 5.8 illustrates these two strategies. Some of the problems and issues that can occur in ATM networks include:

- ATM rerouting requires that protection capacity be provisioned on alternate routes. Insufficient protection capacity can lead to preemption of traffic and increased restoration time.
- Layer 2 failures can occur if the network cannot distinguish cells due to transmission problems or if a link is undergoing BER testing.
- Network and CPE failures are conveyed through network management cells. Receipt of alarm indication signal (AIS) and remote defect indicator (RDI) cells convey faults on along the path of an ATM virtual circuit (VC). Receipt of these cells at both VC endpoints indicates a network problem.
- ATM is known for its *cell tax*, which is a 20% increase in segmentation and reassembly (SAR) overhead that occurs when an ATM switch divides up large IP packets into fixed-length ATM cells.

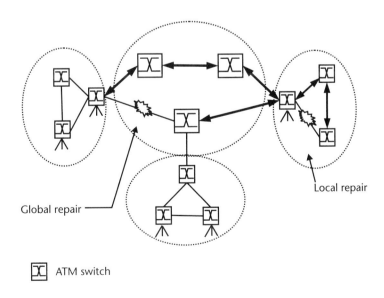

Global repair

Local repair

$\boxed{)(}$ ATM switch

Figure 5.8 ATM reroute strategies.

The last item highlights some of the problems experienced with IP routing over ATM VCs in large networks. Often, a full mesh of VCs must be used to connect routers for better performance. As in the case of frame relay, having an alternate carrier service or connection using ISDN can help circumvent congestion or failures.

5.2.1.3 SONET

When compared to copper, optical signals are not subject to the same electromagnetic interference problems and induced signal issues, such as cross talk, where a signal on one cable induces a signal on the other. In optical networks, maintaining the signal's strength over a continuous, connected link is the main concern. SONET is an American National Standards Institute (ANSI) standard designed for fiber-optic network transmission. SONET is a layer 1 protocol that is the fiber-optic equivalent of the synchronous digital hierarchy (SDH). SONET can be configured in a linear or ring network topology. Linear topologies amount to what is called a collapsed ring, meaning that although the topology physically looks like a string, it is logically configured as a ring. As discussed earlier, ring networks have distinct advantages over linear networks, but cost more to deploy, especially for small numbers of nodes in close geographic proximity.

SONET was developed in the late 1980s and gained popularity in early 1990s in fiber-optic telecommunication networks. This popularity centered on SONET's use of the ring network topology to provide "self-healing" capabilities. SONET/SDH networks were originally designed to provide high-volume backbone transport of connection-oriented voice over long distances. These backbones typically support time division multiplexed (TDM) voice trunking between central offices (COs) in different locations, either for local exchange carrier (LEC) or interexchange carrier (IXC) traffic. In TDM, circuits use a fixed amount of bandwidth whether or not traffic is present. For this reason, SONET/SDH switches, also referred to as add-drop multiplexers (ADMs), are designed so that all ADMs on a SONET/SDH ring operate at the same speed. ADMs typically aggregate traffic from subtending networks for transport over the ring. These networks could consist of other SONET rings as well.

For these reasons, provisioning circuits or upgrading capacity on a SONET ring is quite complex and resource intensive. Not only does it require concurrent upgrade of all ADMs on the ring, but it may require upgrades of some of the interconnected rings as well. Circuit endpoints must first be identified, and then each node on the ring must be configured to pass the circuit through it. Newer SONET provisioning systems do automate and ease some the provisioning complexity. SONET ring networks do require manual engineering to manage traffic and bandwidth utilization. This requires knowing the path and bandwidth requirements of every circuit on the ring. In the end, SONET's inherent design and provisioning complexity are not optimized to meet the growing demand of connectionless IP and Ethernet traffic.

The most positive feature of SONET rings is their restoration ability. In a typical SONET ring, a traffic stream can travel in both a clockwise or counterclockwise direction. One direction is called the *working path* and the other is called the *protection path*. If a link between two nodes (ADMs) on the ring is broken due to a fiber cut or inoperative node, SONET can detect the discontinuity typically within 10 ms. Typically, SONET/SDH frames key off of overhead bytes or calculate BER of the signal payload to identify link failures. All traffic on the ring can be rerouted within

50 ms. This mechanism is referred to as automatic protection switching (APS). The 50-ms interval represents the maximum restoration time that would not disconnect a voice call [15]. Unfortunately, data traffic is susceptible to disruptions of smaller time intervals. When APS takes place, the ring is said to be in a *foldback* state.

APS is illustrated in Figure 5.9 for two SONET ring configurations: bidirectional line switched ring (BLSR) and unidirectional path switched ring (UPSR) configurations [16]. Both can invoke APS in less than 50 ms. In BLSR configurations, APS loopback is performed at the two adjacent ring nodes connected to the disconnected fiber. BLSR configurations require working and protection capacity at every node on the ring, whether or not traffic is present at that node. UPSR configurations are best suited in access networks, where traffic typically takes on a star or hub pattern. In a UPSR configuration, traffic is bridged onto the working and protection paths at the transmitting node and the APS decision is made at the terminating node. UPSR and two-fiber BLSR rings both reserve twice the needed bandwidth, while four-fiber BLSR are inflexible and double the amount of fiber required. BLSR has an advantage over UPSR in that unused timeslots can be used in different portions of a ring when traffic patterns are distributed or meshed.

Assuring successful rerouting during foldback requires utilizing protection paths for each circuit's working path. Several schemes are used and are illustrated in Figure 5.10. *Dedicated protection* schemes guarantee reroute capacity for a given circuit [17]. In this scheme, a traffic stream is transmitted on both the working and a

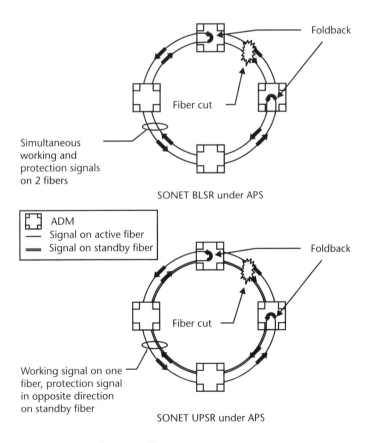

Figure 5.9 SONET APS protection scenarios.

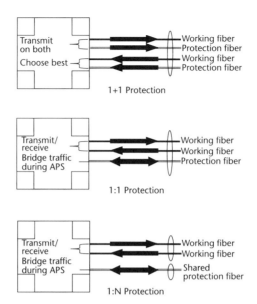

Figure 5.10 SONET protection schemes.

dedicated protection channel at the same time. In a *single ended* or *1 + 1 dedicated scheme*, traffic at the originating node is always transmitted over the two channels. The receiving node selects one of the signals based on purely local information, whether or not the ring is in foldback. In a *dual-ended* or *1:1 dedicated scheme*, the receiving node uses the APS protocol to reconnect working-path traffic on to the protection path.

Dedicated protection schemes can make inefficient use of bandwidth, actively using only half of the available capacity of a ring and adding to bandwidth cost, particularly for distributed traffic patterns. An alternative is a *shared protection* or *1:N* scheme, which offers better bandwidth utilization. In this scheme, one protection channel is shared among *N* working channels. For example, a protection path can be assigned to protect five working paths, assuming that only one working path is likely to fail at one instant. In the event of a failure, APS reconfigures the rest of the ring so that all of the traffic is once again protected.

In SONET, protection mechanisms can be enabled or disabled on a per-channel basis, enabling protection on an individual service basis. This feature enables different services to share the same SONET ring. Most SONET rings are deployed by telecommunication carriers to transport different services for their commercial subscribers. Enterprise system connection (ESCON), fiber channel, Gigabit Ethernet, ATM, frame relay, and other protocols can be carried over SONET networks. As of this writing, SONET networks operating at optical carrier (OC)-3 (155 Mbps), OC-12 (622 Mbps), OC-48 (2.488 Gbps), and OC-192 (9.953 Gbps) rates have been deployed. If an organization is using services that are deployed over a SONET network, they should be aware of the following:

- The SONET/SDH hierarchy is optimized for large-volume long-haul data transport. Depending on how the service is used, it may be less cost effective for localized transport of traffic generated at customer premises.

- SONET requires the determination of the maximum bandwidth required for traffic. A general rule is to add 30% of the required bandwidth for contingency.

- Ethernet traffic does not gracefully map onto the SONET/SDH hierarchy. In most cases, different vendors use different approaches to map such traffic. Consequently, organizations are often confined to using a single vendor's equipment in a ring.

- APS assumes that both the working and protection paths are routed on two diverse fiber paths between ADMs, so that the same failure does not affect both paths. Because carriers have been known to overlook this, knowledge should be obtained of the geographic fiber routes over which circuits are mapped.

- Optical layer survivability does not guarantee survivability at the IP service layer. For Internet networking, the switches at the edge network and access links can still represent points of failure. In the end, for IP routing, all layer 2 networks that support the path of an IP packet should be survivable.

5.2.1.4   WDM

Wave division multiplexing (WDM) systems can convey a signal over a particular wavelength (or color) of light traveling through a fiber [18]. Because light is comprised of many wavelengths, this means that many signals can be transmitted over one light path, unlike a traditional SONET system, which transmits signal using the entire light path of a fiber. WDM in essence multiplies the capacity of a single fiber pair, optimizing the use of fiber in a network. Dense WDM (DWDM) can multiply the capacity of a single fiber even further. At the outset, DWDM is most cost effective when used to offset installation costs of new fiber to increase capacity for high-volume transport.

Use of WDM and DWDM further amplify the need for fiber restoration, as loss of such high-volume traffic links can be catastrophic. As of this writing, many current systems do not yet perform protection or restoration. Efforts are underway to transition traditional SONET networks to WDM optical transport networks (OTNs). Such networks would integrate transport, multiplexing, routing, supervision, and survivability at the optical layer. Instead of on a fiber basis, these networks employ SONET on a wavelength basis, unaware of the underlying WDM layer.

Protection can be accomplished in OTNs in several ways. One method is illustrated in Figure 5.11. DWDM transponder units can send and receive signals from SONET ADMs as well as other types of systems [19]. The transponder units transfer signals into a wavelength for transport over DWDM links. To provide protection at the wavelength level, all wavelengths are routed simultaneously in both directions around the ring. The protection switching occurs at the end nodes of the affected optical channel upon a fiber cut. Each node transmits and receives the signal from the opposite direction.

Efforts are underway to map fast and Gigabit Ethernet directly onto optical wavelengths, avoiding the SONET/SDH layer altogether. This would simplify network architecture and recovery mechanisms. In fact, trends are moving towards

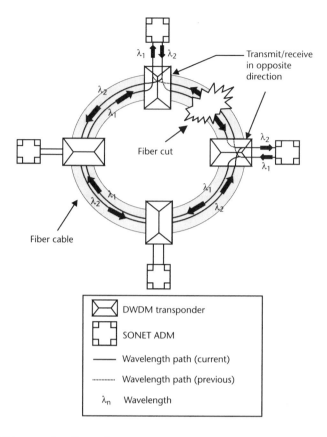

Figure 5.11   OTN protection illustration.

natively mapping IP directly to wavelengths in complete optically switched and cross-connected networks that avoid inefficient conversion to electrical signals.

5.2.2   Routing Methods

As stated earlier, routers are layer 3 devices that are charged with delivering packets of information between layer 2 networks. Traditional routing is done by simply reading the destination address of an IP packet and looking up an appropriate route in routing tables. IP is the standard that routers use to transmit packets and communicate with each other. The current version of IP, version 4 (IPv4), is soon to be replaced by version 6 (IPv6) as of this writing. IPv6 has several major improvements over IPv4. The expanded address space of IPv6 allows trillions of hosts to be connected to the Internet. It also has provisions to ensure that voice and video content can be streamed reliably through networks. But with these improvements, there is still no built-in flow-control or congestion-control mechanisms because these are deployed in layer 4, as in the case of TCP in data networks.

Each router must populate its routing tables using information that characterizes possible routes based on the current state of the network. Within and around IP are many routing protocols to convey such information. Each is designed to incorporate the intelligence to handle specific situations with capabilities that can be used for survivability. There are two classes of routing protocols. One class consists of interior gateway protocols (IGPs) that usually operate within a given network. IGP is a protocol used to exchange routing information between gateways routers within an autonomous network. There are a variety of IGPs used, including the routing information protocol (RIP) and the OSPF protocol. Interior gateway routing protocol (IGRP) and enhanced IGRP (EIGRP) are proprietary IGPs developed by Cisco designed to work on Cisco devices.

The other class is known as exterior gateway protocols (EGPs), which operate between two different networks. They are usually found in routers residing on a boundary between two networks or autonomous systems. They are typically used to route traffic between service provider and customer networks, or across service provider networks. Border gateway protocol (BGP) is the most commonly used EGP, and can be used as an IGP as well.

For survivability and performance, many enterprises multihome to two or more network service providers for either WAN or Internet services. In this environment, there is a need to monitor performance of destinations in another network so that data packets are sent to those destinations providing the greatest assurance of reaching them. In large networks, it is often difficult to control the distribution of traffic to different routes to destinations using an IGP. Although static routes can be defined in Internet access routers to use in case an Internet access link to an ISP fails, it is often difficult to get around failures within the ISP's own network.

Intelligent route control (IRC) is a capability that is being embedded in many routing protocol products to avoid many such situations [20]. They continuously monitor traffic flows and use different schemes to compute statistics that characterize the status of a destination network. These statistics are then used to make routing decisions to those destinations, based on predetermined policies. If a destination's performance does not meet the policy requirements, alternate routes are determined that satisfy those requirements. By using this approach, routing can be better controlled and managed. *Constraint-based routing* (CBR) is an approach that computes routes subject to constraints such as bandwidth or administrative policy. CBR considers more than just network topology to define routes and often can direct traffic on longer but more lightly loaded paths.

5.2.2.1 RIP

Adjacent routers in a network will update each other with information about changes in routes. As the number of routers in a network grows, so does the increase in routing information. (In very large networks, the traffic generated by such information transfers can be significant and affect router performance). RIP computes the best route in a network to use for a given packet and can convey that information to other routers. Routes are compared based on a distance measure called *hop count*, which is the number of routers passed to get to a destination (Figure 5.12). RIP forwards IP packets on the route with the lowest hop count and will broadcast the entire routing table to all routers every 30 sec to keep them informed. For this reason, RIP works best in small networks where routes do not change that often and is inefficient when used in large networks. RIP will discard packets it receives that have traveled through 15 routers (i.e., 15 hops). It will only support one path for the same route. It is also prone to routing loops, which means that a packet can be translated to a route that it has already traversed.

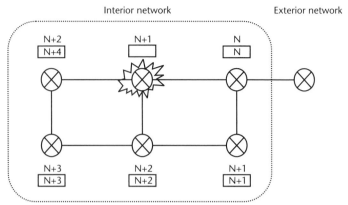

Figure 5.12 RIP example.

5.2.2.2 OSPF

OSPF is an IGP that is favored over RIP. Unlike RIP, OSPF computes a cost metric for each route, which takes link bandwidth into consideration. When a link's status changes, a router using OSPF will broadcast the change to its adjacent neighbor routers (Figure 5.13) [21]. Recent changes in link states are also issued every 30 min. The time required to recalculate the characteristics of each route, often referred to as convergence, is less than that of RIP. When employing OSPF for mission-critical networking, the following features are worth noting:

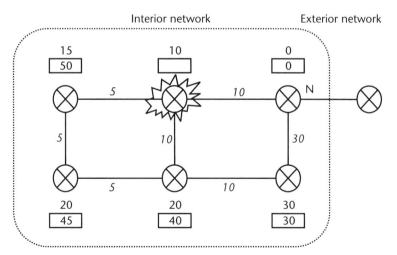

Figure 5.13 OSPF example.

- Because OSPF uses a shortest-path tree to compute routes, it is not prone to loops and can identify multiple paths for a destination, even those with equal costs. OSPF is finding popularity in newly deployed networks, particularly those that have multiple ingress or egress paths.

- Because only route changes are broadcast, the amount of information issued is far less than RIP, making it well suited for large networks.

- If an OSPF router in the network fails, an update is sent to every other router on the network using a multicast address.

5.2.2.3   BGP

BGP is a popular routing protocol used to link networks to one another on the Internet. BGP can operate on both an interior (IGP) device (referred to as IBGP) or on exterior facing devices, as in the case of EGP (referred to as EBGP). EBGP is used to multihome to multiple service provider networks and to provide failover in the event a provider's network experiences problems [22].

BGP enables networks to advertise reachability information to each other for specific IP address prefixes [23]. A loss of route to a specific prefix is detected through the timing out of periodic *keep alive* packets that are sent to a neighboring BGP router. Lost routes are removed from the router's routing table and the secondary routes are used in place. The neighboring routers issue messages to other BGP routers, reporting that the destinations that use the failed router are no longer available. Such messages can propagate to routers across the entire Internet so that each knows which prefixes are still reachable to its connecting network. When the router recovers, this process is repeated. BGP can dynamically characterize alternate routes using a weight metric per route or simply a shortest path. BGP capabilities are illustrated in Figure 5.14. When employing BGP for mission-critical networking, the following precautions should be noted:

- In BGP, traffic is favored on paths through service provider networks showing the best computed path, typically without regard to bandwidth or performance costs. If a link connecting to a service provider's network undergoes congestion, there is no mechanism to overflow traffic to another link connecting to that provider's network. Streaming or other bandwidth-intensive services over such links can be negatively affected.

- When a router goes down and then recovers, it is advertised to every other router. This can create an oscillation in routing advertisements, known as *route flaps* causing routing recompilation throughout the entire network. Loss of synchronized routing information can result between the failed router and other routers. To reduce the effect, BGP uses a mechanism called *route flap dampening* to suppress problematic routes that can result from the flap. This occurs when a recovering router has not rebuilt its routing table and continues routing packets based on information it had prior to the failure [24]. This can create erroneous routes; sometimes they contain forwarding loops whereby an adjacent router continues to route packets to a failed router, bouncing them back and forth across the link.

- Use of BGP can be quite complex and is best suited for organizations with a critical need for multihomed protection at the IP layer. An organization must

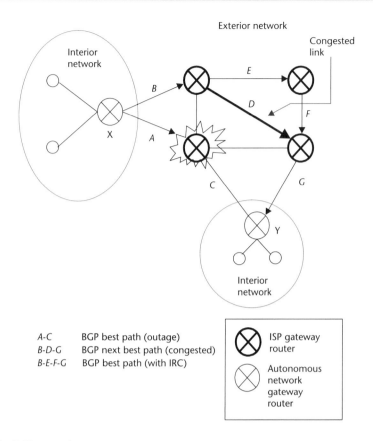

Figure 5.14 BGP example.

typically request that BGP route information be sent to their gateway router. Once in place, the router "officially" joins the Internet and can route all-important Internet traffic. With this responsibility, an organization should only advertise its own IP address block.

- BGP tables, which can include tables for the entire Internet, are stored in active memory. In multihomed situations, multiple copies of the table may be required. This places enormous memory requirements on routing devices.

In all, BGP is most effective when multihoming a router to multiple other routers. Because most Internet failures are those occurring in access links, the benefits of using BGP may be marginal. A simple static routing without any BGP information can sometimes result in a more simplistic approach.

5.2.2.4   HSRP and VRRP

Interior protocols such as OSPF and RIP can identify default gateways and circumvent gateway outages in layer 3 networks. When using routers in conjunction with a layer 2 LAN, any given host must have a primary static gateway router to which to send all IP traffic, representing a single point of failure. If that router fails, then all hosts must be reconfigured to direct their IP traffic to another router. It is often impractical to do this due to the amount of time and effort it would take. The hot standby routing protocol (HSRP), backed by Cisco, and the virtual router redundancy protocol (VRPP) are recently developed layer 3 protocols that are designed to enable the use of a backup router in these situations [25]. They create a virtual default gateway address that identifies a backup gateway router on the same LAN segment or VLAN. As of this writing, VRRP is undergoing standardization. HSRP is a Cisco-developed proprietary design that provides capabilities similar to VRRP.

VRRP enables a router to serve as a primary gateway for an IP subnet and a backup for another subnet having its own primary gateway. This is illustrated in Figure 5.15. The two routers can load-share, with each router acting as a redundant backup for the other. VRRP assigns a virtual MAC (VMAC) address to the primary router. By doing this, hosts do not need to change the default IP gateway address in the event the primary gateway router fails. The primary gateway router sends periodic polling messages to the secondary router. If message sending ceases, the backup router assumes that the primary routers is inactive and activates failover. Because the backup router assumes the VMAC, the failover is transparent to the host systems.

5.2.2.5   Route Optimization Techniques

Multihoming, although effective as a survivability mechanism, results in much unused link capacity to the secondary router. Route optimization products are designed to bring IRC and CBR capabilities within the direct control of an enterprise. They try to utilize idle access link capacity by monitoring information from a variety of sources, including the Internet, on the performance and status of networks behind the access links, so that the right access paths are chosen for outbound traffic. It also allows an organization to assign different types of traffic to links depending on their transmission needs (as in the case of streamed services), Internet service provider (ISP) traffic status, or even cost. Products are now coming to market having such capabilities.

Another approach to route optimization is through the use of routing service providers (RSPs) versus systems (see Figure 5.16). Providers are entering the marketplace with route-monitoring services that include views of Internet, local loop,

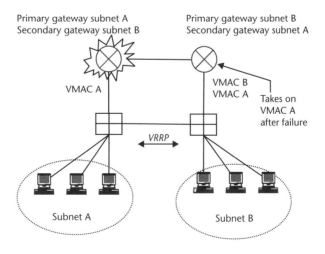

Figure 5.15   VRRP example.

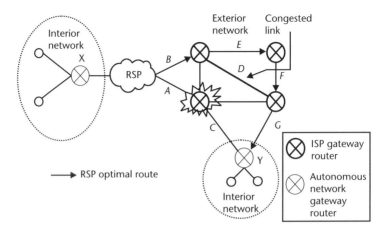

Figure 5.16 Route optimization example.

and CPE performance [26]. These services can make routing changes on behalf of a client organization or simply provide the information to the client's own network managers for their own purposes [27]. In the former case, an access link is required from the client to the RSP's point of presence (POP) over which traffic is routed. The RSP then routes this traffic accordingly to the ISPs. For redundancy, multihoming between the client and RSP is often required. Several factors need to be considered when choosing a route-optimization solution, whether it is a single system or RSP:

- The solution should provide flexibility in specifying how and where performance data is gathered. Some solutions offer proactive monitoring that anticipates where link congestion is likely to happen so that outbound traffic routing can be readily modified.
- The solution should provide enough response time to adequately route traffic to other links if congestion is detected.
- The solution should allow different policies or traffic rules to be used and should support all types of services implemented through data traffic, including virtual private network (VPN) and streamed traffic.
- Detailed reporting on current and projected performance for each access link should be made available.

CPE-based route optimization systems typically provide a more restricted view of Internet traffic status versus an RSP. Use of an RSP also avoids upfront system investment and periodic maintenance costs. However, use of single RSP can result in a single point of failure. Being that such services are relatively new, RSP solvency is usually a key concern.

5.2.3   Multilayer WAN Switching

Integrated switching is the unification of layer 3 routing and layer 2 switching in a single device. Multilayer switching was discussed earlier with regard to LAN environments, with focus on IP and Ethernet technologies. This same concept can be extended to WAN environments as well. Although multilayer switches have distinct advantages, they are plagued by the inherent characteristic of having too much capability in one box, particularly those products that combine layers 2, 3, and 4 in one capability. Many might find overreliance on a single box too confining, complicated, and hence risky.

A multiple-box approach might offer greater flexibility with respect to delivering survivability to a mission-critical network. It also enables a more fluid migration path for each layer's device. As technology evolves within each layer, devices can be upgraded to newer devices more transparently as opposed to a multilayer device.

As discussed earlier, many protocols have some inherent recovery and performance management capabilities. Interactions can sometimes conflict or work together to enhance reliability. Quite often, vendors create products that leverage these interactions in a multilevel device or protocol. The following describes some such capabilities that have gained industry recognition.

5.2.3.1 IP and ATM

Multilayer devices are available that combine IP routing with ATM switching. In many cases, these functions are implemented in hardware for speed and efficiency. Incoming IP packets are mapped onto an explicit ATM VC, creating an express lane for the traffic [28]. To conserve VCs, VCs are merged or pooled together for backbone transport based on traffic priority levels. The devices also allow implementation of both IP and ATM QoS capabilities (QoS is discussed later in this book). VCs can be can be segmented for different application performance priorities and can be assigned guaranteed or burstable bandwidth levels. Such features are targeted towards VPN services (discussed later in this chapter).

5.2.3.2 IP and Frame Relay

Multilayer IP and frame relay devices operate in similar fashion. These devices coordinate IP alternate routing with PVC failover in the event of an outage. Typically, when separate devices are in use, a router could immediately forward packets to an alternate path on a secondary PVC if a primary PVC fails. If the primary PVC recovers immediately, a route flap could occur, causing the router to go back to forwarding packets using the primary PVC. Frame relay is also prone to the *split horizon* problem. This happens when routing updates are received over PVC, but they cannot be forwarded over another PVC.

Multilayer devices have embedded mechanisms to deal with these situations. They can coordinate failover so that routing does not revert back to a primary PVC until a specified time interval has expired, allowing time for routing functions to synchronize. Figure 5.17 illustrates this mechanism. Some devices can be configured not to forward packets on a secondary PVC unless an LMI is received on the primary PVC. They can also be configured to work around split horizon problems.

5.2.3.3 ATM and SONET

ATM cells can be mapped on to plesiochronous digital hierarchy (PDH) circuits, such as digital signal (DS)-1 and DS-3, which are then multiplexed onto SONET/SDH rings. In this scenario, ATM switches are connected in a logical point-to-point fashion with links overlaid onto the physical SONET ring. Typically,

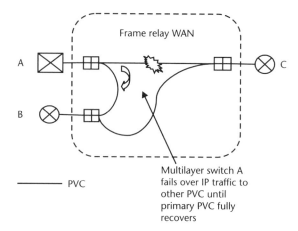

Figure 5.17   IP and frame relay coordinated failover.

ATM switches must rely on the SONET ADMs to guarantee 50-ms APS for all physical link failures. The problem with this scenario is that ATM layer problems cannot necessarily trigger SONET APS. Conversely, ATM switches are not made directly aware of a fiber cut if it happens. ATM networks have been known to malfunction during fiber cuts whereby switches continuously flood a network with broadcast messages regarding the failure of the associated trunks, causing *thrashing* of SONET ring [29].

To get around these types of issues, vendors have offered products with SONET functionality directly integrated into ATM switches. Such ATM switches can then be connected directly in a SONET ring topology, with each node on the ring being an ATM switch, as illustrated in Figure 5.18. This architecture provides additional

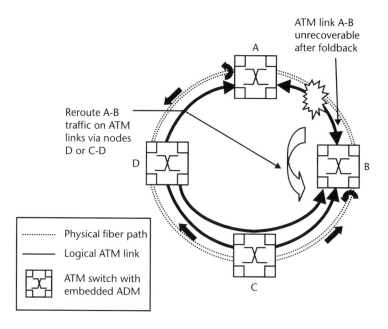

Figure 5.18   ATM and SONET coordinated failover.

benefits in terms of fiber utilization and bandwidth management. Because such ATM devices can cost more than a standard SONET ADM, this higher cost can off-set some of the savings resulting from gains in fiber utilization. SONET rings require excess bandwidth and are inflexible to reallocating bandwidth based on traffic needs. SONET ATM switches can overcome some of these limitations by statistically multiplexing ATM cell traffic on allocated ring bandwidth, while designating remaining bandwidth for circuit-based or non-ATM (e.g., IP and frame relay) traffic. If necessary, the circuit traffic can use standard SONET protection if needed.

5.2.3.4   MPLS

Mission-critical applications require that service survivability be guaranteed when needed. Multipath label switching (MPLS) is a packet-forwarding and routing scheme that attaches a small header (or label) to a packet [30] to convey a service-quality or handling policy for the packet. The header consists of a short, fixed-length value that is typically placed between the IP header and the link layer header. A device called a label switching router (LSR) examines only the label when forwarding the packet [31]. LSRs can make switching and routing decisions faster than conventional IP routers or ATM switches. They make forwarding decisions based on the policy represented by short labels, versus IP address lookups and variable length subnet masking, which involve considerably slower processing. Furthermore, MPLS uses the labels to recognize and rank individual traffic flows so that during failure or times of scarce bandwidth, more important flows can have priority over other flows. Flows are defined as packets traveling between common network endpoints. The ability to identify traffic flows through a network makes the flow deterministic so that it can be assigned to preengineered paths. Service providers are using MPLS not only to speed the transport of IP traffic through their core networks, but also to optimize bandwidth utilization.

MPLS identifies flows in terms of forwarding equivalence classes (FECs). Unlike traditional routing, where routing information can change dynamically via updates from other devices, MPLS calculates routes once for each flow (or FEC). An FEC can be defined by an IP header, the type of interface through which a packet is received, the packet type (multicast or unicast), or even other information describing the service that is involved.

Figure 5.19 illustrates the MPLS forwarding process. The ingress LSR, sometimes referred to as a label-edge router (LER), receives an IP packet, classifies it as a member of an FEC, determines the FEC's route, and prepends a label identifying the FEC to the packet. The next LSR that receives the packet uses the label as a lookup index in a forwarding table. Unlike a router, where the longest subnet mask *match* on an IP address must be performed to look up a route, the index simply points to the next LSR. The entry in the forwarding table indicates how the packet is to be processed. The label is then updated accordingly, and the packet is forwarded to the next LSR. The egress LSR (or LER) removes the MPLS label and extracts the IP packet. The path between an ingress LSR and an egress LSR is called a label switched path (LSP), also referred to as an MPLS *tunnel*.

An LER can use several protocols to define an LSR with adequate transmission capacity for the FEC. They include label distribution protocol (LDP), constraint-based routing LDP (CR-LDP), or the resource reservation protocol with traffic

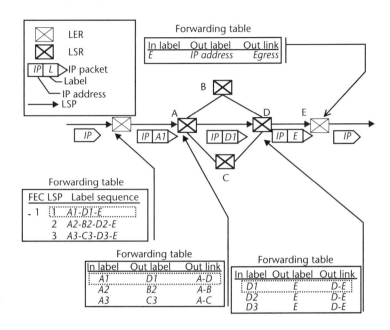

Figure 5.19 MPLS routing example.

engineering extensions (RSVP-TE). LDP and CR-LDP are standardized traffic engineering signaling protocols, transported over TCP, that allow the creation, teardown, and management of an LSP.

MPLS was not initially designed to establish or restore paths based on the type of physical (layer 1) connectivity. For this reason, several mechanisms have been developed to account for this. Generalized MPLS (GMPLS) was devised to enables MPLS to control routers, DWDM systems, ADMs, and other optical devices so that they can be provisioned dynamically in response to network conditions. This involves enhancement to LDP and CR-LDP to convey the type of underlying transport technology, so that LSPs can be defined on the basis of layer 1 and layer 2 attributes. MPLS also has mechanisms for restoring Ethernet in less than 50 ms.

Some routing protocols such as OSPF, IGP, and BGP, have been modified to identify packet routes using MPLS information. LSPs are taken into consideration to build routing tables and forward packets [32]. Frame relay service providers are also devising MPLS-enabled frame relay service to provide any-to-any connectivity for customers, versus the use of point-to-point PVCs. MPLS could provide a less expensive alternative to deploying multiple redundant PVCs between customer sites for reliability.

5.2.3.5   SNA

System network architecture (SNA) traffic is still alive and well in many IT environments. SNA traffic is designed to use connection-oriented transport and is sensitive to delay. If frames are received out of order or lost, SNA views this as an error and reinitializes them. Using a transport mechanism that guarantees packet order can get around this problem. However, many of the widely used WAN protocols such as IP or frame relay are connectionless and cannot guarantee packet order. Packets get disordered while in transit for many reasons, including routing changes, link outages, or equipment problems.

There are several approaches being used to address packet order issues. One approach is to use equipment that can detect disordered packets. However, reordering packets and resending them can cause excess delay or even timeouts. Another approach involves using special devices to separate SNA traffic from IP. Doing so can improve SNA and WAN performance as well as recovery time, but may not necessarily resolve the packet order issue. A popular approach is to identify or trap SNA traffic at the point of origination and route it accordingly. Some routing protocols such as IGRP have mechanisms that can identify SNA traffic, give it priority, and route it over a PVC that is separate from other traffic. Another approach that is currently under industry examination is to develop a standard that uses available bits within MPLS packets to identify the order of SNA packets. In the end, packet order is an issue that should be addressed in SNA implementations.

5.2.4   Design Strategies

Designing WAN connectivity for an enterprise requires consideration of a vast array of factors. PVC performance factors such as committed information rate (CIR) should align with premise LAN speeds to avoid impedance mismatch and having the PVCs become traffic bottlenecks. Topology is another primary concern when designing a mission-critical WAN. Dual connectivity is typically a must for critical WAN locations (Figure 5.20). Traffic should be balanced to and from each router and should be redirected to another link in the event a link or router fails. Each link should be provisioned over a separate physical facility on separate pathways from the premise to the service provider's POP. Although this is achievable through working with the provider, it can be expensive.

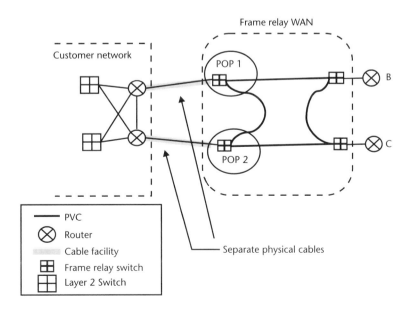

Figure 5.20   WAN access architecture example.

Some WAN providers offer redundancy on a usage basis, which can further reduce costs. When using a redundant WAN circuit or service, critical traffic should be given priority at the time the circuit or service is used. Some organizations even acquire redundant circuits to their most critical customers. The backup service should be established in advance because automatic activation upon outage of the primary WAN service can increase failover time. The time to failover to the other circuit or service should also be taken into consideration. Failover to another PVC, for instance, could require time just to activate the PVC.

An alternative used by many organizations involves use of an entirely different backup circuit or service, such as ISDN, satellite, or microwave [33]. Costs for this redundancy can be contained by using circuits or services that have reduced operating speeds and performance sufficient enough to carry the most critical traffic. Purchasing 0-CIR frame relay or available bit rate (ABR) ATM PVCs on a second WAN carrier backbone can help minimize costs. Some organizations actively use the secondary service for less critical traffic, such as intranet traffic or backhauled Internet traffic. When actively using the primary WAN services, enough bandwidth should be allocated for critical applications at key network locations. In frame relay, this means defining the appropriate CIR and maximum burst rates. Overreliance on TCP rate control should be avoided. Instead, traffic-shaping techniques can be used to ensure QoS for different traffic flows. These are discussed in the chapter on network management.

Care should be taken to assure that the circuit or service is provided over a physical cable and path that is totally diverse from that used by the WAN service. This diversification could even be extended to greater levels. Many carriers will share common backbone providers, fiber, or satellite links, or interconnect with each other at critical network locations, such as Internet network access points (NAPs). It is not uncommon for a major carrier outage to affect other carriers. Thus, when ordering a backup WAN service, it is important to know if there is any interconnection or common network sharing with the primary service provider.

Some WAN service providers offer a 99.99% uptime guarantee, which is almost an hour of downtime. Depending on how an organization values downtime, the question of whether redundancy is even necessary for WAN service is raised. If redundancy can improve availability to 99.999%, then downtime is reduced to about 5 min, which can be significant for those organizations that highly value downtime. In any case, the exact nature and meaning of the uptime guarantee should be carefully reviewed with the service provider. Furthermore, provisions should be made for strict provider penalties if the service guarantee is violated.

5.2.5 VPNs

VPNs are authenticated and encrypted encapsulations of data, sometimes referred to as *tunnels*, which are transported over a public or private network. An IP packet, for example, can be encrypted for message confidentiality and then placed inside of another packet or frame for network transport. Generically, encapsulation is a technique whereby special header information is placed around a data packet. IP packets are often encapsulated in frame relay or ATM WANs using headers, or even encapsulated within other IP packets for transport over the Internet. When security protocols like IPsec are used, encryption converts the encapsulated data from its original or *plain text* to a special encoding or *ciphertext*. This can be decrypted only by use of a decryption key that is associated with the encrypting key. As long as the encryption and decryption keys are kept secret, encrypted data can only be read by the peer key holders and thus can be kept confidential during transport, delivery, and storage. Note that use of MPLS in conjunction with VPNs has yet to be standardized. MPLS VPNs neither encrypt packets nor have them traverse the Internet. Instead, MPLS labels can be used in place of IP encapsulation or encryption to distinguish VPN traffic [34, 35].

Point-to-point tunneling protocol (PPTP), layer 2 tunneling protocol (L2TP), IP security (IPsec), and secure sockets layer (SSL) are the most widely used protocols for VPNs. L2TP and IPsec encapsulate and encrypt the IP packet, which contains the IP source and destination addresses. Header information is added, which is later stripped upon arrival to destination. IPSec can either encrypt the application headers, TCP/UDP headers, and data between a client and a server (called *transport* mode), or encapsulate and encrypt the entire packet (called *tunnel* mode) between two gateway devices. Because transport mode does not encrypt the IP headers, they are exposed and can pose security issues. PPTP is a Microsoft propriety protocol that does not encapsulate or encrypt the IP packet.

A VPN over the Internet can serve as an alternative or backup to a WAN, private line, or long-distance calling. It can also be used as a secondary approach to a storage area network (SAN) for data backup. The VPN can also be used to provide WAN access to locations that do not have a PVC or WAN connectivity. Many firms provide Internet access to their corporate sites through a star-like WAN architecture (Figure 5.21), with all sites connecting to one location in order to access the Internet. Accessing an Internet VPN intended as a WAN backup in this manner often defeats the purpose of having a backup VPN in the first place. Having only one Internet access point on the WAN could be problematic due to the potential bottleneck and single point of failure at that location. A better approach is to provide each office with its own Internet access. VPN software should be installed on only office gateways or routers. Adding redundant WAN links will further reduce bottlenecks and improve WAN survivability.

With mission-critical networks, a key concern is the ability of a VPN server to failover to another server without losing connections and data. In such an event, given that a failed server or gateway will not receive packets, packet transmissions to that device from other systems should be discouraged until failover has been completed. Conveying such information to other devices on the VPN may not be easily done and can take some time [36]. Furthermore, a recovering or backup server would have to reestablish VPN connections, which can consume extra time.

5.3 Metropolitan Area Networks

WANs are often used for interconnecting different customer locations, each with their own LAN, over great distances. WAN bandwidth typically can be deterministically characterized and engineered in service provider backbone networks. Physical, large WAN network topology does not change that frequently such that scalability becomes an issue. In fact, the general networking rule that traffic predictability increases with network size holds true. Enterprise WAN switches aggregate

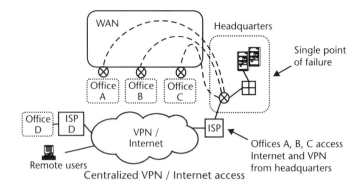

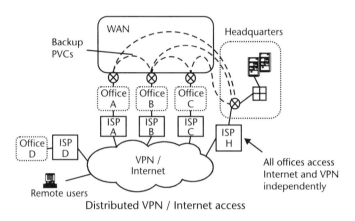

Figure 5.21   Using VPN as a WAN backup.

and consolidate many types of voice and data traffic. As network demands are aggregated among enterprise networks for long-haul transport, small perturbations in local traffic marginally affects the larger network. On the other hand, a WAN topology that supports more localized traffic is subject to greater shifts in topology and usage. WANs at this level are often referred to as metropolitan area networks (MANs). Figure 5.22 illustrates relationships between the WAN, MAN, and LAN.

MAN carriers have a greater need for scalability of their network infrastructure and are always seeking ways to optimize their networks and capacity. These networks, which like WANs reside over physical fiber-optic networks, require more resiliency to support a much wider base of customers, more customer churn, and greater service mix. As layer 2 frame relay or ATM services over SONET may not adequately provide the scalability and efficiency required for MAN operation, carriers are exploring alternatives such as Ethernet and resilient packet ring (RPR) technology. Implicit to these technologies is their ability to offer survivability features comparable to SONET.

5.3.1   Metro Ethernet

Ethernet MAN implementations are also referred to as metro Ethernet. Metro Ethernet service providers use a Gigabit Ethernet over a multiple fiber ring topology so that traffic can be rerouted if link congestion or an outage occurs. Multiple fiber

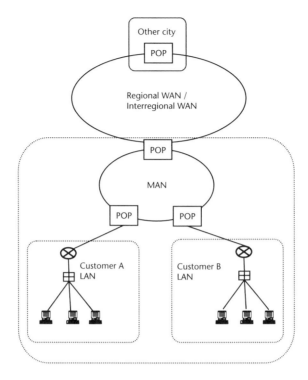

Figure 5.22   WAN, MAN, and LAN comparison.

rings can be defined around a given area so that traffic can route from one ring to another around a given failure (see Figure 5.23). Because it offers the same features of Ethernet, the network is much more resilient to fluctuation in bandwidth demand versus traditional SONET-based services. But like Ethernet, the network operates in a best-effort fashion, where traffic could be discarded during congestion. For networks that cater to multiple clients, each requiring guaranteed services, this can

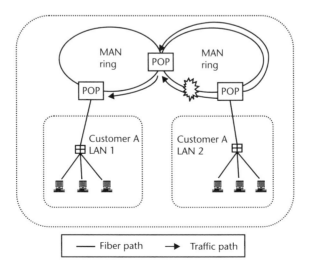

Figure 5.23   Metro Ethernet reroute.

become an issue. Although overprovisioning the network can compensate for this, a more desirable approach is to use traffic engineering and QoS techniques discussed further in this book to build suitably scaled networks.

High-performance switches and systems are being introduced to help alleviate network operations and management when dealing with such large Ethernet networks. Self-testing and self-configuring devices are being developed to assist in updating routing tables, managing fixed and dynamic IP addresses, and recovering from network failures. This last feature is often the subject of debate. A MAN network comprised of Ethernet switches consists of nodes connected by point-to-point links. Ethernet switches must typically route traffic on their immediate links, adding greater delay as traffic accumulates from link to link between intermediate nodes from a source to destination. Such delay can exceed 50 ms to the point where it can adversely impact voice and data transactions.

Some layer 1 technologies, particularly DWDM, are being leveraged to address some of these issues. A more direct, standards-based technique is Ethernet protection switching (EPS) [37]. It is designed to recover metro Ethernet services in 50 ms or less. EPS can work over copper or fiber on various network topologies. EPS uses IEEE 802Q VLAN protocol to segment traffic into flows called path groups. Path groups can contain multiple VLANs and travel in opposite directions on a primary physical path. A predefined protection path is assigned to each group to be used in case failure occurs. If a node detects a link failure, it automatically switches traffic to the assigned protection link.

Rapid ring spanning tree (RRST) is an enhancement to the spanning tree protocol so that ring topologies can be used with traditional Ethernet [38]. It uses rapid reconfiguration protocol, known as IEEE 802.1w, to quickly change between blocking and forwarding port states, resulting in improved recovery times. The protocol divides a large ring network into several smaller ring networks. A root bridge is established within each ring (using spanning tree protocol), which can block traffic, unlike traditional root bridges. It can change a blocking port to a forwarding port if necessary.

5.3.2 RPR

As realized through previous discussion, uses of traditional SONET and Ethernet as MAN technologies are not without their issues. SONET's stranded bandwidth and Ethernet's nondeterministic approach to traffic leaves carriers with less than optimal solutions to providing reliable services that can generate profit. One factor that is fundamental to this gap is the inability of SONET to recognize a layer 2 MAC protocol such as Ethernet. In an effort to fill this gap, a new MAC protocol designed for fiber-ring networks is undergoing standardization by the IEEE 802.17 working group. Called resilient packet ring (RPR), it provides a packet-based fiber optic ring solution with survivability comparable to SONET [39–41].

Many have described RPR as an Ethernet protocol designed for fiber rings, but it differs from Ethernet with respect to layer 2 MAC frame formats and control messages for topology discovery, link and node status, and protection and management. The RPR MAC layer is independent of layer 1 and layer 3 and is designed to handle multiservice networks provisioned over ring topologies. An RPR ring consists of at least four ring channels, each carrying traffic flowing in opposite directions. The ring is somewhat of a giant switch backplane, with each node acting as a port to the backplane. As traffic passes through each node, the MAC removes traffic destined for that node and adds traffic destined for other nodes. Traffic that is not destined for the node passes through [42].

Because of its packet-routing feature, an entire ring capacity can be used at any time without the need for reserved protection bandwidth. Traffic flows can be identified and assigned a quality of service to traverse the ring. Bandwidth is allocated in 1-Mbps increments with bursts up to the data rate of the ring. These parameters are controlled and managed by the carrier and can adapt quickly and easily to changing customer requirements. The MAC protocol regulates bandwidth sharing across the ring so that a single node is not at advantage or disadvantage. Furthermore, each MAC device at a node can convey the utilization on its immediate links to all other nodes on the ring, so that each can adjust the rate at which traffic is placed on the ring. If a certain threshold is exceeded, a node can request upstream nodes to reallocate bandwidth. It can also request transmitting nodes to increase their insertion rates if arriving frames are excessively delayed. RPR frames travel on the shortest path between nodes. Because nodes pass transit traffic through, scalability to higher transmission rates is easier.

A failure of a node or link can be recognized in less than 50 ms via communication among nodes. When this happens, traffic is simply rerouted to a secondary path or in a secondary direction at each node (Figure 5.24). Unlike SONET, where a working and protection ring is defined, RPR can route traffic on either ring, with either ring failing over to the other. RPR uses a default technique called *steering*, which notifies all nodes of the failure and routes traffic away from it. An optional approach is *packet wrapping*, where packets are routed in the opposite direction of travel upon a failure.

Because RPR is independent of layer 1, it can run over SONET or even Ethernet. In fact, RPR is designed to enhance these technologies rather than compete with them. RPR is backward compatible with SONET, enabling SONET payloads to be transmitted in a connectionless data fashion, while traditional TDM traffic is carried in a connection-oriented manner using standard SONET protection. Payload

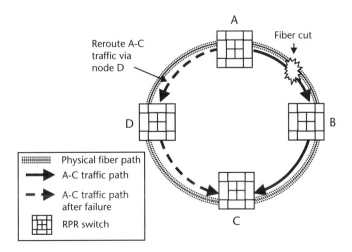

Figure 5.24   RPR traffic failover example.

frames are mapped into the RPR portion of the underlying SONET frame for transmission around the ring. This allows ring operators to migrate from a TDM to a packet-oriented environment while protecting the investment already made in the ring infrastructure.

Resilient packet transport (RPT) is a precursor to the RPR protocol. Viewed as a superset of RPR, it includes RPR capabilities with additional features that include synchronized, stratum-timed transport of TDM circuits and the ability to better support differential services (DiffServ), QoS, and MPLS. If operating over a SONET/SDH ring, it can acquire synchronization from the SONET/SDH framing. It can extend the number of SONET-supported ring nodes from 16 to 254.

5.4   Summary and Conclusions

Ethernet is the predominant layer 2 technology in LANs—popular for its low-cost ease of use. But because Ethernet functions on a best-effort basis, it is subject to frame collisions and discourages multiple links between nodes. It may not inherently guarantee the reliable performance and survivability required for mission-critical network operation.

For this reason, there are several work-around solutions that can be used. Network adapters have the ability to use backoff techniques to minimize collisions. Collisions can also be minimized using architectures that involve switches versus hubs, so that collision domains can be segmented. Transceivers, media converters, and multiple network adapters can be used to establish redundant links. Some newer switch products also have inherent capabilities to establish redundant links. For backbone networks, it is best to use layer 3 switches to provide better link utilization and load sharing across parallel paths. VLAN technology is intended to aid both user and traffic LAN management, but should be used discriminately to avoid creating unnecessary backbone traffic. Finally, because many LANs use dynamic IP addressing, DHCP service must be made survivable. This means using a secondary DHCP host for redundancy.

WANs transport data between LANs and other adjoining networks. The most popular WAN technologies include frame relay and ATM. In frame relay, survivability can be realized through the use of backup PVCs (which must be properly designed and specified), failover to another service such as ISDN, VPN, or a secondary WAN provider. Equal if not greater emphasis should be given to access link survivability. ATM has more inherent traffic-control mechanisms than frame relay, and can provide greater bandwidth. However, ATM is known for its complexity, expense, and management overhead.

SONET is a layer 1 protocol that was originally intended for long-haul transport of TDM circuits. It is has since become the underlying physical transport for most WAN services because of its self-healing ring feature. A SONET ring employs extra protective capacity that is not effectively used all the time. SONET rings are not easily scalable and require engineering of deterministic traffic patterns. For these reasons, they are most suitable for long-haul high volume traffic. WDM technology improves SONET economics by expanding ring capacity, as it operates at the wavelength versus fiber level. It also has similar self-healing capabilities at the wavelength level.

WANs mostly employ layer 3 IP-based routing. IGP and EGP are the two general classes of routing protocols in use. Some protocols handle survivability and redundancy better than others. OSPF is a popular IGP that is useful for networks with multiple ingress and egress points. BGP is a popular EGP that is effective for mulithomed routers. Router redundancy at edge locations can be achieved using the VRRP or HSRP protocols.

Because many enterprise networks access multiple carrier services, it is important to choose a protocol that has IRC capabilities. This provides the ability to identify problems en route to a far end network and to select the best routes to get there. Route-optimization techniques can be used to monitor networks behind access links so that the best access link is chosen. Approaches include using special CPE systems or the services of an RSA.

MPLS technology has been introduced to make routing over WANs more efficient, deterministic, and manageable. Multilayer switches have also become popular in the last several years because of their ability to consolidate routing and switching functions across several layers. Some examples include switches that combine IP with layer 2 functions found in frame relay, ATM, or Ethernet. There are also implementations of ATM/SONET multilayer switching. Yet having too much intelligence in a single box can be risky and impede scalability.

MAN technology is better suited for the less deterministic and lower traffic volumes found in regional networks. Unlike SONET, they can more cost-effectively provide the scalability required to address the traffic pattern changes found at the local level. Metro Ethernet and RTR are two MAN technologies that can also inherently address performance and reliability issues. Like SONET, they both leverage fiber ring technologies for survivability, but they make more effective use of capacity. RPR, in particular, employs packet-based routing around the ring, which makes it more resilient to fiber cuts and congestion.

Although a minimum variety of technologies should be used in an enterprise network whenever possible, having some diverse technologies at the LAN, WAN, and MAN level can aid survivability. An alternative is to select individual technologies that have embedded protective mechanisms. Because cost will always be a limiting factor for most enterprises, the challenge is to select solutions that can cost-effectively protect the most critical portions of the network.

References

[1]   Steinke, S., "Troubleshooting Third Millennium Networks," *Network Magazine*, March 2001, pp. 78–82.

[2]   Sathaye, S., "Turn on Your Network," *LAN Times*, Sept. 28, 1998, p. 41.

[3]   Hochmuth, P., "Users Get Innovative with Virtual LANs," *Network World*, May 27, 2002, p. 20.

[4]   Hochmuth, P., "Managing Backbone Bandwidth Can Pay off," *Network World*, February 25, 2002, p. 20.

[5]   Castelli, L., "Transceivers As Backup Offer Another Port in a Storm," *Lightwave*, November 1999, pp. 72–74.

[6]   Boothroyd, D., "Extending Your Reach with Media Converters," *Cabling Installation & Maintenance*, January 2002, pp. 50–55.

[7]   Furlong, D., "Building Redundant Networks with Media Translators," *Lightwave*, April 1999, pp. 66–70.

[8]   Albin, J., and K. Tam, "Adapting to Ethernet," *Lightwave*, April 2002, pp. 52–54.

[9]   Zimmerman, C., "Fault-Tolerant Servers in the NIC of Time," *Data Communications*, pp. 54–56.

[10] Bourke, T., "The Not-So-Usual Suspect," *Hosting Tech*, January, 2001, pp. 42–46.

[11] Philpot, K., "Maximizing Uptime with Redundant DHCP," *Network Computing*, October 2, 2000, pp. 224–227.

[12] Clark, E., "Building the Fault-Tolerant WAN," *Network Magazine*, August 1998, pp. 62–67.

[13] Taylor, S., and J. Wexler, "Will Convergence Require Additional Disaster Recovery," *Network World*, August 2, 1999.

[14] Woods, D., "Backing up Your High–Speed Local Loop," *Network Computing*, September 18, 2000, pp. 144–150.

[15] Koehler, S., "Protecting the Metro Optical Network," *Network Reliability—Supplement to America's Network*, August 2000, pp. 20S–24S.

[16] Williams, K., and S. Lukes, "TDM-to-IP Evolution in the Metro," *Internet Telephony*, March 2001, pp. 54–59.

[17] Samieian, S., et al., "Capacity Requirements of Ring, 1 + 1, and Mesh Configurations," *Lightwave*, August 2001, pp. 114–132.

[18] Zalloua, P., "Simplicity and Reliability of Multiservice Platforms," *Lightwave*, January 2002, pp. 105–108.

[19] Cvijetic, M., S. Nakamura, and B. Faer, "New Optical-Channel Shared Protections-Ring Architecture," *Lightwave*, June 2000, pp. 123–126.

[20] Torode, C., "Vendors Inject Intelligence into Net Traffic," *Computer Reseller News*, May 20, 2002, pp. 63–64.

[21] Christensen, P., "IP Routing and Product Development Challenges," *Integrated Communications Design*, July 16, 2001, pp. 54–55.

[22] Liska, A., "You Down with BGP?" *Hosting Tech*, December 2001, pp. 59–61.

[23] Johnson, J., "Intelligent Route Control Improves BGP," *Network World*, February 11, 2002, p. 29.

[24] Dobrushin, A., "Network Availability is Vital in the New World of IP," *Lightwave*, May 2001, pp. 148–152.

[25] Kraus, C., "Bringing Redundancy to Layer 3 Switching," *Network World*, December 14, 1998, p. 41.

[26] Allen, D., "Multihoming and Route Optimization: Finding the Best Way Home," *Network Magazine*, February 2002, pp. 38–43.

[27] Cope, J., "Smart Routes," *Network World*, May 27, 2002, pp. 49–50.

[28] Shaw, J., "Using ATM to Enforce IP Net Quality of Service," *Network World*, April 27, 1998, p. 45.

[29] Lisle, S., and P. Havala, "Evaluating ATM Network Topologies, *Lightwave,* April 1998, pp. 56–62.

[30] Ashwood-Smith, P., B. Jamoussi, and D. Fedyk, "MPLS: A Progress Report," *Network Magazine*, November, 1999, pp. 96–102.

[31] Greenfield, D., "Lesson 163: MPLS in Brief," *Network Magazine*, February 6, 2002, pp. 30–32.

[32] Xiao, X., et al., "Traffic Engineering with MPLS," *America's Network*, November 15, 1999, pp. 32–38.

[33] McGarvey, J., "ISDN Fills Disaster Recovery Role," *Interactive Week*, April 27, 1998, p. 31.

[34] Engebretson, J., "The Fatal Flaw of MPLS?" *America's Network*, May 15, 2002, pp. 14–15.

[35] Boardman, B., "MPLS VPNs: The Real Deal," *Network Computing*, June 10, 2002, pp. 105–107.

[36] Greene, T., "Standard Needed so VPN Failures Can be Detected," *Network World*, August 2, 1999, p. 8.

[37] Gimpelson, T., "Metro Ethernet's Third Option," *Network World*, November 3, 2002, pp. 51–52.

[38] Wu, T., "Ring Spanning Tree Brings Resiliency," *Network World*, January 21, 2002, p. 37.

[39] Sharer, R., "Resilient Packet Ring Offers QoS," *Network World*, August 27, 2001, p. 39.

[40] Hawkins, J., "RPR: More than a SONET Facelift," *Telecommunications*, Mid-May 2002, pp. 20–21.

[41] Schuler, J., and K. Lancaster, "Extending Reliability and Survivable Optical Rings to the Access Network," *Lightwave*, July 2001, pp. 120–127.

[42] Vasani, K., and M. O'Connor, "Scalability Issues in Metro/Access Networks," *Lightwave*, January 2002, pp. 74–77.

CHAPTER 6

Processing, Load Control, and Internetworking for Continuity

Until recent years, centralized network architectures using mainframe systems were a staple in many IT environments. They provided vast processing power and gradually acquired fault-tolerant capabilities as well. However, as distributed transaction processing requirements have heightened, mainframes were found to lack the versatility to support today's real-time and dynamic software development and processing environment. An Internet-based transaction, for instance, will often require the use of several independent processors situated at different network locations. The need for scalability in implementing high-performance computing has driven consideration of alternatives to centralized mainframe-based network architectures. This chapter reviews technologies and techniques that can be used to optimize survivability and performance within a distributed internetworking environment.

6.1 Clusters

For mission-critical networks, cost-effective ways of ensuring survivability is always an objective. The concept of *clusters* is designed with this objective in mind. A cluster is a group of interrelated computers that work together to perform various tasks. The underlying principal behind clusters is that several redundant computers working together as single resource can do more work than a single computer and can provide greater reliability. Physically, a cluster is comprised of several computing devices that are interconnected to behave as a single system. Other computers in the network typically view and interact with a cluster as if it was a single system. The computing elements that comprise a cluster can be grouped in different ways to distribute load and eliminate single points of failure.

Because multiple devices comprise a cluster, if one device fails in a cluster, another device can take over. The loss of any single device, or *cluster node*, does not cause the loss of data or application availability [1]. To achieve this capability, resources such as data and applications must either be replicated or pooled among the nodes so that any node can perform the functions of another if it fails. Furthermore, the transition from one node to another must be such that data loss and application disruption are minimized.

Other than reliability, clustering solutions can be used to improve processing or balance workload so that processing bottlenecks are avoided. If high-performance processing is required, a job can be divided into many tasks and spread among the cluster nodes. If a processor or server is in overload, fails, or is taken off line for maintenance, other nodes in the cluster can provide relief. In these situations, clusters require that nodes have access to each other's data for consistency. Advances in storage technology have made sharing data among different systems easier to achieve (refer to the chapter on storage).

Clustering becomes more attractive for large, distributed applications or systems. Clusters can improve scalability because workload is spread among several machines. Individual nodes can be upgraded or new nodes can be added to increase central processor unit (CPU) or memory to meet the performance growth and response time requirements. This scalability also makes it more cost effective to provide the extra computing capacity to guard against the unpredictable nature of today's data traffic.

Cluster connectivity can be achieved in numerous ways. Connecting servers over a network supporting transmission control protocol/Internet protocol (TCP/IP) is a very common approach. Another approach is to connect computer processors over a high-speed backplane. They can be connected in various topologies, including star, ring, or loop. Invariably, in each approach nodes are given primary tasks and assigned secondary nodes to automatically assume processing of those tasks upon failure of the primary node. The secondary node can be given tasks to do so that it is kept useful during normal operation or kept idle as a standby. A reciprocating arrangement can be made as well between the nodes so that each does the same tasks. Such arrangements can be achieved at several levels, including hardware, operating system (OS), or application levels.

Clusters require special software that can make several different computers behave as one system. Cluster software is typically organized in a hierarchical fashion to provide local or global operational governance over the cluster. Software sophistication has grown to the point where it can manage a cluster's systems, storage, and communication components. An example is IBM's Parallel Sysplex technology, which is intended to provide greater availability [2, 3]. Parallel Sysplex is a technology that connects several processors over a long distance (40 km) using a special coupling facility that enables them to communicate and share data [4].

6.1.1 Cluster Types

Categorizing clusters could seem futile given the many cluster products that have flooded the market in recent years. A cluster in which a node failure results in that node's transactions, accounts, or data being unavailable is referred to as a *static cluster*. Dynamic clusters, on the other hand, can dynamically allocate resources as needed to maintain transaction processing across all users, as long as there is one surviving node [5]. These clusters provide greater availability and scalability, typically limited by data access and storage capabilities. Cluster management is often easier with dynamic clusters, as the same image is retained across all nodes. For situations involving large volumes of users, *super clusters* can be constructed, which is a static cluster comprised of dynamic clusters. These types of configurations are illustrated in Figure 6.1.

Each of these cluster types can be constructed in several ways using different technologies. The following list contains some of the most widely used technology approaches, illustrated in Figure 6.2:

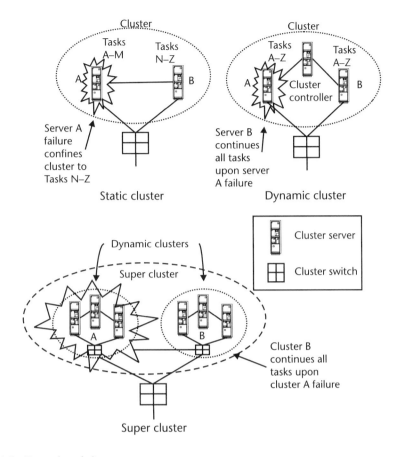

Figure 6.1   Examples of cluster types.

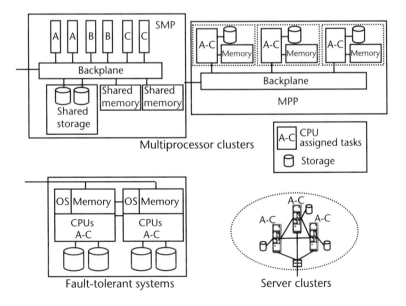

Figure 6.2   Examples of cluster technologies.

- *Multiprocessor clusters* are multiple CPUs internal to a single system that can be grouped or "clumped" together for better performance and availability. Standalone systems having this type of feature are referred to as *multiprocessor* or *scale-up* systems [6, 7]. Typically, nodes perform parallel processing and can exchange information with each other through shared memory, messaging, or storage input/output (I/O). Nodes are connected through a *system area* network that is typically a high-speed backplane. They often use special OSs, database management systems (DBMSs), and management software for operation. Consequently, these systems are commonly more expensive to operate and are employed for high-performance purposes.

  There are two basic types of multiprocessor clusters. In symmetric multiprocessing (SMP) clusters, each node performs a different task at the same time. SMPs are best used for applications with complex information processing needs [8]. For applications requiring numerous amounts of the same or similar operations, such as data warehousing, massively parallel processing (MPP) systems may be a better alternative. MPPs typically use off-the-shelf CPUs, each with their own memory and sometimes their own storage. This modularity allows MPPs to be more scalable than SMPs, whose growth can be limited by memory architecture. MPPs can be limitless in growth and typically run into networking capacity limitations. MPPs can also be constructed from clusters of SMP systems as well.

- *Fault-tolerant systems* are a somewhat simplified hardware version of multiprocessor clusters. Fault-tolerant systems typically use two or more redundant processors and heavily rely on software to enhance performance or manage any system faults or failures. The software is often complex, and the OS and applications are custom designed to the hardware platform. These systems are often found in telecom and plant operations, where high reliability and availability is necessary. Such systems can self-correct software process failures, or automatically failover to another processor if a hardware or software failure is catastrophic. Usually, alarms are generated to alert personnel for assistance or repair, depending on the failure. In general, these systems are often expensive, requiring significant upfront capital costs, and are less scalable than multiprocessor systems. Fault-tolerant platform technology is discussed in more depth in a later chapter of this book.

- *Server clusters* are a low-cost and low-risk approach to provide performance and reliability [9]. Unlike a single, expensive multiprocessor or fault-tolerant system, these clusters are comprised of two or more less expensive servers that are joined together using conventional network technology. Nodes (servers) can be added to the network as needed, providing the best scalability. Large server clusters typically operate using a *shared-nothing* strategy, whereby each node processor has its own exclusive storage, memory, and OS. This avoids memory and I/O bottlenecks that are sometimes encountered using *shared* strategies. However, shared-nothing strategies must rely on some form of mirroring or networked storage to establish a consistent view of transaction data upon failure.

The following are some broad classes of cluster services that are worth noting. Each can be realized using combinations or variations of the cluster configurations and technologies just discussed. Each successive class builds on the previous with regard to capabilities:

- *Administrative clusters* are designed to aid in administering and managing nodes running different applications, not necessarily in unison. Some go a step further by integrating different software packages across different nodes.
- *High-availability clusters* provide failover capabilities. Each node operates as a single server, each with its own OS and applications. Each node has another node that is a replicate image, so that if it fails, the replicate can take over. Depending on the level or workload and desired availability, several failover policies can be used. Hot and cold standby configurations can be used to ensure that a replicate node is always available to adequately assume another node's workload. Cold standby nodes would require extra failover time to initialize, while hot standby nodes can assume processing with little, if any, delay. In cases where each node is processing a different application, failover can be directed to the node that is least busy.
- *High performance clusters* are designed to provide extra processing power and high availability [10]. They are used quite often in high-volume and high-reliability processing, such as telecommunications or scientific applications. In such clusters, application workload is spread among the multiple nodes, either uniformly or task specific. They are sometimes referred to as *parallel application* or *load balancing* clusters. For this reason, they are often found to be the most reliable and scalable configurations.

A prerequisite for high-availability or high-performance clusters is access to the same data so that transactions are not lost during failover. This can be achieved through many of the types of storage techniques that are described in the chapter on storage. Use of mirrored disks, redundant array of independent disks (RAID), or networked storage not only enable efficient data sharing, but also eliminate single points of failure. Dynamic load balancing is also used to redistribute workload among the remaining nodes if a node fails or becomes isolated. Load balancing is discussed further in this chapter.

6.1.2 Cluster Resources

Each node in a cluster is viewed as an individual system with a single image. Clusters typically retain a list of member nodes among which resources are allocated. Nodes can take on several possible roles, including the primary, secondary, or replicate roles that were discussed earlier. Several clusters can operate in a given environment if needed, where nodes are pooled into different clusters. In this case, nodes are kept aware of nodes and resources within their own cluster and within other clusters as well [11].

Many cluster frameworks use an object-oriented approach to operate clusters. Objects can be defined comprised of physical or logical entities called *resources*. A resource provides certain functions for client nodes or other resources. They can reside on a single or multiple nodes. Resources can also be grouped together in classes so that all resources in given class can respond similarly upon a failure. Resource groups can be assigned to individual nodes. Recovery configurations, sometimes referred to as *recovery domains*, can be specified to arrange objects in a certain way in response to certain situations. For example, if a node fails, a domain can specify to which node resources or a resource group's work should be transferred.

6.1.3 Cluster Applications

For a node to operate in a cluster, the OS must have a clustering option. Furthermore, many software applications require modifications to take advantage of clustering. Many software vendors will offer special versions of their software that are *cluster aware,* meaning that they are specifically designed to be managed by cluster software and operate reliably on more than one node. Cluster applications are usually those that have been modified to failover through the use of scripts. These scripts are preconfigured procedures that identify backup application servers and convey how they should be used for different types of faults. Scripts also specify the transfer of network addresses and ownership of storage resources. Because failover times between 30s and 5 min are often quoted, it is not uncommon to restart an application on a node for certain types of faults, versus failing over to another processor and risking transaction loss.

High-volume transaction applications, such as database or data warehousing and Web hosting, are becoming cluster aware. Clusters enable the scaling that is often required to reallocate application resources depending on traffic intensity. They have also found use in mail services, whereby one node synchronizes account access utilization by the other nodes in the cluster.

6.1.4 Cluster Design Criteria

Cluster solutions will radically vary among vendors. When evaluating a clustered solution, the following design criteria should be applied:

- *Operating systems.* This entails what OSs can be used in conjunction with the cluster and whether different versions of the OS can operate on different nodes. This is critical because an OS upgrade may entail having different versions of an OS running in the cluster at a given moment.
- *Applications.* The previous discussion highlighted the importance of cluster-aware applications. In the case of custom applications, an understanding of what modifications are required needs to be developed.
- *Failover.* This entails to what extent failover is automated and how resources are dynamically reallocated. Expected failover duration and user transparency to failovers needs to be understood. Furthermore, expected performance and response following a failover should be known.
- *Nodes.* A number of nodes should be specified that could minimize the impact of a single node outage. An $N + I$ approach is often a prudent one, but can result in the higher cost of an extra, underutilized cluster node. A single system image (SSI) approach to clustering allows the cluster nodes to appear and behave as a single system, regardless of the quantity [12].

- *Storage.* Cluster nodes are required to share data. Numerous storage options and architectures are available, many of which are discussed in the chapter on storage. Networked storage is fast becoming a popular solution for nodes to share data through a common mechanism.

- *Networking.* Cluster nodes must communicate with each other and other nodes external to the cluster. Separate dedicated links are often used for the nodes to transmit heartbeat messages to each other [13].

6.1.5   Cluster Failover

Clusters are designed such that multiple nodes can fail without bringing down the entire cluster. Failover is a process that occurs when a logical or physical cluster component fails. Clusters can detect when a failure occurs or is about to occur. Location and isolation mechanisms typically can identify the fault. Failover is not necessarily immediate because a sequence of events must be executed to transfer workload to other nodes in the cluster. (Manual failover is often done to permit system upgrades, software installation, and hardware maintenance with data/applications still available on another node.) To transfer load, the resources that were hosted on the failed node must transfer to another node in the cluster. Ideally, the transfer should go unnoticed to users.

During failover, an off-line recovery process is undertaken to restore the failed node back into operation. Depending on the type of failure, it can be complex. The process might involve performing additional diagnostics, restarting an application, replacing the entire node, or even manually repairing a failed component within the node. Once the failed node becomes active again, a process called failback moves the resources and workload back to the recovered node.

There are several types of cluster failover, including:

- *Cold failover.* This is when a cluster node fails, another idle node is notified, and applications and databases are started on that node. This is typically viewed as a slow approach and can result in service interruption or transaction loss. Furthermore, the standby nodes are not fully utilized, making this a more expensive approach.

- *Warm failover.* This is when a node fails, and the other node is already operational, but operations must still be transferred to that node.

- *Hot failover.* This is when a node fails, and the other node is prepared to serve as the production node. The other node is already operational with application processing and access to the same data as the failed node. Often, the secondary node is also a production server and can mirror the failed server.

Several activities occur to implement a complete failover process. The following is a general description of the types of events that take place. This process will vary widely by the type of cluster, cluster vendor, applications, and OS involved:

- *Detection.* Detection is the ability to recognize a failure. A failure that goes undetected for a period of time could result in severe outage. A sound detection mechanism should have wide fault coverage so that faults can be detected and isolated either within a node or among nodes as early as possible. The ability of a system to detect all possible failures is measured in its fault coverage. Failover management applications use a *heartbeat* process to recognize a failure. Monitoring is achieved by sending heartbeat messages to a special monitoring application residing on another cluster node or an external system. Failure to detect consecutive heartbeats results in declaration of a failure and initiation of a failover process. Heartbeat monitoring should not only test for node failure but should also test for internode communication. In addition to the network connectivity used to communicate with users, typically Ethernet, some clusters require a separate *heartbeat interconnect* to communicate with other nodes.

- *Networking.* A failover process typically requires that most or all activity be moved from the failed node to another node. Transactions entering and leaving the cluster must then be redirected to the secondary node. This may require the secondary node to assume the IP address and other relevant information in order to immediately connect users to the application and data, without reassigning server names and locations in the user hosts. If a clustering solution supports IP failover, it will automatically switch users to the new node; otherwise, the IP address needs to be reallocated to the backup system. IP failover in many systems requires that both the primary and backup nodes be on the same TCP/IP subnet. However, even with IP failover, some active transactions or sessions at the failed node might time out, requiring users to reinitiate requests.

- *Data.* Cluster failover assumes that the failed node's data is accessible by the backup node. This requires that data between the nodes is shared, reconstructed, or transferred to the backup node. As in the case of heartbeat monitoring, a dedicated shared disk interconnect is used to facilitate this activity. This interconnect can take on many forms, including shared disk or disk array and even networked storage (see Section 6.1.7). Each cluster node will most likely have its own private disk system as well. In either case, nodes should be provided access to the same data, but not necessarily share that data at any single point in time. Preloading certain data in the cache of the backup nodes can help speed the failover process.

- *Application.* Cluster-aware applications are usually the beneficiary of a failover process. These applications can be restarted on a backup node. These applications are designed so that any cluster node can resume processing upon direction of the cluster-management software. Depending on the application's state at the time of failure, users may need to reconnect or encounter a delay between operations. Depending on the type of cluster configuration in use, performance degradation in data access or application accessing might be encountered.

6.1.6 Cluster Management

Although clusters can improve availability, managing and administering a cluster can be more complex than managing a single system. Cluster vendors have addressed this issue by enabling managers to administer the entire cluster as a single system versus several systems. However, management complexity still persists in several areas:

- *Node removal*. Clustering often allows deactivating a node or changing a node's components without affecting application processing. In heavy load situations, and depending on the type of cluster configuration, removal of a cluster node could overload those nodes that assume the removed node's application processing. The main reason for this is that there are less nodes and resources to sustain the same level of service prior to the removal. Furthermore, many users attempt to reconnect at the same time, overwhelming a node. Mechanisms are required to ensure that only the most critical applications and users are served following the removal. Some cluster solutions provide the ability to preconnect users to the backup by creating all of the needed memory structures beforehand.

- *Node addition*. In most cases, nodes are added to an operational cluster to restore a failed node to service. When the returned node is operational, it must be able to rejoin the cluster without disrupting service or requiring the cluster to be momentarily taken out of operation.

- *OS migration*. OS and cluster software upgrades will be required over time. If a cluster permits multiple versions of the same OS and cluster software to run on different nodes, then upgrades can be made to the cluster one node at a time. This is often referred to as a *rolling upgrade*. This capability minimizes service disruption during the upgrade process.

- *Application portability*. Porting cluster applications from one node to another is often done to protect against failures. Critical applications are often spread among several nodes to remove single points of failure.

- *Monitoring*. Real-time monitoring usually requires polling, data collection, and measurement features to keep track of conditions and changes across nodes. Each node should maintain status on other nodes in the cluster and should be accessible from any node. By doing so, the cluster can readily reconfigure to changes in load. Many cluster-management frameworks enable the administration of nodes, networks, interfaces, and resources as objects. Data collection and measurement is done on an object basis to characterize their status. Management is performed by manipulation and modification of the objects.

- *Load balancing*. In many situations, particularly clustered Web servers, traffic must be distributed among nodes in some fashion to sustain access and performance. Load balancing techniques are quite popular with clusters and are discusses further in this chapter.

6.1.7 Cluster Data

Data access can be a limiting factor in cluster implementation. Limited storage capacity as well as interconnect and I/O bottlenecks are often blamed for performance and operational issues. The most successful cluster solutions are those that combine cluster-aware databases with high-availability platforms and networked storage solutions.

Shared disk cluster approaches offer great flexibility because any node can access any block of data, providing the maximum flexibility. However, only one node can write to a block of data at any given time. Distributed locking management is required to control disk writes and eliminate contention for cached data blocks across nodes. Distributed locking, however, can negatively impact I/O performance. Partitioned cluster databases require that transactions be balanced across cluster nodes so that one node is not overloaded. Balancing software can be used to direct I/O queries to the appropriate server as well as realign partitions between nodes.

In shared-nothing data approaches, each cluster node has exclusive access to a static, logical segment of data. This eliminates the need for locking and cache-contention mechanisms, which consume performance. This is why shared-nothing is often preferred in large data warehouses and high-volume transaction applications. On the other hand, shared nothing requires reallocation of data and new partitions if nodes are added or removed from the cluster.

Cluster database solutions must ensure that that all committed database updates made prior to a failure are applied (referred to as a *roll forward*) and that all uncommitted updates are undone during a recovery (referred to as a *roll back*). The roll-forward process is less intensive with greater snapshot frequency of the database.

6.1.8 Wide Area Clusters

Wide area clusters are desirable in enterprises for greater diversity and manageability (see Figure 6.3). These clusters are designed with simultaneous application processing on all cluster nodes at all sites. Applications are operated without regard to the physical location of the platform. If an outage occurs at one site, operations continue at the other site. Concurrent access to the same data image from all sites is achieved through a variety of techniques, including mirroring and networked storage. Coordination is required to manage the data access from each site [14]. All sites are interconnected via a wide area network (WAN). As in a collocated cluster, mechanisms are required to detect failures at the remote site and failover to surviving site. This requires synchronized use of cluster software management, network routing, load balancing, and storage technologies.

To achieve invisible failovers using an integration of multivendor components might be quite challenging. However, vendor-specific solutions are available to

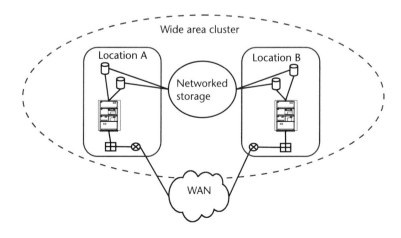

Figure 6.3 Wide area cluster example.

achieve high availability in wide area clusters. An example is IBM's MultiSite and Geographically Dispersed Parallel Sysplex (GDPS) clustering technology. It involves connecting S/390 systems via channel-attached fiber cabling and a coupling facility. This enables controlled switching from one system to another in the event of an unplanned or planned service interruption [15]. A coupling facility is an external system that maintains nonvolatile shared memory among all processors, enabling them to share data and balance the workload among applications.

A Parallel Sysplex cluster can be separated by up to 40 km, where each site is configured with redundant hardware, software, connections, and mirrored data. Standard or user-defined site switches can be executed. If a system fails, it is automatically removed from the cluster and restarted. If a CPU fails, workload on another is initiated. Mission-critical production and expendable workloads can be configured among sites depending on organizational need. For networking, there are several options. Previously, a technique called data link switching (DLSw), a means of tunneling SNA traffic over an IP network, was used for recovery and network failures. Recently, enterprise extender functions have been introduced that convert SNA network transport to IP. The systems can use a virtual IP address (VIPA) to represent the cluster to outside users.

6.2 Load Balancing

Load balancing is a class of techniques used to direct queries to different systems for a variety or reasons, but fundamentally to distribute a workload across some available pool of resources. Load balancing is best used in situations involving large volumes of short-lived transactions or in networks with a large numbers of users accessing a small quantity of relatively static information. This is why it has found popular use in front of Web servers, clusters, and application server farms. It is also used to direct and balance frequent transaction requests among applications involving data or content that is not easily cached.

In the case of Web sites, load balancing can be used to assure that traffic volume will not overwhelm individual Web servers or even individual server farms. It permits distributing load to another site, creating redundancy while sustaining performance. Load balancing is effective on sites where no transactions are involved and when most of the site hits access a small number of pages without the use of hyperlinks to other servers. Data and content between load balanced sites can be partitioned, mirrored, or overlap in some manner so that each site processes the same or different portions of the transactions, depending on the nature of the application.

Load balancers are devices that distribute traffic using a number of different methods. They provide numerous benefits: they can alleviate server system bottlenecks by redirecting traffic to other systems; they provide scalability to add capacity incrementally over time and utilize a mix of different systems; they offer an approach to preserve investment in legacy systems and avoid upfront capital expenditures; they provide the ability to leverage redundant systems for greater availability and throughput; they obviate the need for bandwidth and system memory upgrades to resolve performance bottlenecks; and they can be used to improve operations management by keeping processes running during routine maintenance.

There are a several ways to classify load balancers. Two classifications are illustrated in Figure 6.4. One class consists of *network load balancers,* which distribute network traffic, most commonly TCP/IP traffic, across multiple ports or host connections using a set of predefined rules. Network load balancers originated as domain name servers (DNSs) that distributed hypertext transfer protocol (HTTP) sessions across several IP hosts. They used basic pinging to determine whether destination hosts were still active in order to receive queries. Later, this capability was expanded to measure destination server performance prior to forwarding additional requests to avoid overwhelming that host. With the advent of e-commerce, load balancers were further enhanced with capabilities to monitor both front-end and back-end servers. They direct traffic to back-end servers based on requests from front-end servers and use a process called *delayed binding*, which maintains the session with that server until data or content is received from it prior to making a decision.

Another class consists of *component load balancers,* which distribute requests to applications running across a cluster or server farm. Component load balancers are standalone systems typically situated between a router and internal server farm that distribute incoming traffic among the servers based on predefined rules. Rules can involve anything from routing based on application, server response time, delay, time of day, number of active sessions, and other metrics. Some rules might require software agents to be installed on the servers to collect the required information to implement a rule. Load balancers deployed in front of server farms or clusters can use a single VIPA for the entire site, making the site appear as a single system to the outside world. Because these systems can be a single point of failure between a server farm and external network, failover capabilities are required, including use of fault-tolerant platforms.

Load balancers can be used in this fashion with respect to clusters. Cluster nodes are usually grouped according to application, with the same applications running within a cluster. Load balancers can use predefined rules to send requests to each node for optimal operation. An example is sending requests to the least-used node. It can also be used in innovative ways. For example, load balancers can be used in conjunction with wide area clusters to direct requests to the cluster closest to the user.

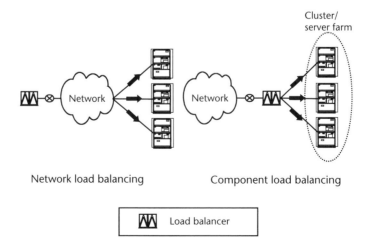

Figure 6.4   Network and component load balancing examples.

Not only does balancing help control cluster performance, it also enables the addition or removal of cluster nodes by redirecting traffic away from the affected node. Furthermore, load balancers can also be used to manage cluster storage interconnects to reduce I/O bottlenecks.

6.2.1   Redirection Methods

Load balancer devices inspect packets as they are received and switch the packets based on predefined rules. The rules can range from an administratively defined policy to a computational algorithm. Many of the rules require real-time information regarding the state of a destination server. Several methods are used to obtain such information. Balancers that are internal to an enterprise often make use of existing system-management tools that monitor applications and platform status through application programming interfaces (APIs) and standard protocols. Some balancers require direct server measurements through the use of software agents that are installed on the server. These agents collect information regarding the server's health and forward this information to the load balancer. The agents are usually designed by the load balancer vendor and are often used on public Web sites. The information that agents collect can be quite detailed, usually more that what would be obtained in a PING request [16]. For this reason, they can consume some of the server's CPU time.

Many Web sites have a two-tier architecture (Figure 6.5), where a first tier of content bearing Web servers sits in front of a back end or second tier of servers. These second-tier servers are not directly load balanced, and sometimes a first-tier server can mask a second-tier server that is down or overloaded. In this case, sending requests to the first-tier server would be ineffective, as the second-tier server would be unable to fulfill the request. To handle these situations, some load balancers can logically associate or *bind* the back-end servers to the first-tier server and query them for the status information via the first-tier server.

One would think that the purpose of using a load balancer is to direct more CPU-intensive traffic to the servers that can handle the load. This is true, but there are occasions whereby other rules may be of more interest [17]. Rules can be classified as either *static* or *dynamic*. Static rules are predefined beforehand and do not

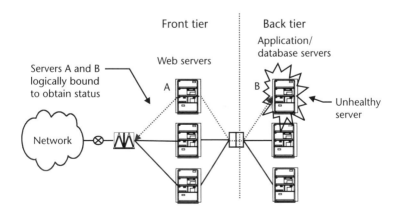

Figure 6.5   Load balancing with multitier Web site.

change over time, while dynamic rules can change over time. Some rule examples include:

- *Distributed balancing* splits traffic among a set of destinations based on predefined proportions or issues them in a predefined sequence. For example, some balancers perform round-robin balancing, whereby an equal number of requests are issued in sequential order to the destination servers. This routing works fairly well when Web content if fairly static.

- *User balancing* directs traffic based on attributes of the user who originated the request. They can examine incoming packets and make decisions based on who the user is. One example is directing requests to a server based on user proximity. Another example is providing preferential treatment to a Web site's best customers.

- *Weight or ratio balancing* directs traffic based on predefined weights that are assigned to the destination servers. The weights can be indicative of some transaction-processing attribute of the destination server. For example, the weight can be used to bias more traffic to servers having faster CPUs.

- *Availability balancing* checks to see if destination servers are still alive to avoid forwarding that would result in error messages.

- *Impairment balancing* identifies when all or some facilities of a server are down and avoids traffic to that server. Although one can connect to a server, a partial failure can render a server or application useless. Furthermore, a server under overload would also be ineffective.

- *Quality of service (QoS) balancing* measures roundtrip latency/delay between the destination and a user's DNS server to characterize network transport conditions.

- *Health balancing* monitors the workload of a destination server and directs traffic to servers that are least busy [18]. Different measurements and approaches are used to characterize this state. For example, some measurements include the number of active TCP connections and query response time. Standalone measures can be used or several measures can be combined to calculate an index that is indicative of the server's health.

- *Content-aware balancing* directs requests based on the type of application (e.g., streaming audio/video, static page, or cookie) and can maintain the connection state with the destination server. Traffic can be redirected based on back-end content as well using *delayed binding*. This involves the load balancer making the redirection decision only until it receives content from the Web server. For streaming video applications, balancers will direct all requests from a user to the video server for the entire session. Content-aware balancing can provide greater flexibility in handling content across Web servers and enables placing different content on different machines.

There are two basic forms of redirection—local and global. Each can use special redirection rules that include some of the aforementioned:

- *Local load balancing.* Local load balancing or redirection involves using a load-balancer device that is local to a set of users, servers, or clusters.

Component or network load balancers can be used. It is used to route requests from a central location across a group of servers or hosts that typically sit on the same local area network (LAN), subnet, or some other type of internal network. They are typically used to route requests across systems residing within a data center. Local load balancing can also be used to distribute requests originating from internal network hosts across firewalls, proxy servers, or other devices. It is often seen as a way to manage traffic across a server complex, such as those that host a Web site.

- *Global load balancing.* Global load balancing involves directing traffic to a variety of replicated sites across a network, as in the case of the Internet [19]. Redirection decisions are made centrally from devices that reside outside a network that intercept requests for content prior to reaching a firewall (Figure 6.6) and direct those requests to an appropriate location. It works best when caches are distributed throughout a network, but each destination does not have dedicated cache.

  Global redirection has become an integral part of mission-critical networking solutions for data-center solutions and Web-based applications [20]. It enables organizations to distribute load across multiple production sites and redirect traffic appropriately following the outage or overload of a particular site.

  Global load balancers come in different forms and can be configured in multiple ways. They can very well be consolidated with local load balancers. They can be configured using DNS or even border gateway protocols (BGPs) (some network load balancers might require configuration with contiguous IP addresses). Throughput can vary among products, but it is important that they can scale up throughput with the volume of requests.

Global and local load balancing can be used in conjunction with each other to distribute traffic across multiple data centers (see Figure 6.7). Each data center can have a local load balancer identified by a VIPA. The balancers would distribute traffic to other centers using the VIPAs of the remote balancers, as if they are local devices. The DNS requests for the Web site hosted by either data center must be

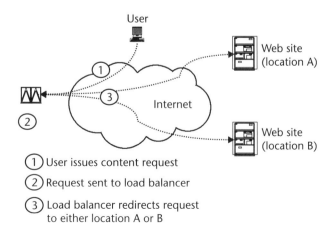

Figure 6.6 Global load balancing example.

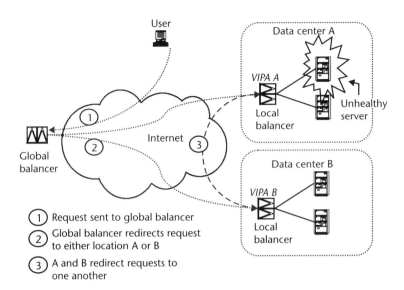

Figure 6.7 Combined global and local load balancing example.

directed to the domain or VIPAs of the load balancers. This requires leveraging DNS and HTTP capabilities to send users to the most efficient and available data center.

6.2.2 DNS Redirection

DNS redirection is commonly used for Web site applications. In this implementation, uniform resource locator (URL) queries traverse Internet DNS devices until the IP or VIPA of the global load balancer is returned. The global load balancer in this case is an authoritative name server. There are several drawbacks to DNS redirection. First, it is time sensitive and can time out if an IP address is not found. Second, the device must have up-to-date information on the status of each destination location. Finally, DNS redirection is at the domain level, so it may not execute redirection decisions at a more detailed rule level, such as those previously discussed.

There are several approaches to DNS redirection. One approach uses round-robin sequencing to direct incoming messages to different Web servers. If a server fails, users remain attached until the server has timed out. Another approach sends incoming requests to a single device known as a reverse proxy that performs the redirection. The proxy server can also interact with the load balancers to obtain status information of their locations. Another DNS approach is called *triangulation*, whereby after a user request is directed to multiple proxy sites, the site having the fastest response is used. It is used to provide DNS devices with higher throughput and greater protocol transparency. DNS redirection is mainly used in conjunction with HTTP redirection, whereby HTTP header information is used to redirect traffic. The approach does not work for non-HTTP traffic, such as file transfer protocol (FTP).

6.2.3 SSL Considerations

Secure socket layer (SSL) traffic can pose challenges to load balancing. SSL encryption/decryption is usually an intensive process for a Web site [21]. To address SSL traffic, a common approach used is to have the load balancer proxy the SSL server so that it can maintain a "sticky" connection to an assigned SSL server. This involves SSL requests remaining decrypted prior to reaching the load balancer. The load balancer retains its own VIPA as an SSL-processing resource. The load balancer then redirects the request to the IP address of the site's SSL servers. The SSL address inside the user's cookie information, whose current value is the balancer's VIPA, is then modified to the address of the SSL server. This forces the balancer to continue redirecting successive transactions to the SSL server for the duration of the session, even if the user requests different Web pages. The balancer can also implement secure session recovery by trying to reconnect with the SSL server if the session is disrupted while maintaining the session with the user.

6.2.4 Cookie Redirection

As previously mentioned, user balancing involves making redirection decisions based on user attributes. Because user IP address information can change and IP headers contain minimal user information, higher layer information is often required. Cookies, data that applications use to gather information from a user, serve this purpose. Cookie-based redirection is designed to make redirection decisions based on technical and/or business objectives. Requests from users that represent good-paying or important customers can be given preferential treatment. Another approach used is to redirect users based on their type of access so that users with slower speed access can be redirected to sites with faster servers. Some balancers allow cookies to be altered for certain applications, as in the case of SSL. For example, a cookie could be modified with a customer service location to initiate a live audio/video session with an agent. Cookie redirection, however, can negatively impact a load balancer's performance, depending on how deep into a cookie it must look.

6.2.5 Load Balancer Technologies

Load balancers are produced using several approaches to implementations, which can be categorized into the following basic groups:

- *Appliances* are devices that are optimized to perform a single function, versus using software installed on a general-purpose server. They are often more cost effective to use, have built-in reliability, and are easier to maintain. Load-balancer appliances are usually placed between a router and switch and are used often for local balancing application. They provide better price performance than server-based balancing because they rely on distributed processing and application-specific integrated circuits (ASICs). These devices are currently quite popular.
- *Software-based* balancing is accomplished by software that is resident on a general-purpose server. Because of this, processing can be slower than hardware-based balancing. On the other hand, software-based solutions provide greater flexibility and can be more easily upgraded. This is makes it easier to keep up with new software releases, especially those where new agents are introduced. They also offer standard programming interfaces for use by third-party or custom applications. Last, they can simplify network topology,

especially if the server is used for other functions. The servers are typically equipped with dual network adapters—one that connects to a router and another that connects to a switch or hub that interconnects with other servers.

- *Switch-based* balancers are just that—a network switch or router platform that has load-balancing capabilities. Because switches typically sit in locations central to users, servers, and the Internet, it makes them a prime candidate for load balancing. Like appliances, they can use ASICs so that balancing is done at wire speed. For example, ports on a LAN switch that connect to a server farm can be designated for load balancing and treated as one network address. The switch then distributes traffic using rules that are defined through configuration.

- *Server switching* is a new approach that recreates the application session management and control functions found in mainframe front-end processors, but it applies these techniques to distributed servers and server farms. The concept delivers three big benefits: it increases individual server efficiency by offloading CPU-intensive chores; it scales application-processing capacity by transparently distributing application traffic; and it ensures high levels of service availability. Server switches achieve application-based redirection by implementing advanced packet filtering techniques. Filters can be configured based on protocols, IP addresses, or TCP port numbers, and they can be applied dynamically to a switch port to permit, block, or redirect packets. They can also be used to select packets whose headers or content can be replaced with application-specific values. By combining load balancing and filtering within server switches, virtually any IP traffic type can now be load balanced. This means administrators can redirect and load balance traffic to multiple firewalls and outbound routers, so standby devices no longer sit idle. Server switches offer this capability by examining incoming packets and making a determination about where they should send them based on source IP address, application type, and other parameters. This is why vendors are trying to avoid having the term *load balancer* applied to their server switch offerings. The issue is not just distributing loads of like traffic across multiple CPUs, but requires distinguishing and prioritizing various types of traffic and ensuring that each one is supported by resources appropriate to the business value it represents. For example, layer 7 switches look at layers 2 thru 7 of the IP packet, recognize cookies, and treat them accordingly. These platforms are application aware, have powerful load-balancing capabilities, and can do geographic redirection as needed.

- *Switch farms* are a manual approach to isolating and balancing traffic among workgroups [22]. It involves connecting servers directly to user's switches to offload backbone traffic. An example is illustrated in Figure 6.8. The network is designed so that traffic is kept local as much as possible. Workgroup application servers are directly connected to the switches that service their users. Core switches support only enterprise services, minimizing backbone traffic and freeing it up for critical traffic. This approach is counter to the concept of server farms, which are usually situated in a central location with the goal of reducing administration. However, server farms can increase backbone traffic and can be a single point of failure. Switch farms do require cable management so that copper limitations are not exceeded.

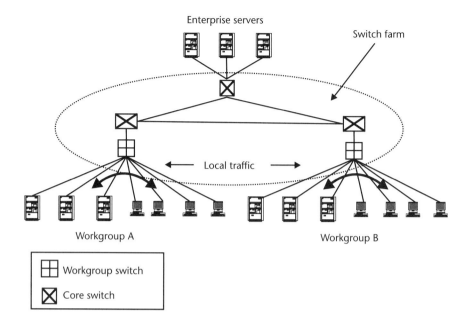

Figure 6.8   Switch farm load balancing example.

6.2.6   Load Balancer Caveats

Despite all of the advantages that load balancing can provide, there are several caveats that have become apparent with their use:

- If not properly managed and configured, load balancers can bring down a site or a system. For example, a mistyped IP address can result in a catastrophic situation. Numerous erroneous requests, say for nonexistent pages or content, can overload a device and bring it to a halt.
- Reliance on single standalone measures of server health can deceive load balancers about the server's status. For example, although a server's HTTP daemon may fail, the server can still appear as alive and respond to PING requests.
- Load balancing works best when transactions are simple and short and data or content is relatively static and easily replicated across locations. Overly complex transactions can pose management headaches and increase operations costs, offsetting the potential savings gained from the scalability load balancing provides.
- Unless balancers are used that can handle multitier Web sites, load balancing is ineffective against back-end server overload.
- Although load balancers can be used to protect against outages, a load-balancer device itself can be a single point of failure. Thus, they should be implemented either on a high-availability or fault-tolerant platform, or used in conjunction with a mated load balancer for redundancy.
- As of this writing, it is still unclear how load balancing aligns or interworks with QoS mechanisms (discussed further in this book). Although QoS can

provide preferential treatment and guarantee service levels for network traffic, it is ineffective if the traffic destination cannot provide the service. The two can be used as complementary techniques—QOS, especially traffic prioritization, can police and shape traffic at ingress points to improve traffic performance and bandwidth utilization, whereas load balancing is generally used to improve transaction rates.

6.3   Internetworking

Today's Internet evolved in a deregulated environment in the span of 10 years. Many of us have firsthand experienced the frustration of trying to access a Web site only to have it take forever to download. The Internet operates using TCP/IP networking, which is connectionless and is designed to slow end systems down as traffic increases. Packet transmission is slowed at the originating end points so that the intermediate nodes and destination hosts can keep up, until buffers are filled and packets are discarded. Yet, many large enterprises are now migrating their business-critical processes to this kind of environment. The next sections describe some practices designed to provide Web-based applications the ability to withstand the irregularities of the Internet.

6.3.1   Web Site Performance Management

Users typically expect to be able to access a Web site when they want it. They also expect a Web site to be viewed and browsed easily and quickly, regardless of where they are and how they are connecting to the Internet. Unfortunately, these expectations are the basis of the frustrations of using the Web. When such frustrations surface, they are usually directed to a Web site's owner or Internet service provider (ISP). Web sites are now considered a "window to an enterprise," thus poor performance as well as poor content can tarnish a firm's image.

Use of the Web for business-to-business 7×24 transactions has even heightened the need for Web sites and their surrounding applications to be available all of the time. Experience has shown that most Web sites under normal conditions can sustain reasonable performance. However, their resiliency to swift, unexpected traffic surges is still lacking. Consumer-oriented sites are typically visited by users who do more browsing than buying.

Business-to-business sites handle more transaction-oriented traffic in addition to buying. The term *transaction* is often synonymous with higher performance requirements. Transactions often require SSL encryption as well as hypertext markup language (HTML) browser-based traffic. From our earlier discussion, we saw that SSL requires more processing and reliability resources. Many times the back-end network situated behind a Web site is often affected when problems arise. The following are some broad categories of problems that are often experienced:

- *Internet service providers.* Losing a connection to a site is one of the leading causes of download failures or site abandonment. Access network connectivity typically consumes about half of the time to connect to a Web site.

- *Site slowdowns*. A general industry requirement for a page download is 8 seconds. Unfortunately, this requirement is gradually moving to 6 seconds. Studies have shown that users will abandon a sites if downloads take longer. Web servers are typically engineered to download a page within 2 seconds. But a single page download often requires the use of a number of resources in addition to the site server. In addition to the issuance of queries to arrive at a site, queries are also issued to other sites for content or data to construct a single page. This includes the DNS query that is usually the first and most critical step. Each query is called a *turn*. The number of turns an average site uses to download a page is close to 50 and is gradually increasing. Further problem arise when users attempt to reload a slow site, further increasing traffic load [23].

- *SSL connections*. SSL operates more slowly than HTML connections. The encryption and negotiation processing consumes server resources. Although many realize that SSL should only be used when necessary, the number of sites requiring SSL is growing. In addition to encrypting only the financial transaction portion of a site visitation, many are also encrypting the browsing pattern as well for privacy.

- *Graphics*. The use of bandwidth intensive graphics and video can degrade performance versus text content. Although use of image compression techniques is catching on, Web pages are being designed with more graphical content.

- *Site design*. Many users abandon sites out of frustration. Much on-line buying is comprised of impulse purchases, implying that closing the deal with a user as quickly as possible can lead to a successful transaction. This includes enabling the user to easily locate the item they want and making it easier for them to buy. Unfortunately, sites with complex designs can make it quite difficult for users to find what they want. Not only does this affect the number of transactions, but it also degrades performance.

- *Distributed applications*. Application architectures are becoming more distributed in nature, meaning that they must access other applications outside of their immediate environment. This not only degrades performance, but it also reduces a site owner's processing control and makes performance monitoring more complex.

- *Browser versions*. Users will access the Web using browsing software, which can vary in nature. Because each browser or browser version is not the same, each can create different types of Web traffic. This makes Web site design and engineering ever the more challenging.

- *Bursty traffic*. Web traffic surges are commonplace and occur during busy hours, major events, or busy seasons. Engineering a network for jumps in traffic is a traditional challenge and involves building an infrastructure that is not expensively overengineered and can sustain adequate performance during traffic surges.

This last item underscores the challenge in devising a Web site infrastructure that adequately fits the traffic access patterns. As the cost of fault tolerance is orders of magnitude greater than best effort, it makes sense to build an infrastructure that can instill a balance between undersizing and oversizing. The best way to do this of course is through scalability—developing the ability to economically add incremental capacity as needed. However, organizations using the Web environment will find that scalability is an acute problem that is achieved by manipulating a limited set of control variables. The following are some broad classes of tactics that can be used:

- *Browser or client.* Characterizing a site's user population can help predict traffic patterns, aid in site design, and produce behaviors that can be used in site testing. Such items include type of user access and bandwidth, browsing pattern, download pattern, and type of browser. Devices are on the market that can read HTTP header information to determine what browser is in use. This helps define what compression algorithms can be best used for large files.

- *The Internet.* The effects of the Internet with respect network and protocol are usually the least controllable. However, some precautions can be used to address the irregularities of the Internet. A Web site should be located as close to the Internet as possible. Reducing the number of hops to the Internet also reduces the likelihood of encountering bottlenecks at peering points (points where private and public networks meet). An unusual slowdown in Web hits is often indicative of a bottleneck or an outage and not necessarily site demand. Site queries are likely to build up elsewhere to the point where they can overwhelm a site once the blockage is cleared. Building in excess site capacity can avoid the expected surge. Routing traffic away from the problem location is easier said than done, as it requires knowledge of the network status.

- *Web infrastructure.* Web applications should be designed to accommodate average peak traffic equivalent to about 70% of the site's capacity in terms of simultaneous users. It is not uncommon for large enterprises to separate Web sites for different lines or facets of their business. This can be used as a load-balancing technique because it can split traffic demand based on the type of visitor. Site designs to handle recurring versus occasional users can be markedly different. Furthermore, special devices called *caches* can be used to store pregenerated pages versus building them on the fly. Caching is discussed further in the chapter. Devices such as *accelerators* or *TCP multiplexers* can pool TCP connections together, minimizing the number of requests for a given page [24]. These reduce connection overhead messaging by acting as proxies to Web site systems, collecting and consolidating TCP/IP requests. They monitor traffic to determine if a connection can be used by another user request.

- *Page components.* Keeping pages as lean as possible is often the wisest approach. Color reduction and text reduction by removing white space in page files can be used if large quantities of graphics are required. Use of compression techniques can reduce page and data sizes. Users having slower access, versus those with broadband connectivity, often notice improvements in download times.

- *Site monitoring.* Use of site performance monitoring and simulation tools can aid in developing cost-effective site enhancements. Software and service-provider solutions are also available that can be used for these purposes. The key to choosing the right solution is knowing what conditions should be monitored and how they should be measured. Measures such as page views per day, average page views per user, page abandonments, and repeat visits can be either useful or wasteful depending on what condition is to be measured. The object of all this is to obtain the ability to reconstruct transactions and identify the root cause of a problem.

Multitiered Web sites pose a challenge to these tools, as they require collecting performance data from each system in each tier and correlating that information across tiers with user sessions [25]. The process flow of a session may have to be tagged and time stamped so that it can be identified across the different systems in the tiers. If several application-monitoring processes are in use, the performance data must be exchanged, collected, and condensed among them. Reporting site performance and problems is still problematic in the industry. There is a need for standards that present monitoring data in a structured fashion.

6.3.2 Web Site Design

Good Web site design can improve reliability [26]. There is a limit to what good site design can achieve, as many variables are outside the control of the site designer. However, skilled designers use techniques to build pages that are less susceptible to Internet-level effects.

6.3.2.1 Site Architecture

Web site architecture will most likely have the greatest influence on performance. A Web site application is commonly comprised of Web site functions, back-end application functions where the business logic and database functions reside. There are two general Web site architecture approaches, shown in Figure 6.9. One approach assigns specific functions or groups of functions to different servers and is often characteristic of multitier architectures. The other approach assigns all functions to

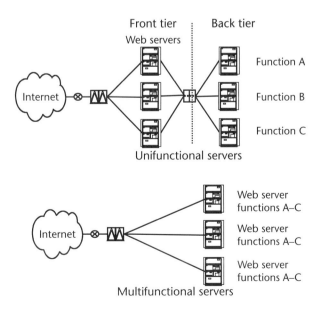

Figure 6.9 Two approaches to Web site implementation.

all servers. These basic approaches can vary and can be used in combination with each other.

In a multitier architecture, each tier can be changed and scaled independently. This is important because adding capacity prior to a seasonal change, major event, or product launch can avoid slowdowns. But multitiered architectures are complex, and interfacing to the back-end systems can be resource and time intensive. They also have more points of failure. Web sites designed for unplanned or planned downtime must either have all functions replicated on some servers or have a failover system for each functional server. Otherwise, they risk losing a portion of their functions during downtime.

6.3.2.2 Site Linking

Today's fast-paced environment has caused many organizations to hastily implement Web sites or site changes without thoroughly testing them. As a result, many sites have a high degree of link and page errors. Page and link errors tend to increase as the number of site pages increases or as the amount of dynamically created content increases. Failure to link to third-party sites, such as those to a credit-authorization site, can cause transactions to fail altogether. One solution is to detect and repair link errors as they occur; the other is to simply test the site thoroughly prior to activating it. Quality error reporting to users can preserve one's image when errors are encountered. Informing the user on recourse could motivate them to return at a later time and discourage them from immediately retrying. Retries on the part of numerous users can potentially overload a site.

Inexperienced Web site designers often use links as subdirectory calls, and fail to realize the magnitude of actively linking to another site during a page download. Links may require going across several networks to a site thousands of miles away, as opposed to accessing a file in a directory on the same machine. Overreliance on linking can spell disaster for a Web site's performance, doing more harm than good. Care should be exerted when inserting links. Each should be inserted with the understanding of the location of the linked site, and proper recourse if the site is unavailable.

6.3.2.3 Site Coding

As Web applications grow in complexity, coding errors will undoubtedly slip through the testing stream. Unlike the traditional system's development process, much testing is being conducted in the production environment, where all of the required data, services, and connectivity are available. Each function must be tested for potential failure and inefficiency. Code validation is a test that typically checks for proper language constructs and syntax. It does not check to see that the output of the application is usable by other processes. Poorly coded Web sites can also pose security issues.

6.3.2.4 Site Content

A lot of graphics, objects, banners, and dynamic content dancing around a screen may do little but dazzle a user—they often provide clutter and confuse the user, making it more difficult for them to find the content they want. Clutter will also lead to a slower site, per earlier discussion.

6.3.2.5 Site Navigation

Failure to understand user-browsing patterns can result in lost transactions and poor performance. Users will access a Web site to find something. The inability for them to find what they want will motivate them to go to another site [27]. When users cannot get the service they want, it is no different than an outage. Getting lost when navigating through a site plus inadequate search tools contributes to the dilemma. Users tend to use primary navigation tools and rarely use supplemental or refined navigation tools [28]. Use of the *back* button is typical when users are lost—they simply backtrack through the pages that got them there.

A common design mistake is to organize content based on how the organization, rather than the customer, views it. For example, a company will arrange product groups in a way that reflects the firm's organizational structure—and not how a customer sees them. Different types of customers will have different browse patterns. Site subscribers will navigate differently than occasional visitors. Subscribers will often require more dynamic content versus casual visitors.

Once a designer understands what types of users will access a site and what the associated navigation patterns are, the next step is to determine how many users of each type will search concurrently, using their most common patterns. This will indicate how much front- and back-end infrastructure will be necessary to handle the transaction load. It should reveal what system components and resources will be mostly used, based on the frequency of the executed patterns.

6.3.3 Web Services

At the time of this writing, the concept of *Web services* is being introduced in the industry. A Web service is software logic or data packaged so that it can be used by other software entities in a network. Similar in concept to APIs, this involves segmenting applications in a way so that other applications in a network can use them, thus turning them into a service for others to use. Web services protocols and languages such as extensible markup language (XML) and simple object application protocol (SOAP) offer a standard way for an application to access software objects on another system. SOAP is a standard object-oriented approach to building software, while XML is the language used to build the messages between objects.

It is too early to tell what impact Web services will have on building mission-critical networks. Web services do imply reliance on other systems and reliable networking for operation—this alone creates multiple points of failure for the application. This book presents many applicable techniques and strategies regarding networking between systems. Needless to say, application developers using Web services should always design logic that heals against situations when a service is unavailable. A recourse or failover strategy should be incorporated, as well as measurement and reporting of the event.

6.3.4 Web Site Recovery Management

Web sites are gradually being subjected to higher availability requirements. Many e-businesses as well as established physical ones depend on them for survival. Problems can arise with site servers, the applications, or the network. Much of this book is devoted to discussion of problems with networks and servers. From the previous discussion, it was evident that application problems due to coding errors, nonexistent links, and inadequate testing are commonplace.

Some of the basic rules discussed throughout this book can be useful in developing Web site survivability mechanisms. First and foremost, always make incremental modifications to a site, and always archive as much of a site and its configuration as possible before you modify it. It is easier to test smaller changes versus wide-scale renovations. All changes, no matter how small, should be thoroughly tested prior to being activated. Carefully controlled distribution of content can also minimize errors. If a failure does occur at a site, page requests and email should be redirected to a secondary site. The secondary site should be at best a mirror image, if possible, of the primary site or should at least be able to provide key functions. Having both sites operational (or load shared) is advantageous and avoids overload. The secondary site should also have an off-line backup image of itself residing somewhere else, usually within the development environment, so that it can be deployed if the secondary site crashes while the primary site is recovering.

The previous discussion on load sharing touched upon how IP address and DNS services can be used to redirect traffic. Avoiding the need to tell users to switch over to another Web site in the event of an outage is of prime importance. There are two strategies (as illustrated in Figure 6.10) that can be used to reroute traffic when a site becomes inactive and that can make the redirection transparent to users. One

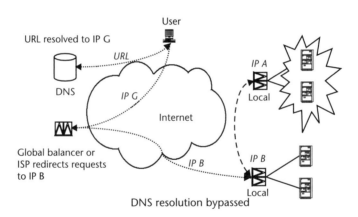

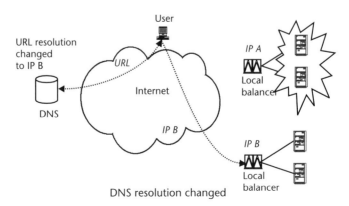

Figure 6.10 Web site recovery examples.

strategy redirects users to the secondary site without changing the IP address and DNS services. The second approach is to change the URL to a new IP address and propagate the change in the DNS services.

In the first approach, applications, load balancers, cluster technology, or even the ISP can be used to redirect users to the secondary site's IP address upon receiving that of the primary site. The second approach involves changing the DNS services over to a new IP address for the URL. Although this seems like a simpler approach, it may be time consuming to implement. The DNS administrators must make the change to a primary and secondary DNS server, through the DNS server hierarchy. This is typically done by alerting the hosting provider of the secondary site. However, the old IP address may still persist in other systems for some time until the new IP address is fully propagated. This is why DNS updates are usually issued only about twice a day.

It goes without saying that the DNS service is critical to both Web site operation and recovery and is a single point of failure. DNS is designed to translate, or resolve, the host name element of URLs to IP addresses so that routing can take place. There should be at least two DNS servers assigned to a domain for redundancy. DNS should preferably be hosted at two locations, by two providers (or a provider and the organization itself). Sole reliance on the DNS process to recover mission-critical sites can be dangerous. DNS services are discussed in more depth in the following section.

6.3.5 Internet Access

Although it is the foundation for e-commerce, the behavior of the Internet grows more unpredictable each day. Protection against such instability needs to be manifested in any Web-based operation [29]. Many ISPs offer their customers service level agreements (SLAs) as an insurance policy against outages. (SLAs are discussed in Chapter 12.) Yet, SLAs with all their loopholes and caveats, offer inadequate protection against the unexpected. To understand why, one must understand how the Internet works.

When a user issues a request to access a Web site, a virtual connection (or session) is set up between a user and a Web server. The Internet decides the routes that packets travel between the two. As packets travel between the two, different routes or paths in different networks can be used at any given time. These paths will vary in bandwidth and traffic conditions. When a network backbone connects to another backbone of greater or lesser bandwidth, an impedance mismatch occurs [30]. These are typically congestion points where packets are often queued, particularly at network egress points. During peak period, buffer overflows can cause packets to get tossed at these points. If this occurs, the TCP protocol resends unacknowledged bytes again, which may constitute multiple packets, creating the potential for more traffic. Selective acknowledgment at TCP minimizes situations where large windows of unacknowledged byte streams are retransmitted. To avoid having congestion conditions cascade into even more congestion, congestion avoidance and control algorithms are implemented. Following detection of lost data, the TCP window size is adjusted radically downward and incremented slowly on each positive acknowledgment until congestion abates.

When a Web request is processed by a Web server, a socket is opened for that user and server resources are allocated for that session. The server acknowledges the request and retrieves the Web site content or data from cache or disk memory. More often than not, the server must connect to other servers to retrieve content, keeping the session alive. The server sends the data back to the user and closes the session. This same process is repeated every time a user requests content. In the end, there is no way an ISP can really guarantee latency due to all of the variables that are outside its control.

Many organizations use the precaution of simply having more than one ISP for their corporate and branch office locations. When choosing an ISP, it is important to understand what type of ISP it is and what survivability capabilities it has. This includes equipment (e.g., router) and network redundancy. ISPs can operate on a national, regional, or local basis. Some cater to only corporate or business-to-business traffic.

6.3.5.1 ISP Access

A regional or local ISP should obtain backbone network access from at least two top-tier national or multinational ISPs to guard against the possibility of outage on one of the backbones. Many regional or local ISPs commonly share the same metropolitan area network (MAN), local exchange carrier (LEC) infrastructure, point of presence (POP), or interconnect site location to access their national backbone provider. A POP generally refers to an access network site of an ISP, where dial, frame relay, digital subscriber line (DSL), and other access servers aggregate traffic. Interconnect or peering sites are where ISPs and multinational providers aggregate traffic onto backbones and relay traffic from one ISP to another. Many ISPs share the same multinational providers as well. In the end, an outage on a multinational provider's backbone or at their interconnect sites will affect many regional or local ISPs.

The architecture of a typical ISP POP site is illustrated in Figure 6.11. The ISP will have line connectivity to a nearby CO with a class 5 switch for dial up access. Remote access servers (RAS) are used to manage the connections. Many POP sites have data centers collocated in the same or nearby building that host the internal and external customer Web sites, including backup and mirrored sites. This is typically done to save the costs of buying or building network access from the hosting site to the POP site. As one can see, the points of failure or congestion are numerous, and an outage at a CO or POP site can be catastrophic.

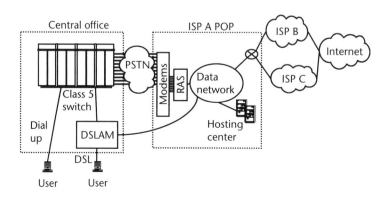

Figure 6.11 ISP access architecture example.

Access from a customer premise to an ISP's POP site should be designed with survivability in mind. Typically, redundant T1 or T3 access circuits are expensive. Redundant circuits should be provisioned along separate physical paths between a customer premise demarcation point and the POP site, as shown in Figure 6.12. Also shown is the case involving circuits to two different ISPs, each also using a separate physical access paths to separate POP sites. In this case, care should be taken to make sure that the POP sites are not collocated in the same building and do not share the same MAN infrastructure, typically a synchronous optical network (SONET) ring.

Organizations should keep the Internet access connectivity separate from their corporate WAN connectivity as much as possible. To save money, many corporations typically backhaul Internet traffic, from the corporate site having direct Internet access, over their WAN. Even at these locations, WAN access should be separate from Internet access. The reason is simple—the Internet can be used as a backup in the event of a WAN outage. VPN service can serve as a backup mechanism. Having separate Internet access connectivity at every corporate location can be a worthwhile approach.

6.3.5.2 Peering and Exchanging

An ISP should have redundant Internet backbone providers and/or redundant connectivity to Internet backbone service providers. Many backbone providers will have their own POP sites that provide connectivity to their backbone networks. Internet backbones exchange traffic at five locations: Metropolitan Area Exchange (MAE) East, MAE West, and Ameritech, Pac Bell, and Sprint network access points (NAPs) [31]. However, there do exist a large number of ghost routes that are used by carriers to bypass the congestion and costs of using the major Internet exchange points. These are accomplished through *peering arrangements*. These are private relationships between ISPs or carriers whereby they exchange each other's traffic at designated locations called *peering points*. These arrangements are also illustrated in Figure 6.12. It should be noted that when establishing redundant ISP access links,

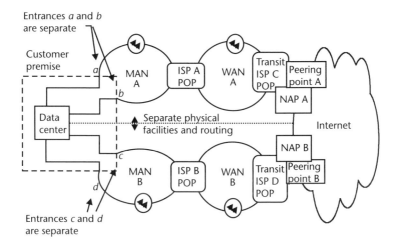

Figure 6.12 Diverse Internet access example.

each ISP POP should reside on different network backbones and connect through different Internet peering points.

Some large ISP carriers will also have *transit arrangements* with smaller ISPs, which are charged for terminating their traffic over the larger ISP's network. Many times, local ISPs may not even have direct access to a national backbone carrier's network and must connect with an ISP who meets the national carrier at either an Internet exchange or peering point. This places a customer's Internet access at the mercy of yet another ISP. These ISPs charge for the traffic they transfer, but often have to transfer much Internet traffic that incidentally flows through their network free of charge. To discourage free traffic, these ISPs often overstate network congestion on their routers at peering locations [32].

6.3.5.3　ISP Performance

To get around these issues, it is important to understand what type of ISP one is using and what type of access they have to a national backbone provider. Some ISPs are demanding SLAs from the transit ISP arrangements to reflect the guarantees made to their clients. Some also seek to minimize the number of exchanges a client's traffic encounters. Furthermore, when choosing an ISP, there are several key performance measures to look for:

- *Initial modem speed*: the negotiated connection speed for dial-up access;
- *Average logon time*: the time to connect to a provider's network;
- *Average download time*: the time for a Web page to download;
- *Average DNS lookup time*: the time for a DNS query to be resolved;
- *Average Web throughput*: the rate at which information is actually transferred;
- *Call failure rate*: the number of times a dial-up connection either cannot be established or is dropped;
- *Average total Web fail/timeout*: Percentage of the time pages fail to download.

6.3.5.4　DNS Services

DNS is a global distributed database that resolves a host name element of a URL to a numeric IP address. Every user's client will send a URL request to its designated DNS server to resolve the URL. A DNS contains literally billions of records and receives millions of updates on a daily basis. DNS functions as a hierarchy of servers. There are currently 13 master root servers at the top end of the hierarchy. The low end includes every domain name registered and advertised by an organization that wants a user-friendly name for its Internet presence rather than dotted decimal IP addresses.

Many large enterprises elect to maintain their own DNS servers internally, so that all queries do not have to always go out to a public DNS on the Internet to obtain an IP address. Depending on how Internet access is configured, this can reduce enterprise WAN traffic and make Internet access faster for users. However, operating DNS services requires care in order to avoid the following potential points of failure [33]:

- Placing all DNS name servers on the same subnet;
- Providing the same IP address for both primary and secondary DNS name servers;
- Failing to list all DNS name servers in domain registration records;
- Locating all DNS name servers on the same physical network segment;
- Placing all DNS name servers behind the same router or single leased line.

Many recommend use of fault-tolerant platforms for DNS servers because of the critical role they play in handling Web traffic. Having secondary failover DNS servers is also wise [34].

6.4   Caching

Caching is a way to improve network access to content or data that has a high frequency of use and that is predominantly static, meaning that it does not change over time. If users retrieve the same static information repeatedly from a Web site, it sometimes makes better sense to place that information in memory on a local server residing as close to the users as possible, even on a LAN or WAN. This avoids sending queries across a network and accessing that content from the main Web site [35]. A popular high-volume site can become overloaded with respect to the access network or server or become less able to retrieve dynamic content. Caching is used to provide relief to busy Web sites and frees them up to handle more dynamic content. Many sites use dynamically generated content (e.g., Java applets, animation, and banners) to make their site more attractive.

Caching is achieved by using special cache devices that intercept user requests prior to reaching their destination. Requests are checked to see if the device can comply with the request and provide the desired content. If not, the request is then forwarded to another site. The basis for making this decision is the currency, or freshness, of the Web object or data. Most cache devices use proprietary algorithms that monitor the hit rate and update frequency for each Web object stored in the cache device. The cache size, performance, and hit rates are continually monitored. Each object is assigned a time to live (TTL), which is a time limit until the object is updated. If a Web object is deemed to be old or outdated, the cache device will perform a *refresh check* to see if the object requires updating with a more recent one. Different vendors use different freshness algorithms. Some solutions always query the main site server to see if the content requires updating, which can cause some delay. Some use algorithms to estimate the life expectancy of an object and refresh based on that time interval. Refresh updates can be scheduled and conducted during off-peak hours, reducing the impact on access response times.

The refresh process involves contacting the main Web site to see if any time or date sensitive Web objects have changed since their last download. This requires objects to be tagged with this information. Because refresh checks can be resource intensive and create latency, they are only done when needed and for tagged objects that exceed a certain size. Although not yet widely used, *explicit expiration*, a process whereby objects are tagged with a definitive freshness date, avoids having the cache device second guessing the freshness of an object. Another approach uses edge side includes (ESIs), which are Web page fragments that are cached and later included with other dynamic objects when the Web site is accessed [36].

6.4.1 Types of Caching Solutions

Caching solutions can come in many forms. Server, appliance, and cluster solutions are available. A cache solution can even be embedded directly into routers or switches. Response time, server performance, hit rate, location, and content currency are key factors to consider when deploying a cache solution. Cache can be deployed in two basic ways (Figure 6.13) [37]:

- *Forward proxy*. Forward proxy caches are intended to minimize redundant network traffic, thereby conserving bandwidth, improving response time, and reducing Web site server load. They are typically placed close to the end users, often on the same LAN, within a workgroup or branch office. They act as a local proxy on behalf of the actual Web site server. They can function in either a *nontransparent* or *transparent mode*. In nontransparent mode, user client browsers and media players are manually configured to point to the cache. The cache proxies the requests and forwards those it cannot satisfy to the Web site. The nontransparent mode is error prone because configuration is done manually. The transparent mode avoids manual browser configuration. In this case, the cache examines all requests going to the Internet and diverts only those it can fulfill. ISPs use it to avoid having to manually configure user

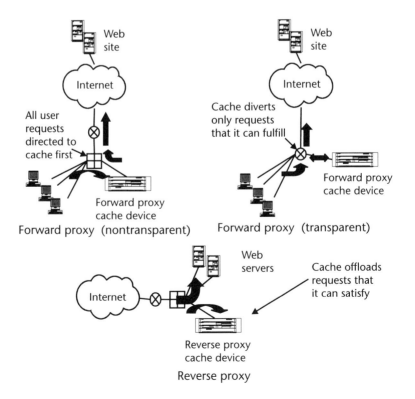

Figure 6.13 Types of cache.

browsers and to save money on Internet connection costs. Transparency is achieved through the use of routers using Web cache control protocol (WCCP) or multilayer switches.

- *Reverse proxy.* Reverse proxy caching, sometimes referred to as *server-side proxy* caching, involves placing a cache device in front of servers in order to improve performance. They are often placed between an organization's Web site and the Internet to improve site performance and protect against traffic surges. Reverse proxy caching can be used in combination with forward proxy caches or even to reduce the number of forward proxies in an organization. A reverse proxy server can act as a front-end processor to a Web site server or server farm. Frequently requested pages are handled by the proxy instead of the Web server. Some proxies even offload SSL encryption processing. Although reverse proxy caching can improve overall Web site performance, it does not necessarily reduce bandwidth consumption.

Load balancing systems can be placed either in front of or behind cache devices. If placed in front, they can be used to redistribute traffic to multiple cache devices. In this case, use of content-aware balancers can significantly improve cache performance. If placed behind, then the caching device will simply process frequently requested pages, reducing requests processed by the load balancer and ultimately the Web site and back-end servers.

6.4.2 Caching Benefits and Drawbacks

Like anything, there are benefits and drawbacks to implementing a caching solution. The following are some benefits:

- Forward proxy caching saves WAN bandwidth costs because traffic is kept close to the network of origin, reducing traffic over a WAN or equivalent network. It avoids the need for backhauling Web traffic to a Web access location [38]. It makes an intranet more scalable and avoids the costs to upgrade network access and bandwidth, as well as procuring additional switches or other types of network equipment.

- Proxy caches on an intranet can act as firewalls or enforce Web access policies. They can restrict outsiders from accessing corporate intranet sites by forwarding only those requests coming from bona fide internal IP addresses. It can also be used to invoke Internet access policies, so that requests for specific URLs can be blocked.

- Caching can help minimize the number of servers required to support a Web site. By offloading Web site servers, server processing performance and utilization is maximized.

- Caching provides a way to efficiently distribute content within a network. Instead of using a data-replication technique to mirror a Web site, caches intrinsically retain a Web site replica for their use. Hot content can be distributed to cache devices when needed to protect the main Web site from a potential traffic surge.

- Web caching improves response and reduces latency in two respects: first, it can reduce the number of turns required of more complex Web sites; second, it avoids having to send requests across a WAN, intranet, or Internet.

- By separating out the static Web objects, caching enables dynamic Web objects with more business value to be developed and implemented more freely without getting bogged with limitations in performance or content distribution.

Aside from these benefits, there are several caveats to be aware of with respect to caching:

- Defective caching can result in data or content that is old or obsolete, or the inability to access Web sites with active content. A chosen cache solution should have proven capabilities to deliver fresh content when desired. Too frequent refresh updates, on the other hand, could overload a Web site. Caching devices that cannot adequately distinguish between static and dynamic objects could end up requesting the same objects repeatedly. Until standard methods of tagging and retrieving objects are agreed upon across different vendors, this is an ever-present problem.

- Some cache devices inadvertently reveal the IP address of a requesting client to the main Web server during a refresh request. Revealing internal IP addresses to the outside world can raise security concerns.

- Sites that have a strong mix of different content, a low cache-hit ratio, or that require streaming may benefit little from using caching. Some vendor solutions do not support all file or data transfer protocols (e.g., FTP) or all streaming media protocols for on-demand or live events. However, some caches can multicast a streamed file to many users.

- The ability to accurately track cached pages can vary among cache-solution providers. Vendors use different methods to capture traffic routed through caching devices, leading to significant discrepancies in traffic reporting. Missed cached pages can result in erroneous Web traffic reports.

- Certain tracking technologies used to observe Web customer/visitor behavior and to customize Web experiences for users may not work as expected when caching is employed.

6.4.3   Content Delivery Networks

Content delivery networks (CDNS) are specialized networks comprised of distributed caches [39]. They provide caching functionality as a service. Customers use CDNs as opposed to purchasing their own cache devices, avoiding equipment costs. Companies use CDN services in the same way that caches are used—to improve user access speeds to content or data [40]. The advantage of a CDN is that cache content can be spread across a geographically distributed network of servers.

A CDN is comprised of a network of servers, caches, and routing devices [41]. In the case of Web access, ISP networks will pull content to the CDN cache server closest to users accessing the Web site (Figure 6.14). Public CDN providers offer their CDN services to enterprise customers, primarily ISPs or content providers. They can

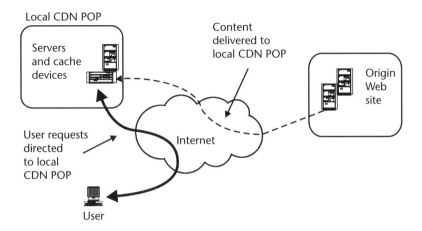

Figure 6.14   CDN example.

use a CDN to distribute content to caching devices in an ISP POP or in one of the CDN's own data centers, instead of the Web site server. Some CDNs deliver content directly to end-user machines as part of the CDN, reducing the capital investment by the CDN provider and gaining as much as 90% customer savings over traditional CDNs.

CDNs can be used in conjunction with load balancers to provide *CDN switching*, an approach that allows CDNs to be used on an as-needed basis (Figure 6.15) [42]. This provides a flexible capability to manage traffic. Using the different types of load balancers discussed earlier, organizations can distribute, in real time, requests to several CDNs based on predefined rules or policies. An organization that hosts a Web site, for example, can lease the services of a CDN for an expected one-time increase in Web site traffic, providing overflow capacity. In another scenario, traffic originating

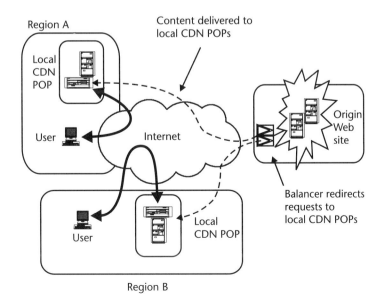

Figure 6.15   CDN switching.

from users in diverse geographic locations can be redirected to the CDN cache that can best provide the needed content based on user attributes. Traffic can also be distributed to CDNs based on CDN or network rate changes.

Public CDNs were initially intended to distribute content to users over the Internet on behalf of Web site owners. This model may not necessarily lend itself to distributing content internally within a private enterprise. As Web traffic and streaming media continues to find its way into enterprise WANs, many organizations are considering creation of their own CDNs to manage and distribute content to remote offices [43]. These are referred to as private CDNs or enterprise CDNs (ECDNs). An ECDN can be used to deliver large streams of content from local caches residing in remote LANs, versus tying up a corporate WAN.

An ECDN is composed of load balancers or content switches, content servers, and back-end application and database servers at the core [44]. Cache devices are then placed in each organization's data center or remote LAN (see Figure 6.16). Cache can even be arranged in a hierarchy so that if content is not found in one cache, another will be searched. The load balancers distribute the load among the content servers as well as providing global redirection. Transparent mode can be used if a caching device is situated in every corporate office, else global redirection can be used. An ECDN will centrally manage, synchronize, and distribute content across all local cache devices and with the content servers [45]. Access and distribution policies can be centrally specified and managed, and can be customized for different corporate locations. Some solutions even allow content created from multiple corporate sites to be centrally consolidated.

CDNs offer themselves as an alternative to investing in expense and resources for content distribution and traffic management and caching. They are used in lieu of overengineering to handle traffic spikes and also eliminate a single point of failure, as multiple copies of content are distributed across a CDN. On the other hand, CDNs can be expensive and suffer from the same general caveats as caches. CDNs can provide access to dynamic content as well as static content. Some CDNs offer content providers special markup languages to assemble dynamic content. The cost

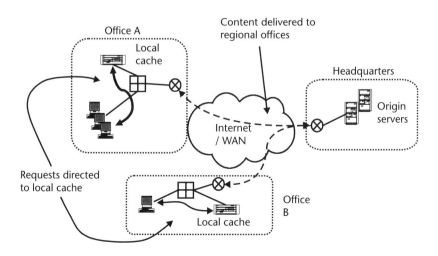

Figure 6.16   ECDN example.

of content delivery to a provider's CDN can be expensive if a high volume of content is being distributed.

6.5   Summary and Conclusions

This chapter discussed various networking and control technologies that can be used to enhance continuity in today's distributed internetworking environment. Discussed were the following:

- Cluster technology is a scalable alternative to centralized mainframe computing networks. Clusters operate on the principal that several redundant servers working together as single resource can provide better performance and reliability than a single system. There are many types of cluster implementations, including multiprocessor, fault-tolerant, and server clusters. Clusters require special software in order to make multiple computers work in unison. Special cluster-aware versions of OS and applications are required for such operation.

  Clusters are designed so that multiple nodes can fail without bringing down the entire cluster. Nodes are given primary tasks and assigned secondary nodes to automatically assume processing upon failure of the primary node. Depending on the type of cluster, cold, warm, and hot failovers can be implemented. For effective survivability, cluster nodes should have access to the same data so that transactions are not lost during failover. Clusters can be implemented at the local or wide area levels. In the latter case, technologies are fast becoming available to geographically disperse cluster nodes at different locations. Clusters also have features that support nondisruptive addition or removal of cluster nodes, OS, and application upgrades.

- Load balancing is a class of techniques used to distribute queries across some available pool of nodes. Load balancing is especially useful in Web sites to assure that traffic volume will not overwhelm an individual site. Load balancers are devices that use predefined rules to redirect requests to different network nodes for optimal operation. There are two basic forms of redirection—local and global. Local redirection is used to route requests from a central location across a group of hosts that typically sit on the same LAN. Global redirection involves directing traffic to a variety of nodes across a network, as in the case of the Internet. Load balancers can be realized through appliance, software, and switch form factors.

- Controlling Web site processing requires a mix of approaches. Good Web site design is the most basic approach to improving site reliability and performance. Multitier and single-tier site architectures can be used, depending on the desired scalability and complexity. Browsing and navigation patterns of visitors will often provide clues as to how much front- and back-end infrastructure will be necessary to handle processing.

  For survivability, a secondary site should be used and situated at a diverse location from the primary site. Both primary and secondary sites should have off-line backup images residing somewhere else, so that they can be deployed if a site crashes. One of two redirection strategies should be chosen in the event one site is overloaded or crashes: send users to an alternate site without changing the IP address, or change the URL to a new IP address and propagate the change in the DNS services. For added protection, many organizations employ more than one ISP. If so, there should be no common infrastructure between the two. Network access to each should be diverse, each ISP should have redundant Internet backbone providers or redundant connectivity to the Internet, and both should not share the same backbone service providers.

- Caching is a way to improve content or data access performance. Caching involves using special cache devices that intercept user requests for content or data that is frequently accessed. Requests are checked to see if the device can comply with the request and provide the desired content. If not, the request is then forwarded to another site. An alternative to using a cache device is to employ a CDN service. These are specialized networks that provide the same function as caches but have the advantage of spreading content across a geographically distributed network of caches. Such content distribution can be used to abate traffic spikes and aid redundancy by distributing multiple copies of content across a network.

References

[1] Chhabria, R., and N. Nandkeshwar, "Availability in Enterprise Computing, *Enterprise Systems Journal*, July 1999, pp. 41–45.

[2] Fletcher, J., "Netplex: A Key to High Availability in the Datacenter," *Enterprise Systems Journal*, March 1998, pp. 30–32.

[3] Bucholtz, C., "Ensuring Uptime," *VAR Business*, January 7, 2002, pp. 35–37.

[4] Fletcher, J., "Parallel Sysplex High Availability with TCP/IP," *Enterprise Systems Journal*, June 1998, pp. 68–70.

[5] Butenko, V., "Dynamic Clustering Provides E–mail," *Network World*, June 18, 2001, p. 43.

[6] Anderson, R., M. Lee, and S. Chapin, "Unraveling the Mysteries of Clustering," *Network Computing*, October 2, 2000, pp. 189–195.

[7] Borck, J., "Clustering Comes of Age," *InfoWorld*, December 17, 2001, p. 32.

[8] Golick, J., "Loading Up: High–Speed Transaction Processing on the Internet," *Network Magazine*, December 1999, pp. 66–72.

[9] Jacobs, A., "Sever Clusters Move Toward Mainstream," *Network World*, March 18, 2002, p. 22.

[10] Neel, D., and T. Sullivan, "Strength in Numbers," *InfoWorld*, December 17, 2001, pp. 28–30.

[11] Pulsen, K., "Clustering Concepts and Technologies," *TV Technology*, August 11, 1999, pp. 34–39.

[12] Connor, D., "Novell Adding Luster to NetWare 5 Clusters," *Network World*, September 13, 1999, Vol. 16, No. 37, p. 1,142.

[13] Stupca, C., "Providing High Availability Through AS/400 Clusters," *Selling AS/400 Solutions*, May/June 1999, pp. 31–38.

[14] Pendry, D., "SANs, Clustered Servers Get Special Hubs," *InfoWorld*, September 14, 1998, p. 12.

[15] Wagner, M., "Data Center Redundancy Keeps Benefits Online," *Internet Week*, November 15, 1999, p. 32.

[16] Chappell, L., "Router Commands," *Network World*, September 4, 2000, p. 48.

[17] Liebmann, L., "Load Balancing: Where the Action Is," *Network Magazine*, March 2000, pp. 60–64.

[18] Schwartz, R., "Simplicity in Load Balancing," *Web Techniques*, November 2000, pp. 58–60.

[19] Liska, A., "Balancing the Load…Globally," *Hosting Tech Magazine*, July 2002, pp. 85–87.

[20] Phifer, L., "Balancing Server Loads Globally," *Internet World*, July 7, 1999, pp. 27–28.

[21] Macvittie, L., "Cover Your Assets, Web Style," *Network Computing*, July 8, 2002, pp. 61–62.

[22] Harman, D., "Switch-Farm Architecture Overcomes Backbone Bottlenecks," *Lightwave*, January 2001, pp. 122–128.

[23] Wilson, T., "Internet Performance Warning," *Internet Week*, October 18, 1999, p. 18.

[24] Susai, M., "TCP/IP Multiplexing Boosts Sites," *Network World*, March 19, 2001, p. 47.

[25] Gomolski, B., "Top 10 Recommendations on Building Scalable, High-Performance Web Sites," *InfoWorld*, January 15, 2001, p. 70.

[26] Wonnacott, L., "Tune Up Your Web Site for Usability," *Enterprise Systems Journal*, December 2001, pp. 51–54.

[27] *The Big Picture—Supplement to Internet Week*, April 3, 2000, pp. S3–S21.

[28] Vonder Haar, S., et al., "Holiday Selling Season: High Hopes, High Stakes," *Interactive Week*, September 27, 1999, pp. 71–74.

[29] Spangler, T., "Racing Toward the Always-On Internet," *Interactive Week*, September 6, 1999, pp. 7–12.

[30] Khan, M., "The Need for Speed," *Communications News*, March 1998, p. 80.

[31] Sweeney, D., "The Coming Crash of E-Commerce," *Network Reliability—Supplement to America's Network*, August 2000, pp. 11S–20S.

[32] Strom, D., "Tech ABC: What's Blocking the Fast Lane," *Internet World*, October 1, 1999, p. 68.

[33] Fontana, J., "DNS Risks Lurk in Corporate Networks," *Network World*, July 1, 2002, pp. 17–18.

[34] Powell, T., "Web Optimization Options," *Network World*, May 27, 2002, pp. 47–48.

[35] Bruno, C., G. Kilmartin, and K. Tolly, "Fast Relief for Slow Web Sites," *Network World*, November 1, 1999, pp. 59–62.

[36] Allen, D., "Lesson 169: Edge Side Includes," *Network Magazine*, Vol. 17, No. 8, pp. 22–24.

[37] Clegg, S., "Cache Panache," *Network World*, April 10, 2000, pp. 57–60.

[38] Riggs, B., and B. Wallace, "Caching is Catching On," *Information Week*, November 1, 1999, pp. 20–22.

[39] Baltazar, H., "Content Delivery Gets Smart," *eWeek*, March 11, 2002, pp. 51–53.

[40] Lyman, J., "Delivering the Goods," *Web Techniques*, February 2002, pp. 28–31.

[41] Margulius, D., "Living on the Edge," *InfoWorld*, March 18, 2002, pp. 52–57.

[42] Skene, B., "CDN Switching Adds Flexibility," *Network World*, January 14, 2002, pp. 29.

[43] Pappalardo, D., "Caching Service Has Int'l Flair," *Network World*, September, 27, 1999, p. 10.

[44] Doherty, S., "Warp Speed Web Content," *Network Computing*, April 29, 2002, pp. 61–79.

[45] Mears, J., "CDNs Taking on an Enterprise Role," *Network World*, February 4, 2002, pp. 17–20.

CHAPTER 7

Network Access Continuity

Network access links are classic single points of failure. Convergence has placed greater importance on access link survivability, as a failed link can mean the interruption of multiple data and voice services. Not only is it important to protect an enterprise location's access links, those connecting to critical customer locations should also be protected. This chapter reviews the survivability and performance features of several voice and data access technologies. We also highlight some of the survivability issues surrounding several wireless technologies that are commonly used to access core networks.

7.1 Voice Network Access

Voice telephone service is considered the lifeblood for many enterprises. A voice outage can place an enterprise at a complete standstill. Nevertheless, many companies pay less attention to the survivability of their voice infrastructure than their data infrastructure. There are sound reasons for this. First, most telephone network outages are relatively small. Voice network infrastructure, from customer premise equipment (CPE) to network switching systems is very reliable. Service provider network transmission and switching systems are designed on fault-tolerant platforms with 99.999% availability. Most systems are designed to minimize the number of active calls in progress that are lost in the event of a failure.

Large-scale data networks, particularly the Internet, have a long way to go with providing comparable availability. Surprisingly, voice and data networks share many common networking elements and infrastructure, and will share even more in the future. A good portion of data traffic, particularly in the access network, is carried on dual-tone multifrequency (DTMF) voice lines, time division multiplexing (TDM) (e.g., DS1, DS3) access circuits, and transported on synchronous optical network (SONET) ring backbones using SONET add-drop multiplexers (ADMs) and digital access cross-connect systems (DACS). Much of this infrastructure was originally intended for voice traffic.

Although voice is a mission-critical service that is designed for high availability, the nature of circuit-switched voice traffic is more forgiving than data. Voice is more resilient to bit errors for the simple reason that the human ear can understand what is being said over the nosiest of lines. But regardless, voice networks are designed with redundancies for high availability of service. Voice networks with redundant TDM links can still have problems during failover to a protected circuit, as a TDM circuit can drop calls and go into alarm.

The same types of precautions one takes with data networking equipment apply to phone switches and associated equipment. Even the software that drives most voice systems requires similar precautions as with application software. Many of the principles and strategies discussed in this book relative to data networks also apply to voice networks, with some exception. The following sections describe some techniques and strategies applicable to enterprise voice networks.

7.1.1 PBXs

Modern private branch exchange (PBX) systems use digital trunks for access to a carrier's central office (CO) from their premise. These are typically in the form of T1 or integrated services digital network (ISDN) primary rate interface (PRI)/basic rate interface (BRI) circuits. Although they are digital circuits, the underlying carrier signals are analog signals (this applies to any type of transmission). Older analog PBXs directly modulate the analog carrier signal using the analog voice signal. In digital PBXs, several voice signals are assigned to time slots and are encoded/decoded over the analog signals. This requires use of digital service unit (DSU)/channel service unit (CSU) devices with T1 interfaces at the customer premise. This capability is sometimes provided in a separate box or integrated within the PBX, depending on the vendor.

Figure 7.1 shows a typical digital PBX configuration, along with some features that provide additional voice service survivability:

- *System redundancy.* Some systems are designed with redundant components with failover to protect the most critical system functions. These are built on

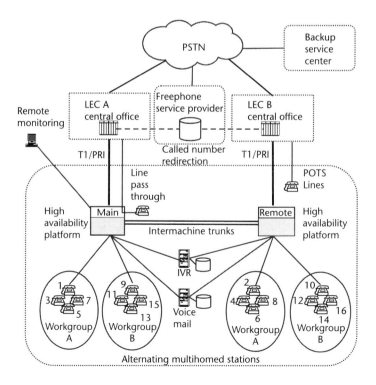

Figure 7.1 PBX with survivability features.

high-availability or fault-tolerant platforms. Consequently, systems with these features are often more expensive.

- *Power protection.* PBXs are software-driven systems. If power is interrupted, all program processes in memory can be lost and require restart. This is particularly true with older PBX systems. Many come equipped with uninterruptible power supplies (UPSs) that typically have enough power to run or shutdown the system until the UPS battery supply is used up or a backup generator resumes power [1]. Some companies have back-up generation or dual-utility feeds to maintain power. Some PBXs system vendors provide generators with their systems.

- *Station failover.* Many PBXs are equipped with the ability to automatically transfer certain stations directly to an access trunk or lines upon failure, bypassing the PBX altogether. It is also wise to have a basic single-line phone connected to dedicated plain old telephone service (POTS) lines, which do not go through the PBX as an extra backup.

- *Remote switch modules.* Organizations and call centers occupying large buildings often use PBXs that are partitioned into multiple switches situated at different locations in the building. This provides switch diversity and enables voice operation to continue in the event a portion of the building experiences a physical disaster or the system simply fails. A two-way trunk is established between the two modules to carry calls from one location to the other. Each module can also be homed to a different building entry point and even to a different carrier CO.

- *Multihomed stations.* Used typically in conjunction with multiple switch modules, this strategy connects every other station to a different switch module, regardless of where the station lies in the building. The purpose of this strategy is to keep each workgroup's operation partially functional in the event one switch module becomes inactive. This approach can result in a cabling nightmare and be expensive to implement, and it requires sound structured cabling techniques. Some techniques are discussed later in this book.

- *Call reroute.* If a PBX is down altogether, organizations must rely on their local exchange carrier (LEC) to reroute inbound calls to either another predetermined location, live operator, or an announcement. Organizations that subscribe to 800 or a freephone service can have reroutes performed in almost real time through the freephone provider's software system. Because freephone numbers translate to POTS numbers for routing, based on North American numbering plan (NANP) format, the routing number can easily be switched or calls rerouted to an alternate destination's POTS number. This makes the outage transparent to callers. On the other hand, if an enterprise uses POTS numbers for access, then they must arrange with the LEC beforehand to revise the switch translation to another destination. Some LECs do provide this service to subscribers. Some enterprises may also use another LEC as a backup, requiring transfer of the client's POTS number to their switches. The interexchange carrier (IXC) will also have to make changes as well, particularly in those cases where enterprises bypass the LEC access network. In any case, the transfer is not likely to be automatic and may consume precious time. Local number portability (LNP) services are being deployed in selected areas that can make this transfer process easier.

- *Informative announcements.* In the event of an outage, announcements can be used to inform callers about what has taken place, what is being done to correct the situation, and what recourse is available. Doing so not only preserves one's image, it can also discourage retries and avoid call overload once service resumes.

- *Remote diagnostics.* Many PBX providers can remotely dial into and perform troubleshooting, provided they have access to the PBX. Although this feature is quite effective, it can make one's system more susceptible to hackers.

- *Spare components.* Having spare replacement components available or readily delivered reduces recovery time. It is important to know how to reactivate the component. Quick access to replacement components for important nonredundant system components is always wise. Depending on the type of service agreement in place, some PBX vendors will guarantee delivery and installation of a completely new system if a catastrophic event took place.

- *Backup storage.* Switch configuration settings, programming, and databases should be backed up using some of the storage techniques discussed in this book. This also includes voice mail and intelligent voice response (IVR) settings and data as well.

- *Alternative backup services.* There are a number of ways to provide continuous voice services if an outage does take place and no failover location is available. One way is to contract with a telemarketing or service center companies to maintain call services during the event. Centrex service, which is not optimized for redundancy but is highly reliable, can also be used. Some enterprises use private lines or even virtual private network (VPN) services as an option.

7.1.2 IP Telephony

The next generation PBXs will be Internet protocol (IP)–based systems that transport voice as IP data packets, as opposed to circuit-switched modulated signals. IP-PBXs (iPBXs) operate as communication servers on a data network. In this regard, they are subject to many of the same protection strategies that are discussed throughout this book. Supporting mission-critical voice IP services over nonmission-critical, best-effort IP networks is still problematic.

As a best-effort transport, IP is designed to route packets with no guarantee that they will arrive at their destination. When packets do arrive at a destination, they are assembled in their original order. Any missing packets are requested and inserted once they are received. This process can cause latency (or delay) in delivering a data stream. Packet order and proper time arrival is critical for voice service. Delays exceeding 250 ms are not uncommon, depending on the type of call. Calls are transported over the public Internet are subject to its inherent instability, making it even more difficult in assuring performance and availability.

One way to get around this issue is to transmit a stream of packets in a continuous stream through a network in the proper order. Protocols such as H.323 and session-initiated protocol (SIP) are designed to support this. However, total protocol compliancy and full interoperability is still not assured among vendor products.

Because the current voice infrastructure is still predominantly circuit switched, many vendors offer hybrid solutions comprised of circuit-switched systems with IP interfaces. As voice over IP (VoIP) matures and is supplemented with multipath label switching (MPLS) or other quality of service (QoS) mechanisms, VoIP transport will no longer be best effort, requiring organizations to pay a premium for priority service for streaming data and voice.

7.1.2.1   Softswitch Technology

Next generation telecom switching systems will be created using softswitch technology. Softswitches are designed to switch packet-based traffic [2]. Traditional CO call processing and switching functions are unbundled and spread across different modular software components. Third-party software can be deployed to provide special or customized services. Some systems are built using a network of servers, each performing a specialized function or service. The switch can handle traffic on legacy interoffice trunking and interface with the service provider's operational support systems (OSSs).

Softswitches can be deployed in mated pairs for reliability (see Figure 7.2). As with a traditional CO switch, signaling system 7 (SS7) interfaces are provided to handle call-processing messages [3]. SS7 is a packet-based service originally intended to handle these messages between switches and other network elements in a circuit-switched environment. SS7 operates on a network separate from the circuit-switched network. In a typical call-processing scenario, an originating switch would perform global title translation (GTT) on called party phone numbers to identify the SS7 network address, known as a *point code*, of the destination switch.

Traditionally, a single point code was defined for each signaling element. Because a failure of an SS7 link could potentially stop call service, switches and other signaling elements were typically homed to two signaling transfer points (STPs), which are switches designed to transfer SS7 traffic. Softswitches enable support for a single code to point to multiple network elements so that traffic can be rerouted to another destination using the same point code.

Softswitches, much like servers, can be deployed in mated pair configurations for redundancy. They can also be used to provide a redundant path for call traffic using Internet call diversion (ICD). This is a feature that uses the Internet or an IP-based network to divert traffic. Call traffic originating from an SS7-based network is switched to a gateway device, from which the calls are then routed to a softswitch.

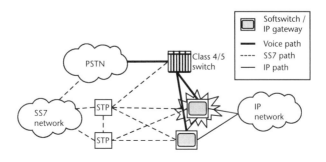

Figure 7.2   Softswitch mated pair example.

7.1.2.2   VoIP Gateways

VoIP gateways are devices that that convert traditional circuit-switched voice to IP packets for transmission over a local area network (LAN) or an IP network. These devices typically attach to a PBX station (FXO) port or trunk (FXS) port. The former is typically used for inbound calls, and the latter is used for outbound calls. Station (or extension) numbers are mapped to the IP address of a user's IP phone or *softphone*.

7.1.2.3   IP Phones

A key component of any VoIP implementation is the IP phone. It is essentially a telephone designed to operate on a transmission control protocol (TCP)/IP network, typically over an Ethernet LAN. Although they have the same look and feel of a traditional phone, they are in essence a personal computer (PC) disguised as a telephone. A *softphone* is a software application that emulates a telephone on a PC. Because of this feature, they are portable and can be used to facilitate a recovery operation by allowing users intercompany dialing from a remote location. Unlike an IP phone, a softphone can share the same data port as the host PC. Like any Ethernet device, both IP and softphones require use of dynamic hierarchical configuration protocol (DHCP) services.

7.1.2.4   VoIP Architectures

Figure 7.3 depicts a VoIP architecture that is often used for recovery and high availability. This approach uses a wide area network (WAN) and PSTN as redundant networks. Traffic can split over the multiple networks or the WAN can be used as an alternate voice network in the event a public switched telephone network (PSTN) carrier's network experiences an outage. Furthermore, phones can be ported to another corporate site having WAN access.

7.1.3   Intelligent Voice Response and Voice Mail Systems

Intelligent voice response (IVR) and voice mail systems are often overlooked critical components of a mission-critical operation. Voice mail, for example, is becoming

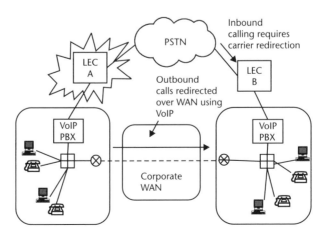

Figure 7.3   Call redirection using VoIP.

more and more mission critical in nature. Having a backup voice-mail server is always prudent, but one should consider network-based voice mail as an alternative to a backup server. Many voice carriers have network-based voice-mail services. Establishing network-based voice mail after a voice-mail system crashes is usually too lengthy. It is best to have the service already set up and be able to readily forward calls to the service when necessary.

IVRs have become the voice portals of many enterprises and often see high volumes of traffic, much of which involves little, if any, human intervention. In the event an IVR fails, a failover IVR should be made available. In addition, staff may have to be freed up or an answering service may have to step in if an IVR fails completely or the backup IVR has less capacity to handle the usual call volumes. Furthermore, IVR failures are known to trigger rolling disasters. Because user input is often written to databases, database corruption is possible. Precautions should be taken to protect the data, such as those discussed in the chapter on storage.

7.1.4   Carrier Services

Many voice telecom carriers had aspirations of becoming all-in-one carriers to clients. This means they would be a one-stop shop for local and long-distance services, and in some cases wireless, cable-TV, and Internet access. In an attempt to achieve this, the Telecommunications Act of 1996 aimed to deregulate local and long-distance services and allowed newer carriers to provide these services [4]. In many cases, incumbent LECs (ILECs), namely the regional Bell operating companies (RBOCs), had intentions of entering the long-distance market, while IXCs targeted the local market for their services. To jumpstart competition in the local markets, the Federal Communications Commission (FCC) granted newer competitive LECs (CLECs) the ability to enter local markets served by ILECs and buy local-loop infrastructure from them. In return, the ILECs were allowed permission to enter the long-distance market. At the time of this writing, many of the newer carriers have gone bankrupt, and ILECs have limited long-distance offerings. The FCC has all but acknowledged that perhaps its model of a deregulated telecom environment was flawed.

LECs terminate their access lines (usually twisted pair) and circuits at their COs. There, calls over the lines or circuits are switched using electronic switching exchange systems, which are typically comprised of fault-tolerant computers managing a switch fabric. Local networks are logically constructed as a hierarchy. COs direct incoming calls to local subscribers or to an access tandem office, which then direct calls to other COs, other tandems, or a designated IXC. Although COs in and of themselves are very reliable facilities (hence use of the term *carrier class* to mean high availability), using them and the service provider's capabilities for survivability requires careful planning. The following are some strategies that should be considered:

- *Carrier commonality.* In many cases, carriers lease parts of their network from an assortment of other carriers, including other LECs, cable-TV operators, or even an IXC. When choosing a voice service provider to support a mission-critical operation, it is imperative to know what facilities they lease, from who, and how those leased facilities are used in their network. True carrier diversity requires that neither carrier owns, maintains, or leases any infrastructure or services from the other. Their networks should operate in parallel from end to end, with no commonality between them.

- *Call reroute capabilities.* Rerouting was discussed earlier in this chapter in relation to PBX problems. In addition to a PBX becoming inactive, other situations can arise that require redirection of calls. Situations can arise whereby call volume exceeds the number of engineered lines to a location. In this case, a service provider could overflow calls to another location. Another situation can involve blocked calls when a T1 or PRI fails, or lines drop due to facility outage. In these situations, a service provider should be able to reroute calls to another location. This means that calls to a dialed number will no longer be terminated to their usual lines—they will be switched to other lines or trunks connecting to another destination having another telephone number. Typically, switching exchanges can translate POTS numbers to other numbers. The carrier should allow call redirection to be initiated almost instantaneously either automatically or through manual activation.

- *Wire center diversity.* Diverse paths to the customer location can be established via two separate COs (see Figure 7.4). Both paths would require separate physical paths to the location [5]. COs typically do not have alternate paths to other COs, except through access tandem offices. If the path through one CO could not be used, the access tandem could reroute calls through the other CO. COs should subtend to multiple access tandems for better reliability and call throughput. Some very large firms will maintain PBXs that are almost equivalent to a CO switch. In some cases, these systems can be configured as an end office in an LEC's network. If the company location is large enough, the LEC can even designate a single exchange (i.e., an NXX with a single block of numbers) to that location. Thus, one of the diverse COs is the actual customer switch. Load sharing across the diverse routes could be achieved several ways. Routing outbound calls on one path and inbound calls on the other is an approach that is often used.

- *Path diversity.* Diverse paths to a customer location using diverse carrier COs can provide ideal protection. If 800 service is being used, then inbound calls can be distributed between the two paths, effectively load sharing them. COs situated on SONET rings provide even better protection. There are several access scenarios that can be used, illustrated in Figure 7.5. Some LECs will

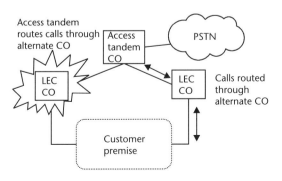

Figure 7.4 Wire center diversity.

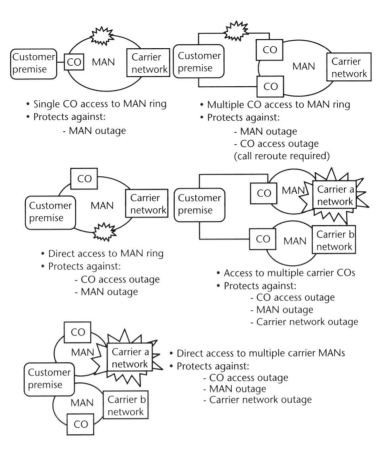

Figure 7.5   Carrier network access scenarios.

even situate a customer's location directly on their SONET ring and will install an ADM at their location.

- *Grade of service.* Grade of service (GOS) is the percentage of calls that are blocked after trying all paths to a destination. Service providers should guarantee a GOS, typically about .005. Most carrier networks are designed for oversubscription, meaning that there are many more subscribers than there are lines (or trunks). The ratio of customers to lines is typically in the range of six to one to eight to one.

- *Survivable switch architecture.* Electronic switching exchanges should have a distributed system architecture so that a switch module failure can be isolated and calls in progress can be transferred to another module, with little if any lost calls. Transmission facility transport and customer line provisioning should be located in separate portions of the switch. This keeps trunk or line configurations independent of the transmission circuits. It is important to also know what type of network monitoring, technical assistance, and disaster-recovery services the carrier has with their switch and transmission equipment vendors.

- *Problem resolution and escalation.* Redundancy using multiple local and/or long-distance voice carriers can provide a false sense of security if not correctly planned. A good voice service provider should have well-defined problem escalation procedures that can guarantee resolution within a specified time period. They should be able to customize call processing and routing services according to a customer's need during a crisis. It is also important to know if they had contracted with the National Telecommunications and Information Administration (NTIA) to see if they have priority treatment in event of a national crisis.

- *Telephony infrastructure.* Many of the newer local providers overlay voice services on nontelephony infrastructure. It is important to understand what type of plant is in use. For example, voice and Internet access services can be overlaid on cable TV or data networking plant. Voice frequencies are susceptible to interference, so any type of shared infrastructure with cable TV or high frequency digital subscriber line (DSL) could be problematic. Cable TV infrastructure was never originally designed to the same level of standards as telephony. This also carries over to cable TV network operations and management.

- *Line quality.* Faulty or noisy phone lines are becoming more and more unacceptable. Noise on a line can disrupt modem communications. Likewise, data communication using DSL can affect voice using the same line. When noise is suspect, use of a line tester or standard telephone could be used to verify the line quality or the ability to make or receive calls. High-frequency line noise may not be perceptible to the human ear, however. The service provider should have the ability to readily test, repair, or replace a faulty line. DSL modems, which utilize the higher end of the frequency spectrum, can cause line noise and often cannot be used simultaneously with lines that use a modem. Use of line filters or lower speed communication can be used as alternatives to work around such problems.

- *IXC redundancy.* Many of the previously discussed concepts lend themselves to IXCs as well. Multiple IXCs should be retained in the event one of them experiences an extensive outage or call overload. This has been known to happen. Arrangement should be made with the LEC to readily redirect long-distance calls to the other carrier in such instances. If a subscription to 800 or freephone service is in place with a carrier, arrangements should be made to either transfer or activate the same numbers with another IXC or redirect callers to another freephone number using an announcement.

7.2   Data Network Access

Carrier access issues were discussed earlier in this chapter with respect to voice access to a LEC. More often than not, voice and data access links are configured over copper loop using time division T1/T3 access. These links typically terminate on a DSU/CSU device at the customer premise. These devices function as a digital modem and are often placed between the line and a router. Provisioning an access line is often a time-consuming and resource-intensive process—one that was further aggravated by telecom deregulation.

When configuring access links, particularly for data, matching the DSU/CSUs and routers at the CPE with those on the far end of the WAN connection can avoid many problems down the road. Consistency in equipment manufacturer, models, software, and configuration settings can help eliminate some of the hidden problems that often bug network managers. Obtaining as many components as possible from the same dealer can also minimize the number of customer service representatives that need to be contacted in case of problems.

7.2.1   Logical Link Access Techniques

Having multiple diverse physical connections into a location is only effective if equipment can failover, switch over, or actively use the other link at the logical level. If one of the access circuits fails, failover to another circuit is required, accompanied by swift recovery of the failed circuit. Physical link failures or problems (*hard outages*) can lead to logical link failures (*soft outages*), unless physical connectivity can be reinstated immediately, as in the case of SONET. For example, physical line noise can cause bits to drop, causing transmission errors. Logical links can drop without necessarily involving the physical link, due to transmission, traffic, software, or equipment problems. Problems in the core of a network may also lead to logical link access problems as well.

A logical access link can be viewed as a single point of failure. If a redundant link is implemented, steps must be taken to assure redundancy across all levels. As discussed earlier, if multiple carriers are being used, there should be no commonality in access and core network facilities or operations. Customers are usually not informed if their circuit is provisioned on the service provider's own fiber or if the fiber is leased. If a fiber cable is cut or if a DACS unit fails, logical access links can fail as well.

Data centers designed to perform centralized processing for an enterprise should have redundant access links and be located as close to the network backbone as possible. The links should be connected through different points of presence (POPs) on different backbones. This avoids the possibility of core network and local loop problems affecting access links. If the links involve Internet access, then each POP should also connect to different Internet peering points.

Redundant link solutions can be expensive. Use of a backup link of less capacity and speed can be an attractive alternative to providing redundancy or upgrading an existing access link. Many frame relay access devices (FRADs), DSU/CSUs, or routers have features to detect failures and dial up using several preconfigured connections over a redundant ISDN or switched 56 service. Failover may not necessarily be instantaneous, so the possibility of losing traffic exists. Using such links to protect only the most critical traffic can avoid performance issues while in use. These links could also be used to handle overflow access traffic from a main access link during overload situations.

Access between the premise LAN and the link access devices, such as the routers or FRADs, should also be protected. Redundant routers should connect to critical LAN segments, especially those connecting to servers or clusters. Use of hot standby routing protocol (HSRP), virtual router redundancy protocol (VRRP), and LAN link protection methods discussed earlier in this book can be applied.

7.2.1.1   Link Aggregation

Link aggregation is the ability to configure many point-to-point links to act as one logical link. Point-to-point links can be consolidated to travel over a single path or multiple paths [6]. Link aggregation technologies include inverse multiplexers (IMUX), IMUX over asynchronous transfer mode (IMA), multilink point-to-point protocol (PPP), multilink frame relay, and router/switch load-sharing. Such technologies can be used to establish multiple access links to a carrier or core network.

- *Inverse multiplexing.* Inverse multiplexing is a technique that aggregates many access links into an aggregate link overlaid across different paths, hence the term *inverse multiplexing*. If one link fails, others can take over (see Figure 7.6). An IMUX system can combine diverse channels and spread the traffic across multiple DS1 channels, sometimes referred to as *NxDS1s*. The clock rates of the individual channels are synchronized in a manner such that frame alignment is maintained across the channels. Transmission delays among the channels are corrected. If one DS1 fails, the others continue to operate. Use of an IMUX in this fashion avoids having a DSU/CSU for each individual DS1 channel. NxDS1s can be grouped together to allocate bandwidth or improve utilization for specific types of traffic. For example, one DS1 can be allocated as a backup channel while the remaining DS1s are combined into a primary channel.

  IMA and multilink frame relay accumulate diverse access links for transmission over multiple channels through a network. Several ATM switch platforms provide IMA capability. Unlike the previous case, IMA segments data into ATM cells traversing over parallel DS1 channels, without the need for clock synchronization. Frame alignment is maintained through the use of IMA control protocol (ICP) cells that are inserted on each link. The ATM segmentation ICP cell insertion makes IMA more processing intensive than an IMUX system. Some switches and routers come equipped with IMUX or IMA capabilities.

- *Load-sharing routers.* Routers connecting parallel WAN links can use open shortest path first (OSPF) equal cost multipath (ECMP) capabilities to distribute packets across the links (Figure 7.7). In the case of equal-cost paths, packets are sent across the parallel links in a predefined order, round robin or some other means, depending on the router's capabilities. On approach involves sending all traffic destined to the same location on one link, referred to as *per-session load sharing*. Uniform traffic across all links may not be as easily achieved using routers versus using IMUX or IMA solutions. On the other hand, routers do provide greater flexibility.

  In layer 2 networks such as Ethernet, multiple paths from a single switch to a router would have to be established without violating the spanning-tree

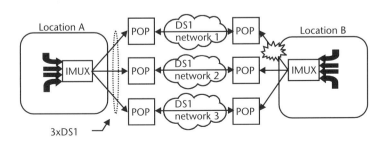

Figure 7.6 Inverse multiplexing example.

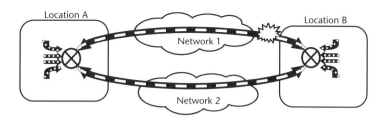

Figure 7.7   Load-sharing router example.

algorithm. The two links must be provisioned so that the switch views them as a single aggregated link. Traffic is then distributed over the links using various techniques, including load balancing or multiple adapters. Some switches have inherent capabilities to distribute traffic over multiple links. An edge device such as a router or switch is considered a single point of failure, so dual devices or fault-tolerant devices would be required if a high degree of survivability is desired. Dual routers would require use of VRRP or comparable protocol [7].

- *Multilink PPP.* Multilink PPP (MLP) is a router-based solution for use with links having PPP, an encapsulation protocol typically used on dial-up lines. PPP frames are divided up, assigned a new header, and transferred across multiple parallel links, usually in a round-robin sequence. PPP frames are reassembled at the receiving end. Frame segmentation and assembly can add significant processing overhead.

Several rules should be followed when implementing multiple parallel access links. First, line speed, latency, and technology should be matched across the links to the extent possible. Systems with built-in buffering for frame synchronization and latency correction are a good choice. Second, both ends of the access link should be configured identically, using the same equipment wherever possible. These rules apply especially to geographically diverse links, which are always highly recommended.

Backup links and link aggregation can serve different purposes and have different requirements. Unlike aggregated links, backup links are not necessarily matched with the primary link in speed or technology. Backup links are typically separate standby links designed to carry only critical traffic at minimum service levels during relatively short outages. In the case of frequent outages, upgrade of the primary link may be required versus constant reliance on the backup link. Because they are activated upon an outage, backup links may require failover process, resulting in some lost traffic.

On the other hand, link aggregation is a load-sharing technique usually done to enhance performance, especially in cases where a primary link upgrade is not achievable or desirable. It is also done to sustain reasonable performance during outages. Link aggregation is more costly and complex than having a simple backup link. Both techniques can be used together to fortify mission-critical access.

7.2.2   Physical Access Techniques

There has been much discussion thus far in this chapter regarding local access. This section will reemphasize some of the points discussed earlier and will introduce some additional points. As stated earlier, logical link access is predicated on physical access. The service provider is usually responsible for physical access. Physical access is typically the copper, fiber, or wireless infrastructure that carries the logical circuits between a customer premise and a carrier premise. This portion is typically referred to as the local loop. The carrier premise, usually referred to as a POP, represents the point where the circuit is connected to another network, either the carrier's own network or that of someone else. The local loop is traditionally one of the most vulnerable parts of a network.

Some local loops are divided into a feeder and distribution networks. The latter is a high-capacity network, usually made of fiber, which carries circuits from a POP (or CO) to a central location called a remote terminal (RT). From there, higher bandwidth circuits are demultiplexed into lower bandwidth circuits that extend to a customer premise. The Telecom Deregulation Act of 1996 allowed carriers to lease out portions of the local loop to other competing carriers needing that infrastructure to provide service. This not only includes physical portions (e.g., cable, fiber, and conduit) but also logical portions (e.g., channels). This environment has further exacerbated the situation of many providers sharing the same loop and POP site locations while using a mix of technologies [8].

Redundant physical access links are most cost effective when protecting access for high volume (i.e., large failure group) or mission-critical access traffic. Access links established on fiber or copper cables from two different carriers should be contained in different conduit, routed on different paths to two different POP sites on different backbone networks. This concept can be modified depending on the type of logical links in use. Redundant links that are aggregated parallel links might use the same transmission technology in all redundant links. If one link is a backup link, two different transmission technologies going to the same POP or different POPs might be used.

Many carriers will convert a circuit at a POP or CO to a transmission technology different from the access technology for backbone transport. Many carriers offload traffic as soon as possible to the network that can best transport it. This is done to avoid nodal congestion. For example, carriers will offload dial-up data access and DSL data traffic to data networks to avoid class 5 switch congestion, sometime referred to as *preswitch offload* [9].

It is important to know exactly where and how the circuits are actually routed in the carrier's network. Quite often, fires or disasters affecting a telecommunication carrier's network site will affect several other carrier networks. If both links are routed over fiber, they should reside on completely different ring networks and use different dense wave division multiplexing (DWDM) systems. SONET backbone protection can provide a false sense of security if access to the POP where the carrier's ADM is located is not redundant.

For added protection, each link should enter the customer building in two different locations. Figures 6.12 and 7.5 both illustrated access scenarios. Dual entrances are especially needed if the premise is situated directly on a SONET fiber ring. Fiber should enter the building and terminate at two different locations such that if one side of the ring fails, traffic can enter through the other side. It is also important to note that the transmission systems (e.g., multiplexers or DSU/CSU) on which the links terminate should share a different power supply or have a backup power architecture in place.

Diverse routing of access links to two different POPs does little to protect against a widespread physical disaster. In densely populated cities, two POPs located within a couple of blocks from each other can be affected by the same disaster. Choosing POPs that are significantly distant from each other can protect against this situation, but this may be expensive and not possible to implement. An alternative is to have a geographically distant recovery site. After all, a widespread disaster that brings down two redundant POP sites will most likely bring down a customer's site as well.

7.3 Wireless Access

Wireless services are fast becoming an effective alternative to protect against network access and backbone outages. There are many different types of wireless technologies. Many of the techniques and strategies discussed thus far lend themselves to at least the core portion of wireless networks. This is because wireless networks are access networks connecting to larger scale wireline networks that are supported by wireline technologies.

Wireless, for the most part, involves sending a radio frequency (RF) signal through air. One finds the same kinds of fundamental transmission components across most of the wireless technologies: a modulator, amplifier, filter, and antennae. The modulator takes a digital or analog signal and encodes it onto an RF analog carrier signal. The amplifier adds lots of power to the carrier so that it can radiate from an antenna. Before it gets to the antenna, some filtering is done to eliminate any noise that was accumulated during the processing of the signal. An outage in any one of these components can disrupt service.

Cellular or personal communication services (PCS) usually have extensive redundancy and protection mechanisms embedded within them so that an outage does not significantly disrupt an entire network. At the time of this writing, public expectation regarding wireless network performance and reliability is still relatively low. Wireless solutions, such as cellular/PCS, wireless LANs, and microwave, offer the potential to provide backup for those instances in which a wireline network experienced an outage. Experience has shown that wireless solutions work best in smaller scale isolated situations where local protection is required.

Wireless networking is often used to back up or supplement wireline networking. It is critical that a wireless backup option can indeed operate and support critical traffic when required. The remaining discussion in this chapter focuses on the suitability of some popular wireless services and technologies for this purpose.

7.3.1 Cellular/PCS

Cellular and PCS services are accessed through cell sites that are geographically distributed. Cell sites provide the ability to reuse frequencies that were allocated to the service provider by the FCC. The exact area and coverage footprint of any cell site is a function of many factors. Cell sites are in effect a distributed antenna system that can receive RF signals from a mobile user's device. Many cell sites connect to a base station controller (BSC) via a wireline network. A BSC then connects to a mobile switching center (MSC) also through a wireline network. Some vendors combine BSC and MSC functionality within the same device. The MSC switches traffic to a wireline public or private network depending on the type of service.

A mobile user's device, such as a cell phone, can often transmit to more than one cell site. Usually, the site that offers the best signal strength is used, but many times a user is in range of at least two cell sites, particularly in urban areas. Because many cell sites home to a BSC, and many BSCs home to an MSC, a failure of either system could significantly disrupt service. For this reason, many service providers home every other cell site to the same BSC (Figure 7.8). If a cell site or BSC fails, users could still communicate with another site, although coverage at times can be spotty or of less quality.

Due to the human ear's ability to comprehend words even under the worst circumstances, this strategy may work well for voice services, but it could be unsatisfactory for data services. Many mobile voice operators that now use circuit-switched technology are planning to move to IP-based data network in the future, both in the wireless and wireline portions of their networks. This in effect raises the performance and survivability requirements of their networks. Some will leverage IP packet routing capabilities that make networks more reliable by rerouting packets from one cell site to another based on prevailing circumstances, such as load and frequency.

Heavy call volume is another threat to cellular and PCS services. Typically, mobile subscribers compete for a limited number of frequencies in a cell site. If a carrier network outage occurs or overloads, such that coverage in an area or service is disrupted, it is not uncommon for traffic to surge in other parts of the carrier's network or even in other carrier networks. In such instances, the reduced number of available channels makes it even more difficult for users to access service. Sometimes, carriers will work out roaming or overflow arrangements with other carriers, hence the increase in volume on their networks. Even an outage in a public wireline network can induce heavy volume in wireless networks, as many enterprises will resort to wireless service as a backup.

To cope with such situations, many wireless operators will use spare cell sites. These are either fixed or are on wheels, often referred to as cells on wheels (COWs). Spare cell sites are used either in response to an outage or to support an expected surge in call volume in response to an anticipated event.

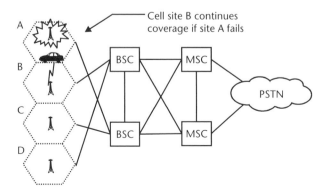

Figure 7.8 Multihomed cellular example.

7.3.2   Wireless LAN

Wireless LANs (WLANs) involve mobile users transferring data through common access points (i.e., antennas) or directly to each other in a peer-to-peer mode. If an access point fails, users may be able to communicate to each other or through another access point. The current and most popular wireless LAN standard is IEEE 802.11. Data is encoded over RF signals using a *spread spectrum* approach, which codes channels over many frequencies to provide greater bandwidth and better security.

Two spread spectrum techniques are used in WLANs. One technique, called direct sequence spread spectrum (DSSS), transmits a signal simultaneously over a group of frequencies, creating a virtual *channel*. Another technique, called *frequency hopping*, transmits a signal sequentially over a group of signals. A channel operates in a half-duplex mode, meaning that data can be either transmitted or received at any one time. To enable full duplex transmission, DSSS devices use multiple channels to create simultaneous bidirectional communication, providing there is an adequate supply of channels. Peak bandwidth capacity can be used in either direction in half-duplex mode, while half of the peak bandwidth is used simultaneously in each direction under full-duplex mode.

As in cellular and PCS, users compete for a finite set of channels. For example, the IEEE 802.11b standard for DSSS defines only 14 channels. The number of channels that are actually offered can vary by region. As in an Ethernet LAN, packet collisions will occur on a WLAN, though more frequently due to channel scarcity. WLANs have mechanisms to discourage collisions that lower the transmission throughput to reduce collision occurrence. In a large WLAN with many users, this can result is significantly less channel bandwidth.

WLANs are gaining widespread popularity at the time of this writing. One may say that the jury is still out as to whether this technology can offer the performance, reliability, and security of traditional wireline LANs. A key concern is that current WLAN deployments are over unlicensed frequency spectrum. This poses concerns over security and quality in terms of the possibility of interference, further aggravated by the use of spread spectrum *and* the overutilization of the spectrum assigned in the unlicensed bands. Some service providers are deploying WLAN Internet and VPN access services in public *hot spots*. Because several carriers can simultaneously offer services in these locations, they can become highly congested.

7.3.3   Microwave

As was stated earlier, wireless can be an effective way to provide redundancy to wired network access. Microwave has long been a popular, reliable, and cost-effective way to provide a transmission path in cases where a physical wireline path could not be established or was cost prohibitive. In fact, many older backbone transmission networks traversing rural networks were built solely using microwave because placing wireline fiber or copper cable was not possible at the time or too costly. In those days and even now, service providers will not install fiber along a right of way unless there is a community of subscriber interest that can eventually pay the cost. Many public agencies and broadcasters created their own microwave backbone networks over the years, as it was more cost effective and secure than leasing carrier services.

This is not to say that microwave is a replacement for fiber-optic transmission. Microwave, even digital microwave, has bandwidth and line-of-site limitations. Bandwidth capacity comparable to SONET/synchronous digital hierarchy (SDH) OC-3 rates can be achieved with increase in carrier frequency, but distances between microwave radio transceivers (sometimes referred to as *hops*) can grow less, which in turn can increase infrastructure costs. This is why microwave and even broadband digital microwave is best used to supplement fiber-optic networks. It can be used to back up fiber ring segments or provide redundant tributary ring access (Figure 7.9). Most vendors offer SONET ring solutions that can integrate digital microwave links as part of the ring.

Many point-to-point microwave backbones are stringlike in topology. A failure along any span can isolate pieces of the network. Although a network can physically resemble a string, it can be overlaid with a different logical link topology, such as star or ring network. A backbone supporting a collapsed ring or star topology has even a greater risk of service disruption in the event of a failure. A couple of effective, but expensive, strategies can be used to get around this. One strategy is to use microwave systems that have high reliability capabilities integrated within them. Many vendors offer architectures with redundant transmitter and receiver systems (Figure 7.10). Another strategy is to convert the physical microwave string into a physical ring by connecting the two endpoints to another network. Thus, a "break" in the string would not isolate parts of the network.

The key to using microwave effectively is engineering the links to adequately carry traffic. When using microwave for a backup access link, a general rule is to engineer the links to support at least 80% of the operating capacity of the primary link. A number of factors must be considered when choosing and engineering a microwave solution, including:

- The number channels and channel bandwidth required for the service application must be considered.
- The spectral efficiency of the modulation scheme, typically measured in bits/hertz. Different modulation schemes will have greater capacity to transfer information in a frequency cycle than others. Phase shift key (PSK),

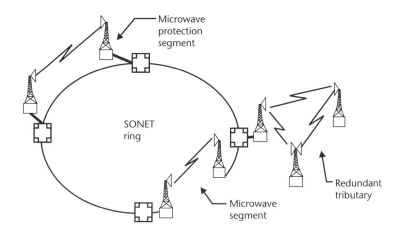

Figure 7.9   SONET interoperability with microwave.

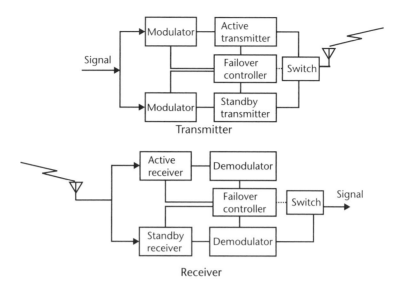

Figure 7.10   Redundant microwave transmitter-receiver system.

quadrature PSK (QPSK), and quadrature amplitude modulation (QAM) are popular modulation schemes.

- Forward error correction (FEC) is a technique used to correct transmission errors along a path. This technique, although quite effective, can consume extra bandwidth.

- Clear line of site between spans is critical. A microwave provider will often conduct field surveys of an entire microwave path to assure that no physical elements interfere with the line of site between hops.

7.3.4   Free Space Optics

Free space optics (FSO) use low-powered infrared laser beams to transmit data through the air between transceivers that are typically placed atop or inside buildings [10]. The transceivers must be accurately aligned with each other and have clear line of site. These systems can provide bandwidth in the range of 2 Mbps to 1.25 Gbps in distances ranging up to 4 km. The achievable bandwidth for a given distance depends on numerous factors, including atmospheric conditions. Phenomena such as building sway, fog, temperature variations, precipitation, and aircraft interference all can impact the signal reliability and achievable bandwidth for a given distance [11].

Many mechanisms are built into these systems to overcome many of these effects. They include parallel transmitters, beam dispersion, tracking systems to maintain endpoint alignment, and even RF backup systems if the primary laser entails problems. For further reliability, such systems are deployed in mesh architectures versus point-to-point or multipoint architectures. The mesh arrangement allows signals to be distributed among several paths across different buildings in the same area.

Although studies have demonstrated link reliability above 99%, the relationships between signal quality, reliability, stability, distance, and attenuation for these systems are still somewhat vague and lack standard guidelines. For these reasons, they are best used as a backup medium for last mile network access and critical backbone links, particularly for short link distances in dense urban areas where lines of sight are available [12].

7.3.5   Broadcast

At the time of this writing, broadcast television is undergoing a significant revolution with the emergence of digital TV (DTV). A conventional analog TV signal experiencing out-of-range or channel interference usually results in a bad or "snowy" picture. The forgiving human eye, like the human ear, has enabled viewers to tolerate poor picture and audio quality. User expectations of conventional TV have traditionally been lower than that of voice telephone, which is considered a lifeline service.

Unlike analog TV, a DTV receiver requires all of the data from a signal to recompose the picture. This means that any disruption of the signal from interference or being out of range will produce nothing or produce a picture with "tiles" that is less viewable than a poor analog picture. Thus for a DTV operation, the reliability requirements are orders of magnitude greater than for conventional TV. Broadcast network facilities have thus become mission-critical facilities.

The digital nature of DTV and the fact that a DTV signal can carry data as well as video and audio are transforming broadcast facilities into complex data networks. Broadcast facilities are designed so that all measurements, signal testing, and confirmation of quality are performed before the signal is encoded for broadcast using a transmitter or stream server. Webcast operations often test, measure, and correct audio and video quality after encoding. For these reasons, many of the techniques discussed in this book lend themselves to application in broadcast facilities.

7.3.6   Satellite

Very small aperture terminal (VSAT) systems have been used in mission-critical data networking applications for quite sometime and are growing in popularity [13]. Like microwave, satellite links are often used as redundant access or backbone links. Broadcast networks use satellite links extensively to feed programming to stations. Satellite has also found popularity in WANs for multicast data transmission through VSAT stations. Satellite service providers offer layer 2 and layer 3 WAN protocols (e.g., ATM and frame relay) that are compatible with typical enterprise data networks.

Relative to microwave, there are significant differences in power, frequency, modulation, and antennas to drive a directed signal to a satellite. Earth stations need very precise engineering because of the distance factor. Error of a fraction of a degree can miss a satellite by hundreds of miles. Satellites use a *spot beam* approach to send signals back to Earth. They divide their coverage area or *footprint* into cells and broadcast back to the cells in different frequencies.

In satellite transmission, latency (delay) is a major criterion in choosing a service. This is why satellite transmission has been popular for data and TV, but not for real-time voice conversation (you can usually tell when you are talking to someone on the phone over satellite). Low-Earth-orbit (LEO) satellites orbit the Earth at distances of about 1,000 miles or less, so the distance and consequently the latency is less. Geosynchronous (GEO) satellites orbit more than 20,000 miles above the Earth and impose at least 260-ms delay. Medium-Earth-orbit (MEO) satellites occupy orbits that range in between LEO and GEO. Some LEO service providers advertise latency of about 100 ms, which is considerably lower than standard terrestrial latency of about 200 ms. Reduced latency has dictated new reliability requirements for satellite infrastructure.

GEO systems are still the most predominant systems used. Although they impose the greatest delay, GEO satellites can be deployed in smaller and simpler constellations than MEOs and LEOs systems. In spite of the delay imposed by higher altitude, they provide consistent reception because of their stationary position relative to the Earth. Consequently, they require less network management than LEO or MEO systems. LEO and MEO systems require handoffs and switching between satellites because they orbit at speeds greater than the Earth's. Their higher altitude affords them with a wider footprint. Uplink transmission to a GEO satellite can be more expensive because it requires greater power to drive a signal to such high altitudes. Because they are deployed in mesh-like constellations and in large quantities, LEO and MEO systems are touted as having greater reliability. Traffic can be switched around a failed satellite just as a data network can switch traffic around a failed node.

An Earth-bound communications network, sometimes referred to as the ground component, supports most satellite networks. The ground component resembles a traditional telecom or data network and is often the less reliable portion of the network. In fact, many broadcast networks use satellite links extensively because of their reliability. Satellite providers typically design to an end-to-end availability of 99.75%. Satellite links are subject to numerous forms of interference, many of which are corrected for in many satellite systems. Weather conditions, shadowing, electrical activity in the ionosphere, refraction, and solar flares can affect transmission quality and reliability. However, when subscribing to a satellite carrier for a mission-critical operation, it is important to know how availability is estimated, whether such items have been factored in, and whether the unavailability is one time or accumulated over time.

Because a repairman cannot be sent into space very easily, most satellite systems use redundant infrastructure within the ground and space components to assure availability. $N + K$ sparing logic is applied to on-board components such as transponders and power supplies. Much of the reliability in GEO satellites is based on sparing of components. In MEO and LEO constellations, whole satellites can be spared within a constellation. Because spares are limited, they are allocated at different times to different users, enabling them to be shared. MEO and LEO satellites are designed to be replaced at intervals of about eight years, while GEO satellites can last up to 15 years.

7.4   Summary and Conclusions

This chapter reviewed survivability and performance issues surrounding voice and data access. Convergence has somewhat compounded access link survivability issues, as a failed link can interrupt multiple data and voice services. Voice and data networks share many common networking elements and infrastructure. Voice network infrastructure, although traditionally very reliable, requires similar precautions to data networking. PBXs require system redundancy, power protection, and failover capabilities similar to data systems. Call rerouting must also be planned in conjunction with LECs and IXCs so that traffic can be redirected to a firm's working location in the event of an outage. Calls to a dialed number must be switched to other lines or trunks connecting to another destination having another telephone number.

Next generation VoIP introduces additional concerns with respect to performance, as it can potentially reduce voice to a best-effort service. As a best-effort transport, IP is designed to route packets with no guarantee that they will arrive at their destination, unless streaming protocols such as SIP are used. As carriers are gradually employing softswitch technology in their networks, there are greater opportunities to enhance survivability and use IP networks as a backup resource. At the enterprise level, VoIP gateways and IP phones can help facilitate survivability and recovery.

Using a service provider's capabilities for survivability requires careful planning. Diverse paths to the customer location must be established via two separate COs, to different carriers if possible. Individually, carriers should have incumbent capabilities to avoid service outages. Distributed switching exchanges should be used so that a switch module failure can be isolated and calls in progress can be saved. A good voice service provider should also have well-defined problem escalation procedures that can guarantee resolution within a specified time period.

When configuring access links, particularly for data, matching CPE systems with those on the far end of the WAN connection can avoid problems. Multiple connections into a location are effective only if equipment can failover, switch over, or actively use another link. Redundant access links, at a minimum, should protect the most critical traffic. Link aggregation is a capability that makes parallel access links act as one logical link. Inverse multiplexing, load-sharing routers, and multilink PPP are several known approaches to link aggregation. Regardless of the approach used, parallel access links should be matched as close as possible with respect to technology and capacity.

Many carriers share common infrastructure that can present single points of failure—a problem further exacerbated by telecom deregulation. Many providers share the same loop and POP site locations using a mix of technologies. It is important to know exactly where and how circuits are actually routed in a carrier's network, and that there are no commonalities with another carrier. Outages and problems affecting one carrier's network will likely affect another as well.

Wireless services and technology has fast become a popular alternative to protect against network access and carrier backbone outages. Many of the techniques and strategies applicable to wireline networks also apply to wireless networks, as wireless is an access technology to wireline networks. Cellular/PCS and WLAN services have become accepted backup resources for voice and data networking, respectively. Microwave has long been a popular, reliable, and cost-effective way to provide redundant transmission paths, especially in cases where a physical wireline path cannot be established. FSO technology has found use as a backup medium for last-mile network access and critical backbone links, particularly for short link distances in dense urban areas.

Like microwave, satellite links are often used as redundant access or backbone links. In satellite transmission, latency is a key criterion in choosing service. Satellites at higher altitudes often provide more consistent reception because of their stationary position relative to the Earth, but they can incur more latency because of the distance. On the other hand, lower altitude systems deployed in mesh-like networks can offer greater reliability and reduced latency.

References

[1]   Bodin, M., "When Disaster Comes Calling," *Small Business Computing & Communications*, February 1999, pp. 54–61.

[2]   Llana, A., "Fault Tolerant Network Topologies," *Network Reliability—Supplement to America's Network*, August 2000, pp. 10S–14S.

[3]   "Circuit to Packet," *Lucent Technologies Solution Note*, May 2001.

[4]   Barrow, C., "The Impact of the Telecommunications Act on Business Continuity Plans," *Disaster Recovery Journal*, Winter 1998.

[5]   Smith, M., "Central Office Disaster Recovery: The Best Kept Secret," *Disaster Recovery Journal*, Spring 2002, pp. 32–34.

[6]   Jessup, T., "Balancing Act: Designing Multiple-Link WAN Services," *Network Magazine*, June 2000, pp. 50–59.

[7]   McPherson, D., "Bolstering IP Routers for High Availability," *Communications Systems Design*, April 2001, pp. 21–24.

[8]   Dillon, M., "Carriers May Foil Arrangements for Network Redundancy," *ISP World*, May 29, 1999, p. 29.

[9]   Williams, K., and S. Lukes, "TDM-to-IP Evolution in the Metro," *Internet Telephony*, March 2001, pp. 54–59.

[10] Willebrand, H.A., Ghuman, B.S., "Fiber Optics Without Fiber," *IEEE Spectrum*, August, 2001, pp. 41–45.

[11] Hecht, J., "Free-Space Lasers Shining as Obstacles are Overcome," *Integrated Communications Design*, November 12, 2001, p. 25.

[12] Buckley, S., "Free Space Optics: The Missing Link," *Telecommunications*, October 2001, pp. 26–33.

[13] Sweeney, D., "Pies in the Sky," *Network Reliability—Supplement to America's Network*, August 2000, pp. S8–S13.

CHAPTER 8

Mission-Critical Platforms

Many organizations entrust survivability and performance to their platforms. The evolution of platform technology has seen a decline in system costs and life cycles, an increase in platform "churn," numerous new features, operating system and application diversity, and consequently more topological and administrative complexity. There has been a shift from large mainframe systems towards server-based and appliance architectures. Systems have become more disposable in nature as salvage costs exceed the costs of new systems.

Manufacturers of fault-tolerant (FT), fault-resilient (FR), and high-availability (HA) platforms usually integrate commercial off-the-shelf (COTS) components for economic reasons. This has given rise to standards for chassis, boards, software, and other components. However, few COTS components have been designed for FT/FR/HA capabilities. Manufacturers must typically bring these components together and add the FT/FR/HA features. FT/FR/HA is more often a *soft* upgrade than a hard one—building resilient systems on general-purpose hardware does not necessarily improve the availability.

FT/FR/HA platforms are typically more expensive because of these additional features. They should be placed in locations where they are most needed—usually at single points of failure that can expose a network to greater risk, such as the edge and access portion of the network. A classic example of selective placement of FT/FR/HA equipment is the telephony world's use of expensive, yet effective, FT platforms for local and toll telephone switching exchanges. These systems are designed with internal redundancies and hardened operating systems (OSs) that can withstand almost any system fault or outage.

Yet, history has demonstrated that sole reliance on FT/FR/HA platforms do not protect against other adverse situations, such as a fire or flood, which can cause an outage of the *physical* facility. Many of the practices discussed in this book are directed towards providing survivability mechanisms that can protect against such circumstances. Use of FT/FR/HA platforms in conjunction with such practices can improve survivability and performance overall. This chapter explores platform capabilities and their relevance in providing mission-critical protection.

8.1 Critical Platform Characteristics

There is no single component within a system that ensures availability and performance. These are achieved by optimally blending system software, hardware, configurations, applications, and management. The following are some of the key characteristics that one should look for in today's products. These characteristics should form the basis of general criteria for choosing mission-critical server and networking platforms [1, 2]. They are presented in no particular order:

- *Simplicity (versus complexity).* "The simpler, the better" is a classic general rule that is apropos when it comes to platforms. Platforms with small numbers of components (and moving parts) are less likely to break. The counterpositive to this rule is the need for redundancy (discussed later). Redundancy in components makes platforms more reliable yet increases the numbers of components. For this reason, one should look for platforms that have an optimal *mix* of different components.

- *Equipment density.* This is an attribute that can drive many other platform attributes, including unit size, weight, and racking options. Small system footprints with central processing units (CPUs), storage, and memory that do not take up much space are becoming more desirable. System density is often measured using the number of processors per unit area as a general rule. This rule is not necessarily universal and is usually more applicable to rackmounted CPU systems. This rule implies that dense systems will likely have more distributed functionality per CPU than a single large CPU system. This is driven in large part by the nature of the software applications that a system will be used for. The rule also implies economy in backplane slots, enabling greater functionality to be incorporated in a limited space [3].

- *Economy.* Cost is always a major factor in platform selection. The adage, "you get what you pay for" prevails when buying a platform. A trade-off between features, availability, and cost should be evaluated when selecting a platform. Some use cost per unit processor as a measure of system economy. Others break it down further into cost per million instructions per second (MIP) or cost per CPU cycle. Such measures can be useful, but they should be used in conjunction with other attributes. As platform components fast approach commodity prices, any investment should be weighed against the value the system can bring in improving application performance and availability.

- *Equipment cooling.* Numerous processor cards or *blades,* large multiple power supplies, and disk drives generate considerable amounts of heat. The more components crammed in a rack space or card cage, the greater the heat density that is generated. As density is becoming more characteristic of fault-tolerant systems, they will most likely require more cooling. Fans are used to increase airflow, producing forced positive pressure throughout a platform assembly. They reduce the temperature by circulating cooling air throughout the system and venting it out. Failure of a single fan can cause the temperature inside the assembly to increase dramatically, leading to CPU and drive failures. Duplicate fans are frequently used and should be monitored by the system. Some systems compensate for a failed fan by increasing the speed of the remaining fans. Fan capacity is usually measured in cubic feet per minute (CFM).

- *Equipment ventilation.* In general, air should be exchanged several times per second, typically on the order of five to seven times. Fans alone do not reduce the temperature. Although they push air through the cabinet or cage, the air must have an inlet and be vented out. Some designs use side inlets and vents, enabling airflow for cards mounted close together in the rack. Some designs have inlets and vents in the rear, so that cabinets or racks can be placed in rows side by side. Having inlets at the bottom and vents at the top creates airflow with a chimney effect, accommodating the tendency for hot air to rise. It is also important to have an air conditioning system that can remove hot air from the room and can keep ambient temperatures to reasonable levels, rather than having it drawn into the systems again. Many systems come with a filter to remove accumulated dirt, which can be corrosive to components.

- *Serviceability*. Systems with good serviceability can significantly reduce mean time to recovery (MTTR). All systems will require servicing at some point in time. How easily a system can be serviced depends on several characteristics. Some of those just discussed, including density and complexity, are contributing factors. Accessibility to components and ease of installation and removal are desirable. One should be able to easily gain access to interior components and not have to remove or impact other components to fix an offending one. Good cable management around a rack or cabinet can aid in servicing. Newer designs use cable assemblies and connectors that enable component replacement without ever touching a cable.

- *Swappability*. This is a coined term used to convey the quickness that components can be replaced and assimilated within a system. *Hot swappable* components are those that can be replaced while the system is operational, without the need to shut down the system during the repair. Components such as fans, processor cards or blades, power supplies, or drives can be hot swappable. Newly added components should automatically notify the platform of their presence, without requiring master processes to be restarted [4].

- *Memory*. There should be enough cache and dynamic random access memory (DRAM) to accommodate background and application processing. When memory utilization approaches capacity, memory channel adapters can drain power from the CPU and degrade performance. This is why investment in additional memory can prolong platform lifetimes in addition to improving performance and transaction rates. Memory protection and volatility is of prime importance when dealing with mission-critical applications. When faced with an outage, information stored in memory should be preserved and recoverable to the extent possible.

- *Drivers*. Drivers are software components that various suppliers use to interface their products with a system's OS. Board-level components, peripheral devices such as printers, and network interface cards (NICs) usually require drivers for operation.

- *Redundancy*. Component redundancy identifies the secondary components that automatically engage when primary components fail. Secondary protection components can range from hardware such as power supplies, power feeds, processors, chassis, backplanes, fans, I/O channels, controllers, network adapters, and disk host adapters. They can also include software components at the OS, middleware, and application levels.

- *Failover*. The use of redundant systems implies that the platform will have failover mechanisms so that operation can continue during failure and repair. Failover involves cutting over to another CPU for continued processing.

Whether application states and memory are preserved during this process depends on the platform's level of availability. Failover will often involve dynamic reconfiguration of the platform and reallocation of resources [5].

- *Openness.* Platform openness is the degree of reliance on COTS components. The ability to run commercially available or custom applications, peripherals, or other components reduces dependency on proprietary implementation. Openness is usually manifested in the form of interfaces, OSs, and middleware. But FT/FR/HA platforms, although integrated with many standard components, usually require substantial proprietary development to differentiate them from general-purpose platforms.

- *Certifications.* Adherence to standards is just one form of certification. Use of components, be they hardware or software, provides some assurance of functionality. Certifications go a step further as assurances that a platform has passed special testing, usually in the areas of safety, electromagnetic interference (EMI), and fire prevention. Such certifications include those from Federal Communications Commission (FCC), International Standards Organization (ISO), Underwriter Laboratories (UL), and Telcordia.

- *Quality.* ISO certifications are also used to convey quality. But quality can also mean workmanship. When shopping for a car, buyers typically will slam the door shut on a showroom vehicle to see if it is solidly made. When shopping for a mission-critical platform, using the same pedestrian approach will not hurt. Workmanship and attention to detail are often overlooked but are critical factors in determining potential availability and life cycle. A rigid chassis made of steel or aluminum that does not sag in the middle can mitigate potential card or blade failures. Slot connections that enable cards or blades to fit snug are critical. Many systems include clamp down mechanisms to ensure that cards are secure and stable. Many vendors perform *burn-in tests* to evaluate if and how hardware components will fail during a period of time. Some vendors will quote an *infant mortality rate,* which is an estimated rate at which components will fail after turning the platform on for the first time.

- *Functionality.* This attribute conveys the extent that a platform satisfies the networking, application, and operational needs of the organization. Each organization will look for something different in a platform. A written requirements document conveying functional specifications can be a highly effective tool to evaluate platform functionality.

- *CPU.* Processor speed is often a specified platform characteristic. Many organizations usually want platforms with fast CPUs. But having a fast CPU will do nothing if the rest of the platform components are not designed accordingly. CPU time should be protected and used in an optimal fashion so that critical operations are executed expediently.

- *Storage.* Much discussion on storage is available in another chapter of this book. Storage comes in many shapes and forms. Drive capacity and type, media, and I/O channel capacity are usually the focus of attention. Storage device failures are not typically the major cause of application interruption. Yet, attention should be paid to how a platform's storage capabilities can fit within an enterprise's storage protection program.

- *OS.* Off-the-shelf OSs are becoming more the norm. Open source OSs, such as Linux, are also fast becoming popular. Vendors will also fortify OSs with special features for specific uses typical of FT/FR/HA platforms. It is important to know which OSs have been tested and certified for use with a platform's CPUs. This assurance will aid compatibility and enable support from the OS software vendor if needed.

  OSs used for FT/FR/HA platforms should have the ability to manage all system resources with continuous nonstop processing. The platform should exhibit overall stability and be impervious to transient OS faults or application interruptions to the extent possible. Quite often, third-party applications can destabilize an OS. Some systems can hang or freeze as a result of an unstable application or too many applications trying to draw system attention. Data access and network connectivity should also be continuous and unaffected by faults. The OS should also enable hot swapping of components. It should recognize new devices while running without the need to restart or reboot. Secure OS evaluations should also be part of the selection criteria. OSs with poor vulnerability track records and low security ratings often undermine FT/FR/HA. OSs that are demonstrably more resilient to attacks, more readily *hardened*, and for which there are choices in OS-level protection software should be favored.

- *Options.* Many platforms come with optional add-ons, which are typically expensive and often underutilized. Special software applications, development suites, and utilities can make custom application development on the platform much easier. However, incorporation of platform-specific utilities into custom applications could adversely affect their portability to another platform.

- *Ability to upgrade.* Overall, a platform should allow improvements to be made without requiring major overhauls or replacement. Scalability, as discussed previously in this book, is highly desirable. The ability to add new chassis, circuit cards or blades, additional memory and storage, power supplies, peripherals, and interfaces should be made with minimal effect on the core platform. Software upgrades in certain situations may require updating to support additional hardware, depending on how tightly they are coupled with platform features. Such upgrades should involve a single-step download process. Upgrades should be conducted in a phased approach, one upgrade at a time.

8.2   Platform Tolerance

Principles of tolerance were discussed earlier in this book. These principles have implications on platform design and operation. The types of services a platform supports, and where the platform lies within the network architecture, can influence platform tolerance. Voice call processing may require higher degrees of availability while high-volume data transactional service might require greater performance. Platform availability may not be as critical in this case as throughput. Ultimately, platform tolerance levels alone do not drive overall service availability. It must be achieved through a conscious combination of the system's physical environment, storage, network architecture, operations, and policies.

8.2.1 Fault Tolerance

Platform FT is achieved at both hardware and software levels. FT hardware involves redundant CPUs, memory boards, network interfaces, input/output (I/O) subsystems, host adapters, buses, power supplies or feeds, cooling fans, disk storage, and other components. Figure 8.1 illustrates a generic FT platform architecture. Some systems feature redundant processor boards, each having redundant CPUs. Software FT involves implementing redundant software processes and software to execute failover and recovery mechanisms during faults. Process or application replications, and all of the platform resources required to support them, must be redundant as well. Some platforms enable redundant components to operate simultaneously to facilitate performance, failovers, or upgrades. Current state-of-the-art failover times are typically in the range of 10 to 15 ms. To facilitate failover, data must be made available to the redundant processor either from memory or a storage device so that service can continue.

Hardware and software FT is not mutually exclusive—a true FT platform will have both levels of FT. Much of the recovery intelligence is achieved at the software level, effectively eliminating the need for complex circuitry at the hardware level. FT systems typically use partitioning at the hardware and software levels so that a fault in one component can be isolated and not affect other components. *Loosely coupled* architectures favor this kind of modularity. CPU and memory are duplicated on another board (sometimes referred to as a *side*) so that application states, network connectivity, and I/O data are preserved in the event of a fault. A mirroring process, similar to that used in data replication, is employed where a process from one side is replicated to the other. Both CPUs on each side intermittently poll each other and compare their information for consistency (Figure 8.2).

Tightly coupled architectures, on the other hand, use strict coordination and consistency between processor states. Use of *lock-step* synchronization between CPUs is popular for these architectures. This involves identical OSs and applications running simultaneously on each CPU, executing each program instruction at the same time with the same data on a clock-cycle basis [6]. This is done to further assure that memory resident data and application states are preserved to the highest

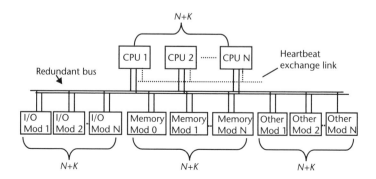

Figure 8.1   Generic FT platform architecture.

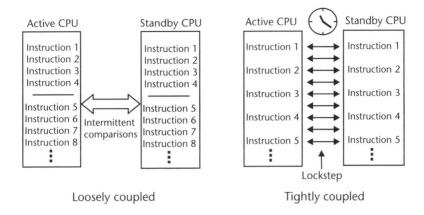

Figure 8.2 Loosely coupled versus tightly coupled architectures.

degree possible. If a discrepancy between the systems is detected, correction is attempted. If a CPU fails, it is taken out of service while the other continues processing. When corrected, all memory and register states are copied from the working CPU. This architecture is popular in telecom applications where call state and signaling information needs to be preserved during failover. Overall, FT architectures involving duplicate sides invite opportunity for passing corrupted memory and data from one side to another in the event of failure. Bad hardware, a software error, or security intrusion could corrupt data in both sides.

FT systems achieve availability by rapidly detecting, isolating, and correcting faults. Active services should continue to process in light of any hardware or software fault, at least in theory. It is most likely that there is no single FT platform that can absolutely guarantee no transaction loss. It is best to acquire a platform that can sufficiently satisfy stated service availability and performance requirements that are acceptable to the organization.

The high degree of replication adds to the system cost. The embedded investment in these platforms compels organizations to hang on to them as long as possible. Depending on the system architecture, many protective components may not yield immediate operational value, remaining idle in the system for long periods of time. Although use of standardized components (e.g., CPUs, I/O subsystems, and NICs) is commonplace, the platform integration and development surrounding their use can be highly proprietary in nature. This can drive life cycle costs even higher, as proprietary implementations are limiting and are characterized by long and expensive feature development cycles.

Classic FT platforms came with two identical redundant sides that were each, in effect, entire systems having a single CPU. Advances in circuit and software design have led to systems that can scale to many CPUs in a single platform. Having more than two CPUs solves a fundamental problem with a two-sided platform: when a side (or CPU) becomes inactive, the platform is tentatively not operating at an FT level. One counter measure is to use an $N + K$ CPU sparing arrangement. ($N + K$ strategies are discussed in the chapter on principles.) Another measure is to cluster systems together in a network to behave as a single entity. This can greatly improve reliability and create quantum increases in performance but can be quite expensive.

In addition to performance and reliability, FT systems favor instances where personnel are not immediately available or required. They are often found in remote locations having unmanned operations but require high availability. Many systems have remote monitoring capabilities where they automatically contact a network manager or support center, even by phone, in case of a fault.

FT platforms are not typically used as general-purpose systems, but tend to be used as point solutions for specific applications such as telecom, air traffic control, scientific control, and process control. They are also used in Internet data centers, e-merchant *critical paths*, and other transaction-based network services. Work group servers or clusters typically do not have to be FT systems.

8.2.2 Fault Resilience

Like FT systems, FR systems come with many of the component redundancies found in FT platforms. On the surface, their architecture can resemble that of an FT platform. They have redundant components and subsystems that can failover to one another, but there is no guarantee that transactions will not be lost. This does, however, result in lower platform costs and yet provides availability comparable to an FT platform. The reason for this is that CPU failures are most often caused by excessive heat. As long as cooling fans are duplicated so that CPU cooling is protected, the integrity of the CPU is further assured. This partial tolerance has found acceptance as long as other components are redundant and hot swappable.

At one time, FR platforms having only one CPU were common. Today, products can vary in architecture. Some are an amalgamation of hot-swappable components, while others are comprised of rack-mounted arrays of single board computers (SBCs) that are whole computers in and of themselves. If an SBC fails, it can be swapped out with another at the expense of losing transactional information and disrupting service. Some platforms come with a segmented high-capacity passive backplane so that CPUs can continue processing if one fails. Software resiliency to faults is achieved by separating application processing from platform I/O processing and storage. This isolates application faults from I/O faults. They also have monitoring and alarming subsystems similar to FT platforms.

8.2.3 High Availability

HA platforms do not provide the same level of survivability as FT/FR platforms. Availability is collectively achieved at the system level through the integration of redundant configurations of low-cost off-the-shelf components. Unlike FT/FR platforms, HA platforms are designed to withstand component failures or faults, versus proactively attempting recovery. Enough redundancy is provided so that failure of one component has little effect on others. For this reason, HA systems will usually exhibit a higher fault rate than FT/FR platforms. They also cost less than FT/FR platforms.

Some HA products elevate their availability through the use of drivers that can recognize when a component is malfunctioning. Like FT/FR platforms, some use highly customized real-time operating systems to perform failover mechanisms to other platforms, as in the case of clustering [7]. Clusters comprised of symmetric multiprocessing (SMP) systems, discussed earlier in this book, will often use this type of capability. Some systems are being designed with hot-swappable components.

8.2.4   Platform Comparisons

Table 8.1 provides a general summary comparing the three kinds of platforms just discussed. Some products will deviate from the comparison. For instance, some FR platforms may not necessarily require secondary CPU initialization upon failover, depending on the system architecture. It is evident from the table how the lines of distinction are growing blurred between them.

8.3   Server Platforms

Server platforms are the computing elements of a mission-critical network. They provide the processing that enables services to be delivered to users. In general, there are various types of server platforms. The following list details some of them. As always, some multifunctional server products will transcend these categories and provide a mix of capabilities:

- Transaction-based servers, designed for the purpose of performing real-time transaction processing;
- Application servers, designed to run specific applications such as accounting and customer-relationship management (CRM);
- Web servers, which handle Web pages or run applications such as email;
- Special-purpose servers, such as security servers (i.e., firewalls and antivirus gateways), storage servers, print servers, fax servers, Internet protocol (IP) telephony, and core Internet services (i.e., domain name servers and email) servers;
- Gateway or host connection servers used to provide connectivity between systems or networks;
- Management servers, designed to provide monitoring and administrative functions;
- Communications servers, designed to manage and control communications, such as remote access servers (RAS);
- Streaming servers used to manage and control video/audio or voice services;
- General-purpose servers designed to support desktop operations and run a variety of applications.

Table 8.1   Platform Comparisons

Characteristics	Fault Tolerant	Fault Resilient	High Availability
Failure detection	Yes	Yes	Yes
Failure recovery	Yes	Yes	Yes
No single point of failure	Yes	Yes	Yes
CPU failover mechanism	Immediate	Requires warm initialization	Requires cold initialization
Availability	99.99% to 99.999%	99.99%	99.5% to 99.9%
Transactions preserved during failover	Yes	No	No

Server survivability and performance in a network environment are usually achieved in a couple of ways. One way is to add redundant components to the platform, such as CPU and memory. Another way is to simply add more platforms to the network, perhaps even assigning them specific functions, like those presented in the previous list. These approaches can also be combined. The following sections will discuss characteristics of server platforms relative to mission-critical operation.

8.3.1 Hardware Architectures

A discussion of server hardware architecture should be prefaced with the caveat that regardless of what type of platform is used, whether FT/FR/HA or other, proper network design will always be the mainstay for service availability. Good server hardware architecture should facilitate restoration in a mission-critical environment. It might have become evident in previous discussion that different levels of tolerance can be achieved through various configuration and arrangements of components. For example, within the class of HA platforms, different levels of availability can be provided when active and standby CPUs are arranged using $N + K$ sparing logic.

Server hardware architectures have undergone a major transformation in recent years. Server systems evolved from personal computers (PCs) with motherboards, where most of the main processing and circuitry resided, to systems with numerous processors connecting into a large passive backplane with many slots. Scalable performance, redundancy, and cost were some of the key reasons for a backplane approach. Backplane systems have classically found use in telecom and control environments. Motherboard systems, on the other hand, lend themselves for use in applications requiring a smaller, economical system footprint.

8.3.1.1 Backplanes

Backplanes have evolved from single centralized chassis with active circuitry to passive backplanes that are segmented, or split, so they can support multiple separate computing systems [8]. They have also progressed into continuous backplanes with numerous slots that can support different bus types [9].

A system using a passive backplane relies on other components to provide processing. The backplane itself is comprised of connectors into which other components can be inserted. These are often referred to as *slot cards*. Because the backplane is passive, the likelihood of a malfunction is reduced. Yet, additional reliability can be introduced by partitioning the backplane into segments, often with their own individual power supplies. This enables several independent systems to occupy one chassis (Figure 8.3). The power segmenting enables powering down a defective system in a platform without impacting the others. This is sometimes referred to as a *power-sequenced segmented backplane*.

SBCs are becoming popular in use as the processing elements that plug into a backplane. An SBC typically occupies one or two slots. It usually has its own CPU, memory, and I/O ports. Because the SBC is independent of the backplane in terms of its capabilities, it can be upgraded by simply replacing it with another card. Many SBC boards are evolving with highly integrated features, such as local area network (LAN), small computer system interface (SCSI), universal services bus (USB), and video interfaces [10].

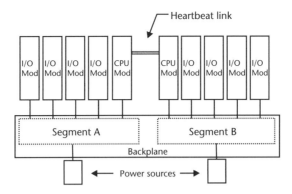

Figure 8.3   Segmented backplane example.

In many configurations, a single CPU can control components in each backplane segment. If a CPU fails, it turns over control to another CPU. Some architectures allow CPUs to exchange heartbeat information with each other through the backplane, or via a serial or network connection. This preserves CPU communication in light of any backplane problems.

The need for high-density compact systems has spawned passive backplanes that can handle numerous expansion slots. Some systems, such as SMP and asymmetrical multiprocessor (AMP) systems segment the backplanes further into groups of four or five slots, each configured as an independent system.

8.3.1.2   Buses

A backplane is a physical circuit that enables communication among the components that plug into it. A bus, as discussed earlier in this book, is a circuit (or link) that is shared among different hosts. It is in effect a logical and physical network that can be realized on a backplane. There have been several major types of bus technologies used in computing platforms:

- *ISA*. The Interactive Services Association (ISA) bus has been used for many years and is still popular in many PCs. Made to process 16-bit words at 8 MHz, the ISA bus has a theoretical maximum transfer rate of 16 Mbps, which is not fast relative to newer technologies. Its use prevails in applications that do not require high bandwidth, such as voice processing. Because it occupies an embedded base of systems, its use is still prevalent. Segmented ISA bus backplanes have been in use as well.

- *PCI*. The peripheral component interconnect (PCI) bus operates at 32-bit words and runs at either 30 MHz or 33 MHz. The theoretical maximum transfer rate is 132 Mbps, providing greater bandwidth. This makes it useful for broadband applications, such as video, and high-volume transfers of data. The PCI Industrial Computer Manufacturers Group (PICMG), a 500-company industry consortium, produces the standard for PCI passive backplanes. PCI is built on the PICMG 2.13 standard, which was intended as a standard for removable redundant CPU systems. PICMG 2.13 posed some electrical limitations affecting PCI's use in mission-critical platforms. Failure or short circuit in a board could fail the entire bus, making it a single point of failure and creating the potential for losing transactions. Failover would have to be implemented using platform-management software to transfer processing to an operational segment of the backplane.

The PICMG standard allows the PCI and ISA bus to be supported on the same backplane. Called *concurrent PCI*, it enables both PCI and ISA buses to simultaneously transfer data over the same backplane. This allows use of legacy ISA boards and enables smooth migration to a PCI framework.

- *CompactPCI.* CompactPCI (cPCI) is an upgraded version of the PCI bus. It also operates at 33 MHz but uses a 64-bit word producing a maximum transfer rate of 264 Mbps. cPCI is electrically similar to a PCI bus but is mechanically different. It is designed using more rugged and reliable connectors, based on the EuroCard standard. EuroCard is a family of boards with pin-plug connectors rather than edge connectors. This enables the boards to be held in place more firmly.

  cPCI supports eight slots while PCI supports only four. cPCI better supports failovers by allowing boards to be hot swapped without affecting others on the bus. This makes it much more attractive for use in FT/FR. Some cPCI implementations divide the backplane into several mirrored segments, each having a group of device component boards, controlled by a central CPU on another segment. This enables survivability of different platform functions, such as I/O. Others dedicate a slot in each segment to a CPU and failover control board. But because cPCI also follows the PICMG 2.13, it suffers from some of PCI's electrical limitations.

- *CPSB.* The compact packet-switching backplane (CPSB) arose out of the PICMG 2.16 standard. This standard allows boards to be pooled together so that failures are contained within a pool. The standard also allows boards to communicate using a packet-switched approach. This makes it attractive for use with newer packet-based Ethernet and Internet protocol (IP) backplanes. Board control is achieved using redundant packet links, enabling the configuration of centralized and decentralized multiprocessing architectures. This allows redundant systems to function with failover on the same backplane.

- *VME.* The versa module europa (VME) bus has been in use since 1981 in high-end real-time telecom, military, and industrial control platforms. Built on the EuroCard standard, it uses a 32-bit wide data path. Because of its extensive use, it has grown to support numerous types of protocols and real-time OSs. It uses an efficient interrupt scheme that allows numerous boards to be hot swapped without system disruption. VME has grown to support more off-the-shelf boards and has gained some popularity for use in lower end applications.

- *Infiniband.* The Infiniband bus has been viewed as a long-term replacement for the PCI bus. However, the embedded invested base of ISA and PCI in systems coupled with an ailing economy (at the time of this writing) has slowed industry adoption. Because it can achieve gigabit-level speeds, it has been positioned for use in platforms supporting data-intensive applications, storage networks, specialized computing, and direct high-speed communications between platforms. Infiniband is further discussed in the chapter on storage [11].

8.3.1.3 Form Factors

Server systems are taking on new shapes and sizes, breaking with some of their traditional form factors [12]. Some of these form factors include the following:

- *Motherboard*-based products come equipped with a full-size motherboard, typically with PCI and/or ISA slots. Traditionally found in PC products, this form factor has been used in networking devices such as routers and switches, even in rack-mount configurations.

- *Pizza box* is a space-saving form that is flat and rectangular, as its namesake implies. They provide cleaner modularity than a motherboard-type system and easier upgrade management. They are finding widespread use in communications networking and hosting devices, where numerous boxes are rack mounted in hosting center. Expansion cards, referred to as *riser cards*, are mounted parallel to the motherboard, versus the perpendicular approach found in true motherboard systems. The flatness of the pizza box affords only two to three expansion slots in a typical system. An important characteristic of the pizza box form factor is that it conveniently fits into standard equipment racks used by large enterprises and service providers.

- *Shoebox* is another space-saving form that also can be rack mounted. It too can support several PCI or ISA slots in a single system.

- *Rack-mount* systems can be comprised of many of the aforementioned forms. The 19-in rack mount is found in many data networking environments, while the 23-in rack mount is found in telecom environments. Although rack-mount systems were commonly found in central office (CO) and point-of-presence (POP) facilities, they have been gaining widespread use in enterprise data centers and telecom closet (TC) spaces. *Blade servers* have become popular residents of rack-mount systems. Preference for rack-mount systems is driven by the need for processor density and the ability to easily service and upgrade systems.

8.3.1.4 Platform Failover

Multiple CPU systems are preferred in mission-critical environments. Individual host systems could be spread redundantly across a network using techniques such as clustering and load balancing. Multiprocessor systems with passive backplanes are also, in effect, small self-contained networks. Any infrastructure that can bring together disparate components and interconnect them in some way for a common purpose, either through a backplane or a network, should be designed, engineered, and managed as a single system. Network nodes comprised of multiprocessor platforms can be leveraged in many ways. Because they can individually and dynamically scale to different network conditions, they can provide the extra computing power that is often needed to handle overload traffic or traffic that is redirected from a failed node.

Multiprocessor systems consist of a blending of CPUs, backplanes, buses, and other components to realize an FT/FR/HA platform. They can range in form from two entirely redundant platforms with failover capabilities, as found in many traditional FT systems, to platforms comprised of numerous SBCs, I/O, memory, and interface boards running in parallel. HA systems typically have a redundant CPU in a cold standby state that is activated upon a malfunction of the primary CPU. Failovers can be automatic, transparent, or require manual intervention. Figure 8.4 illustrates some general differences in CPU failover between platform tolerance levels [13].

Partitioning is a technique that is common in many multiprocessor platforms. It involves dividing a multiprocessor system into smaller systems called *domains*. Each domain is a self-contained system having its own OS, CPU, memory, I/O, network interfaces, and in some cases disk drive. Hence a mission-critical multiprocessor platform can be comprised of multiple domains, providing redundancy to facilitate FT/FR processing. Platforms that use domains perform *dynamic resource allocation* if a domain experiences a malfunction or if processing conditions change. Dynamic resource allocation is the process of reassigning resources, such as CPU, I/O, and memory. Some platforms use redundant CPUs that can access each other's domain, so that a failed component in one domain can be backed up by the same component in another domain (Figure 8.5).

Earlier in this book, we discussed the use of multiprocessing platforms in conjunction with clusters. In SMP platforms, a pool of CPUs operates simultaneously on the same program thread, sharing common memory, I/O devices, and other resources. Only one instance of the OS and application is in memory at a given time. AMP platforms use a single master processor to coordinate slave processors, each

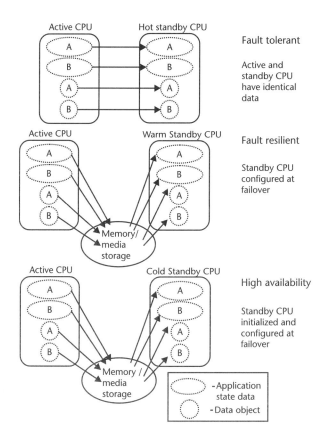

Figure 8.4   Failover examples.

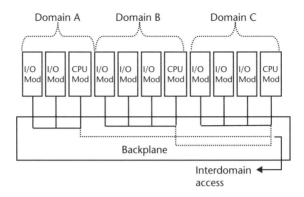

Figure 8.5   Domain partitioning example.

independently performing individual tasks. Although the CPUs are processing the same program, they are not necessarily processing the same program thread, as in the case of SMP. AMP platforms are often used to improve application response time.

8.3.2   Software Architecture

Mission-critical software begins with the OS. Platform vendors have come to realize that an OS is just one of the many components in a platform. Because software can malfunction and experience faults, it is subject to the same if not more exhaustive platform governance mechanisms as other components. To this end, many vendors have recognized the existence of a governance level above that of the OS. This is sometimes referred to as the platform abstraction layer (PAL) [14]. It is usually tied directly to the platform vendor but enables use of commercially available OSs. It is designed to manage such items as event handling, hot swapping, device trees, fault management, interhost communications, failovers, heartbeats, checkpoints, and other platform aspects, depending on the capabilities of the platform OS. Figure 8.6 illustrates a generic software architecture.

Although the presence of a PAL might make a platform more complex, incorporating high-availability support functions into the OS can result in longer restart times and potential loss of platform state information during restart. Longer restarts are also characteristic of OSs stored on a hard drive versus flash read-only memory (ROM).

A CPU will normally divide its time between system tasks and application tasks. When more applications are running, system and application processing time are further consumed. As more jobs are run, application processing time consumption peaks and starts to decline. Meanwhile, the OS continues to increase time devoted for system tasks in an effort to manage the heavy workload and resulting resource contention. Many systems are known to freeze or lock up in those instances when too many application-processing and system-level tasks demand the attention of the OS.

This fundamental behavior must be kept in mind when planning networks. For example, as network speeds increase to gigabit levels, a host system processing an incoming data stream from a network link may be unlikely to keep up with the influx of information. Overwhelmed, the system might exhibit unstable behavior.

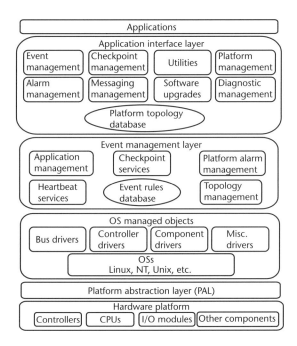

Figure 8.6   Generic platform software architecture.

Although network rate control mechanisms, like TCP, can try to compensate for such behavior, the available network bandwidth is not fully leveraged.

A good OS (or PAL) coupled with high-quality platform hardware are the ingredients for a solid mission-critical platform. Platform topology management should isolate platform hardware and software components such that they collectively exhibit stable behavior under normal and extreme circumstances. The OS should support dynamic loading and unloading of the component *drivers* and be able to reallocate system resources among them when needed. Drivers are software applications that manage a hardware or software component and communicate with the OS. Some systems use *hardened drivers* that are designed with more smarts to manage the component and provide notifications. The most fundamental situation where this is normally useful is component repair, either preplanned or hot swapped. They can alert the OS to the impending removal or insertion of a component to aid in restoration.

Stability is the most important mission-critical characteristic one should look for in an OS. The ability to predict system behavior is the key to mission-critical platform success. Additionally, well-designed systems also mask much of the underlying infrastructure tasks from the manager. OSs, particularly those found in FT systems, are often customized for these purposes. Features such as lockstep and partitioning are added to the software by the vendor. Classic real time OSs (RTOSs) such as Unix RTR, found in many local/toll telephone switching systems, can tolerate software faults to the extent that telephone calls in progress will not be lost. Vendors produce enhanced versions of UNIX descendants, such as Linux and Solaris, with many additional survivability and performance-enhancement features like hot swapping and clustering.

8.3.2.1   Fault Management

Automation of the fault-management process is key to improving availability. Different vendors use different failover and fault-recovery strategies. Data/transaction loss and failover or fault recovery times will vary. There are no industrywide failover standards. This has impeded multivendor solutions for failover management software and support. Much of this implementation is left up to the platform vendor; hence the high degree of proprietary platforms. Some systems are designed with a commercial off-the-shelf OS with the option of using third-party fault-management software. Even in this framework, it is incumbent on the system operator to develop the platform management, procedures, and configuration information that will drive the fault management.

Vendors will have their own interpretations of a platform tolerance level. An FT platform can often withstand an OS failure and keep the platform operational. HA platforms can tolerate CPU and component failures, but not necessarily an OS or application outage. It is best to test out such scenarios prior to acquiring the platform and routinely afterwards, if operations permit. Test reports obtained from the vendor's lab can reveal to what extent a platform satisfies the advertised tolerance level and how it will react in specific situations.

Often times, a misbehaved application can cause OS instability. Many systems have a software process called an *application manager* that manages applications as a component. The proliferation of shrink-wrapped software in today's computing environment has unfortunately given rise to an abundance of poorly designed and tested software. An application manager treats problematic software as a component and as such will attempt to recover and restart it if necessary. Application software should be certified by the vendor (either the platform or application vendor) for problem-free operation on a given platform. The same holds true for device or component drivers.

The types of faults, failover controls, and their location(s) usually dictate fault-management procedures. Control can be handled by the PAL, OS, device driver, an application, or even an external device such as a network-management system—or combination there of. The fault-management process involves diagnosing, locating, isolating, and executing failover and recovery procedures. The following functions are usually involved:

- *Configuration management.* This function defines and maintains the configurations of all platform components. Configurations are updated as components are repaired, upgraded, added, or removed from the platform.
- *Event management.* Many systems use a software process called an *event handler* to provide governance over fault and failover procedures. They are designed to implement a set of actions in response to events such as software faults, hardware failures, or other situations. The job of the event handler is to execute actions that return the platform to a stable operating state. The actions and procedures that are carried out are usually specific to the platform and are typically programmed by the platform vendor. Some platforms will also allow the system operator to specify and customize the procedures—usually in the form of scripts—for their own purposes. The higher the tolerance a system has, the more automated the procedures come, requiring less manual specification.

- *Alarm management.* Alarms are the notifications made to the OS and system operator of a condition, such as a fault or failure. Alarms can be triggered before, during, or after a condition has taken place. The most tolerant systems have an alarm infrastructure that is logically and electrically distinct from the rest of the system so that they can continue to operate during adverse situations. Alarming systems can monitor hardware conditions such as system temperature and fan speed. The Telcordia Network Equipment and Building Specifications (NEBS) advocates use of alarms at low levels of a platform, including switchboards, network connections, and software components. Alarms should be accurate in reporting platform conditions. Persistent false alarms cannot only be a nuisance, but can eventually mask a true fault when it occurs. Telecom platforms typically require 95% alarm first-time accuracy in fault identification.

  Alarms can be manifested through various means, such as ringing/buzzing, light emitting diodes (LEDs), simple network management protocol (SNMP) notifications, page alerts, and even e-mail. Many systems can notify a service center and/or network manager from a remote site either over a data network or dial-up modem connection.

- *Checkpoint management.* This function handles the communication of platform state changes to an internal or external standby system. Many FT platforms have checkpoint communication between an active and standby processor. Checkpointing involves the use of a *heartbeat protocol* that is comprised of periodic messages from an active CPU conveying its operational state. A standby CPU might respond with its own state information as well, conveying its ability to pick up processing in the event a problem occurs. State information received from an active CPU is used to continue processing from the state the active CPU was in prior to a failure. If a standby processor is activated due to a problem in the active processor, it alerts the OS.

- *Failover management.* This function coordinates the activities across different platform components to implement failover and recovery procedures. In doing so, it must ensure the coherency and consistency of data in transferring control from one component to another in an effort to preserve the system state and prevent transaction loss to the extent possible. In tightly coupled components, this transfer is virtually seamless. In loosely coupled components, state and transaction information could be lost during failover.

- *Diagnostic management.* This function is used to execute the correct sequence of diagnostic tests to determine if a device is in the correct state and is capable of continued operation. This function will often collaborate with other management functions to perform the diagnostics.

- *System logging.* This function obtains, records, and reports status information upon the occurrence of platform events. A mission-critical platform should come with software tools to analyze log information so that potential or even past faults can be identified or evaluated. In the case of a past fault, such information can give clues as to what type of repair is required. The information can also be used to identify, predict, and prevent unplanned faults. A gradual increase of logged corrected errors in a component could signify that failure is imminent. Such is the case with disk read/write errors that succeeded on subsequent tries or contiguous bit memory errors indicating potential memory failure.

The level of a platform's *failover granularity* will often convey the nature of its fault management. In an *all or nothing* redundant arrangement, granularity is at the backplane segment level. A component error on a backplane segment will shut down the entire segment and revert operation to another. Such is the case in systems with the PCI bus. Granularity at the component level will revert processing to a corresponding component in another backplane segment. Granularity results in the definition of *fault zones*, which are the working levels in a platform whose operations are monitored and reported on (Figure 8.7).

Different levels of granularity across different platform elements can ultimately be blended together into a hierarchical fault-management framework. Hierarchical fault management applies fault management first in fault zones of lower granularity, and then escalates to higher levels if necessary. If a problem cannot be resolved within a component or a backplane segment, it is then elevated to a higher platform level. In the end, if a platform itself cannot internally resolve the problem, it is escalated to a system operator or to network management.

8.3.2.2 Dynamic Reconfiguration

Dynamic reconfiguration is the ability of a platform to reconfigure while continuing operation. It enhances physical hot swapping by logically detaching and attaching the swapped components without application interruption and loss of data. Upon logical detachment of a component, the component's state, memory contents, and logical interfaces are transferred to a corresponding component elsewhere on the platform. The information is then imported back to the component upon logical reattachment to the platform.

8.3.2.3 Restart/Masking

Today's hardware components are manufactured with such quality that hardware failures are gradually becoming rare. The result is greater system availability. When

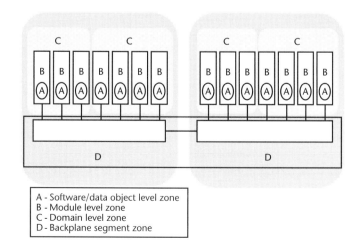

Figure 8.7   Platform granularity example

hardware does malfunction, either at the platform or component level, a restart (or *reboot*) of the component is usually attempted. OS crashes have been known to result in OS reboots. The most widely used retry model is quite simple: if the first retry does not succeed, a second attempt is made. If the second attempt succeeds, than the fault was likely a transient one. If the second fails, the fault was likely permanent and the component has failed [15]. A failover procedure is hence initiated. The failure of a component, even the OS, may not necessarily trigger a restart. A failed component is temporarily detached logically from the platform until it can be repaired. Dynamic reconfiguration would kick in to reallocate system resources in light of the change in platform configuration.

Masking of component faults is a misleading term in that it does not imply concealing faults from the system operator. Faults, even corrected ones, should be logged for reasons stated earlier. Masking simply hides much of the work involved in automatically detaching, diagnosing, and correcting a failed component so that it can be done without manual assistance. If a problem is at a high-severity level, then operator assistance is requested via alarm notifications.

8.3.2.4   I/O Handling

I/O bottlenecks can happen within a platform for many reasons. Heavy data traffic, hardware failure, or software faults could result in such bottlenecks. For one thing, there are a limited number of board slots on any backplane. I/O is the process a component uses to send and receive data to and from an external system. A backplane slot is typically occupied by components performing some kind of I/O: hard drive and tape drive controllers, various network cards, and cards supporting special peripheral devices. Heavy activity in one slot can create an I/O bottleneck. Because a backplane is in essence a network, impedance in one node, like a host, can slow down activity in the entire network. In fact, many systems detecting such circumstances will adjust bus activity accordingly. There are several ways to avoid I/O bottlenecks or at least minimize their impact on the overall system:

- *Alternate pathing* involves automatically redirecting I/O traffic to another predetermined I/O channel if the primary channel becomes inoperative [16]. A backplane typically consists of many paths, sometimes called *traces*, between the I/O modules and processors. Alternate pathing sets up redundant I/O channels (or *paths),* between modules or between modules and processors. If the primary path experiences trouble, then the platform automatically transfers all of a channel's I/O operation to the other redundant path. This technique is commonly used in communication platforms to establish redundant paths between a CPU and the I/O modules that connect to a termination card or network interface (Figure 8.8).

- *Domain redundancy* can be used with multiple I/O modules, as with other components. This involves logically configuring an I/O module on redundant system domains. If a problem occurs in a domain, the corresponding alternate domain can still utilize the module.

- *Module redundancy* can introduce additional reliability. If problems occur in one module, a secondary module can be used. A pair of modules residing

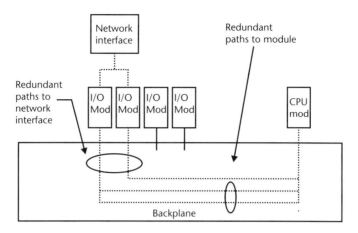

Figure 8.8 Alternate pathing example.

in one domain can be homed to a secondary domain as well. Use of redundant modules might also require redundant connections to the same external resource, such as a disk, or redundancy in the external resource. Redundant modules can be made to run in parallel in loosely or tightly coupled configurations. Parallel operation can be used in conjunction with a disk mirroring application by performing simultaneous writes to two disk systems over separate channels. Once a disk write is committed and reported on a channel, the system will commit a write over the other channel.

- *Retry* is a general approach that can be used in the manner discussed previously. In an I/O operation on a component that has failed, at least one retry on the operation is performed. At the operation level, having more than one retry could eventually lead to a successful one. However, using many retries can degrade system performance and can indicate that potential failure of a device is imminent [17].

- *Failover* requires transferring an I/O module's register states, memory contents, cache contents, and other critical information to another in a seamless fashion so that information is not lost during the transfer. This also includes preserving the connection state with the external device that the I/O component was communicating with at the time of failure.

- *Preemptive shutdown* of certain I/O modules is a precautionary action that is used if a platform undergoes a problem. In such cases, the platform must give preference to those components and processes to sustain service during the situation, at the expense of those I/O modules that are less critical.

- *I/O partitioning* is an architecture approach that places I/O tasks on processor modules apart from the CPU modules. This offloads I/O processing work from the CPU and makes application processing impervious to I/O problems, such as I/O contention. On the flip side, I/O processing can continue if a CPU is problematic and is undergoing failover. It also supports preemptive shutdowns. Redundant I/O modules can be homed to one or multiple CPUs, depending on the platform tolerance level.

8.4   Network Platforms

The trend in network convergence is having far-reaching effects. In addition to the merging of voice and data traffic over a common network, it also implies consolidation of platform form factors. For the first time, we are seeing switches that look like servers and servers that look like switches. Many of the principles just discussed pertaining to servers can also apply to network platforms.

Network switching and routing platforms contain many of the same elements as a server: OS, CPU, memory, backplane, and I/O ports (in some cases, storage). But network platforms typically run a more focused application aimed at switching or routing of network traffic. They also must have more versatile network ports to accommodate various networking technologies and protocols. Because they have a more entangled involvement with a network than they do a server host, they are subject to higher availability and performance requirements. Routers, for instance, were known to have availability that was several orders of magnitude lower than carrier class requirements would allow. Much has changed in the last several years. Convergence of voice, video, and data onto IP-based packet-switching networks and e-commerce, and a blurring of the distinction between enterprise and service provider, have placed similar high-availability requirements on data networking as on telephony equipment, to the point where now these devices must have the inherent recovery and performance mechanisms found in FT/FR/HA platforms.

Switches and routers have become complex devices with numerous configuration and programmable features. Unplanned outages are typically attributed to hardware failures in system controller cards or line cards, software failures, or even memory leaks. Some of the most major outages in recent years were attributed to problematic software and firmware upgrades. The greater reliance on packet (IP) networks as a routing fabric for all services has unintentionally placed greater responsibility on IP-based network platforms. As was seen in our discussions in this book, problems in an IP network device can compound across a network and have wide-scale effects.

8.4.1   Hardware Architectures

The service life of a network platform will typically be longer than that of host server platform. With this comes the need for reliability and serviceability. Network platform reliability has steadily improved over the years, due largely in part to improvements in platform architecture. The differentiation between carrier class and enterprise class products has grown less and less. For one thing, architectures have become more simplified, with fewer hardware components, added redundancy, and more modularity. On the other hand, many network platform adjunct devices, such as voice mail systems and intelligent voice response (IVR) systems, are often built on or in conjunction with a general-purpose platform. Switch-like fabric of some type is interconnected with the server, which maintains all of the software processes to drive the platform.

Figure 8.9 illustrates some of the generic functional components in a network platform [18, 19]. A physical network interface would provide functions for interpreting incoming and outgoing line signals, such as encoding/decoding and multiplexing/demultiplexing. Protocol processing would perform media address control

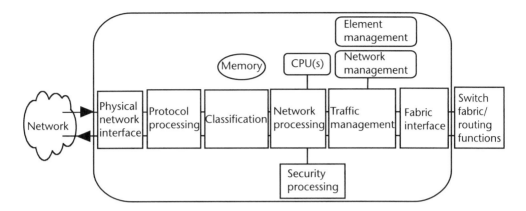

Figure 8.9 Generic network platform functions.

(MAC) layer processing and segmentation/reassembly of frames. The classification function would classify frames based on their protocol. The network processor would provide the network-layer functional processing. The security function would apply any needed encryption or decryption. The traffic-management function applies inherent or network management based traffic controls. The fabric portion manages port-to-port connectivity and traffic activity. Depending on the networking functions (e.g., switching, routing, and transmission) functional architectures will vary and can be realized through various software and hardware architectures. Many of the platform attributes already discussed in this chapter apply.

The following are some general principles regarding network platforms for use in mission-critical networks. Because of the variety of form factors on the market, they may not apply to all platforms. Many of the principles discussed with respect to server platform hardware architecture will in many places apply:

- *Modularity.* Modularity, as talked about earlier in this book, is a desirable feature of mission-critical systems. Modular architectures that incorporate externally developed components are steadily on the rise. This trend is further characterized by the ever-increasing integration of higher level processing modules optimized for a specific feature versus lower level generic components. For example, one will likely find systems having individual boards with firmware and on-board processing to provide T1/E1, SS7, voice over IP (VoIP), asynchronous transfer mode (ATM), Ethernet frame relay, and integrated services digital network (ISDN) services. This kind of modularity reduces the number of components, board-level interconnects, EMI, power, and ultimately cost. It also enables more mixing and matching of features through various interface configurations. Additionally, it lengthens the platform-technology curve by enabling incremental board-level upgrades, versus costly quantum-platform upgrades.

- *Backplanes.* Redundant components with failover capabilities are becoming the norm. This includes redundant backplanes. Alternate pathing is often used to keep modules uncoupled so that if one fails the others are unaffected. Path redundancy, particularly between I/O modules and switch-fabric modules, allows concurrent support of bearer traffic (i.e., data and voice) and network-management instructions. Additionally, establishing redundant bearer paths between I/O and switch modules can further enhance traffic reliability.

• *Dual controllers.* The presence of redundant controllers or switch processors, used either in a hot or cold standby configuration, can further improve platform reliability. This feature can also improve network reliability, depending on the type of network function the platform provides. For example, failover to a hot standby secondary controller in a router is prone to longer initialization times, as it must *converge* or in other words learn all of the IP layer routing and forwarding sessions. This process has been known to take several minutes. To get around some of these issues, vendors are producing routers in which routing sessions and packet forwarding information states are mirrored between processors, thereby reducing extended downtime. Such routers, often referred to as *hitless routers*, are finding popularity as edge network gateway devices, which are traditionally known to be a single point of failure.

• *Clocking.* Clocking places a key role in time division multiplexing (TDM)–based devices but also has a prominent role in other network applications. Having dual clocking sources protects against the possibility of a clock outage and supports maintenance and upgrade of a system-timing source. Improper clocking, particularly in synchronous network services, can destroy the integrity of transmitted bits, making a service useless. Recently, many systems have been utilizing satellite-based global positioning system (GPS) timing for accuracy. Regardless, it is imperative to use a secure source where reliability and survivability are guaranteed.

• *Interface modules.* Traditionally, network devices have often used a distributed approach to the network-interface portion of the platform. Line cards and switch port modules have been a mainstay in switching equipment to support scalable subscriber and user growth. Interface modules, however, are a single point of failure. A failure in one processor card can literally bring down an entire LAN. Simply swapping out an interface card was usually the most popular restoration process. However, the higher tolerance mandated by today's environment requires better protective mechanisms. $N + K$ redundant network interface boards with alternate backplane paths to the switch or routing processor can provide added survivability. Use of boards with multiple different protocols enables diversity at the networking technology level as well. Edge gateway devices, in particular, are being designed with individual processor cards for the serial uplink ports and each user interface port. Some designs put some routing or switching intelligence inside the port modules in the event a catastrophic switch fabric or controller failure takes place.

• *Port architecture and density.* Port density has always been a desirable feature in network platforms. The more channels that can be supported per node (or per rack unit/shelf), the greater the perceived value and capacity of the platform. Dense platforms result in less nodes and links, simplifying the network. But one must question whether the added density truly improves capacity. For one thing, real-time processing in a platform is always a limiting factor to platform capacity. Products such as core switches are typically high-end devices that have a *nonblocking architecture*. In these architectures, the overall bandwidth capacity that the device can support is equivalent to the sum of the bandwidth over all of the ports. Lower end or less expensive workgroup or edge switches have a *blocking architecture*, which means that the total switching bandwidth capacity is less than the sum across all of the ports. The bandwidth across all user ports will typically exceed the capacity of an uplink port. These switches are designed under the assumption that not all ports will be engaged at the same time. Some devices use gigabit uplinks and stacking ports to give the sense of nonblocking.

- *Hot swapping*. As discussed earlier, components that are hot swappable are desirable. This means not only that a component can be swapped while a platform remains powered and running, it also means nondisruptive service operation during the swap. Network interface modules should have the ability to be swapped while preserving all active sessions (either data or voice) during failover to another module.

- *Standards compliance*. High-end carrier grade equipment is usually subject to compliance with the Telcordia NEBS and/or open systems modification of intelligent network elements (OSMINE) process. NEBS has become a de facto standard for organizations, typically service providers, looking to purchase premium quality equipment. NEBS certification implies that a product has passed certain shock, earthquake, fire, environmental, and electrostatic discharge test requirements. Equipment will usually be required to comply with Federal Communications Commission (FCC) standards as well. Some systems may also require interface modules to satisfy conformance with communication protocols. In addition, many vertical market industries have certain equipment standards as well, such as the American Medical Association (AMA), the Securities and Exchange Commission (SEC), and the military.

8.4.2   Operating Systems

An OS in a networking platform must continuously keep track of state information and convey it to other components. In addition to items such as call processing, signaling, routing, or forwarding information, administrative and network management transactions, although not as dynamic, must also be retained. During processor failover, such information must be preserved to avoid losing standing transactions. A platform should also enable maintenance functions, such as configuration and provisioning, to continue operation during a failover.

Quite often, provisioning data is stored on a hard drive device. As in previous discussions, there are numerous ways to protect stored information (see the chapter on storage). Networking platforms using mirrored processors or controllers may also require mirrored storage as well, depending on the platform architecture. Configuration or subscriber databases typically require continuous auditing so that their content is kept as consistent as possible and not corrupted in the event of an outage. Some appliance-based network products, in order to stay lean, offload some of this responsibility to external devices.

As discussed earlier, router OSs have classically been known to take extended amounts of time to reinitialize after a controller failure. Furthermore, the ability to retain all routing protocols and states during failover can be lacking, as the standby processor was often required to initialize the OS and undergo a convergence process. This not only led to service disruption, it also required disrupting service during upgrades.

Routing involves two functions. A routing engine function obtains network topology information from neighboring routers, computes paths, and disseminates that information. A forwarding engine uses that information to forward packets to the appropriate ports. A failure of the routing engine to populate accurate routes in the forwarding table could lead to erroneous network routing. Many routers will assume a forwarding table to be invalid upon a failure, thus requiring a reconvergence process. Additionally, system configuration information must be reloaded and all active sessions must be reestablished. Before routing sessions can be restored, system configurations (e.g., frame relay and ATM virtual circuit mappings) must be loaded. A failed router can have far-reaching effects in a network, depending on where it is located. Router OSs are being designed with capabilities to work around some of these issues.

Furthermore, many switch and router platforms are coming to market with application programming interfaces (APIs) so that organizations can implement more customized features and functions that are otherwise unavailable in a platform module or module upgrade. APIs enable configuration of modules using available software libraries or customized programming. The platform OS will encapsulate many of the platform hardware functions, making them accessible through the APIs. Use of APIs can reduce time to implement system or service-level features.

8.5 Platform Management

Manageability is a vital quality of any mission-critical server or networking platform. The platform should enable monitoring and control for hardware and software fault detection, isolation, diagnosis, and restoration at multiple levels. The platform should also enable servicing through easy access to components and well-documented operations and procedures. Some systems come with modules, software, and procedures for emergency management. Serviceability, or lack thereof, is typically a common cause of many system outages. Human errors made during software or hardware upgrades are a result of complex system and operational processes. Such situations are avoided through a user-friendly element management system (EMS) with an easy to use graphical user interface (GUI).

8.5.1 Element Management System

An EMS integrates fault management, platform configuration, performance management, maintenance, and security functions. A mission-critical EMS should come with redundant management modules, typically in the form of system processor cards each with SNMP (or comparable) network-management agents and interface ports for LAN (typically Ethernet) or serial access to the platform. LAN ports, each with an IP address, might be duplicated on each management board as well for redundant connectivity to the platform.

Many network management software implementations are centered on SNMP. Other implementations include common management information protocol (CMIP), geared towards the telecom industry, and lately Intel's Intelligent Platform Management Interface (IPMI) specification. These solutions are designed primarily to interface with platform components in some way in order to monitor their vital signs. These include such items as temperature, fans, and power. Much discussion has been given to monitoring thus far. In all, any effective monitoring solution must provide accurate and timely alerts if there is malfunction in a component, anticipate potential problems, and provide trending capabilities so that future problems are avoided.

Hardware alerts are usually provided by an alarm board subsystem. As discussed earlier, such systems have interfaces so that generic alarms can be communicated through various means, such as a network, dial-up modem, or even a pager. Alarm systems come in many different forms, ranging from an individual processor board to an entire chassis-based system. Alarm systems should have, as an option, the ability to have their own power source and battery backup in case of a power outage. Many will have their own on-board features, LEDs, and even some level of programmability.

Alarm communication to external systems is usually achieved using various industry-standard protocols or languages. In telecom applications, Telcordia's Man-Machine Language (MML) protocol is widely used, while enterprise networks commonly use SNMP. To the extent possible, alarm communication should kept out of band so that it can persist during a network or platform CPU failure.

8.5.2 Platform Maintenance

There will come a time during the service life of any platform when some type of preventive maintenance or upgrade is required. Upgrades usually refer to the process of modifying the platform's hardware, such as adding or changing processors, memory, NICs, or even storage. It also refers to software modifications, such as installing a new or patch version of an OS or application. Some general rules should be used with respect to the upgrade process in a mission-critical environment.

First, other network nodes should be unaffected by the node undergoing an upgrade. Availability goals may warrant that upgrades are performed while a system is in an operational state, actively providing service. This requires upgrading without service interruption. Many of the types of platform characteristics, such as redundancy and failover, can be leveraged for this purpose. A good user-friendly GUI can help minimize manual errors, which are quite common during the upgrade process.

If availability requirements permit a platform to be taken out of service for an upgrade or repair, it should be taken off line in the off hours or during a time when the least disruption would result from the shutdown. Network level redundancy techniques, many of which were discussed earlier in this book, can be leveraged so that another corresponding device elsewhere in the network can tentatively provide service during the upgrade.

Once the repair or upgrade is completed and the system is reinitialized, it should be in the identical state as it was prior to the shutdown, especially with respect to transaction and connection states. Its state and health should be verified before it is actually placed on active duty. In some situations, an upgrade that has gone awry might require backing out of the upgrade.

It is recommended that a service agreement be in effect with the system vendor to provide on-site repair or repair instructions by phone. Quite often, difficulties arise during the startup process versus shutdown. Retaining backup copies of configuration data and applying those configurations upon restart will ensure that the platform is in a state consistent with that prior to shutdown. Sound configuration-management practices should include saving backup copies of configuration files and keeping them updated with every configuration change, even the most minor ones.

Some multiprocessor platforms can operate in *split mode*, which permits the upgraded environment to be tested while the preexisting environment continues to operate and provide service [20]. This allows the upgrade to be tested before it is committed into service, while the platform is in an operational service mode. This minimizes service interruption and improves availability. Split mode in essence divides a platform into primary and secondary operating domains, each served by a CPU and at least one I/O component (Figure 8.10). The primary domain retains the preexisting system and continues to actively process applications. It keeps the secondary domain abreast of application states and data, so that it can eventually transfer service after testing. Some applications from the primary domain can participate in the testing of the secondary domain if administratively specified.

Maintaining onsite spare components for those that are most likely to fail should be part of any maintenance program. Of course, this also requires having trained personnel with the expertise to install and activate the component. However, keeping replacements for every component can be expensive. Platform vendors will normally ship needed components or send repair technicians, especially if it is part of a service agreement. Replacement part availability should be a negotiated clause in the service agreement.

This last point cannot be emphasized enough. A platform vendor can be a single point of failure. If a widespread disaster occurs, chances are good that many organizations having the same platform and similar service agreements will be vying for the same replacement parts and technician repair services. Component availability typically diminishes the more extensive a widespread disaster grows, regardless of the terms in a service agreement. One countermeasure is to use a secondary vendor or component distributor. If a particular component is commonly found in platforms across an enterprise, another strategy that is often used is to maintain a pool of spares that can be shared across company locations. Spares can be stored centrally or spread across several locations, depending on how geographically dispersed the enterprise is.

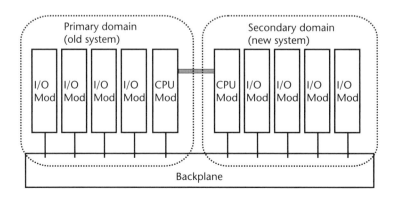

Figure 8.10   Split-mode operation example.

The use of *fixed spares* requires having a spare per functional component. An alternative is the use of *tunable spares*, which are spares that have most of the underlying native capabilities for use but require some tuning to configure and prepare them for their service function. For example, many types of I/O components may share the same type of processor board. All they would need is last-minute configuration based on their use in the platform (e.g., network interface or device interface). This can include such things as installing firmware or software or flipping switches or jacks on the board. Figure 8.11 illustrates the concept. Thus, a pool of universal spares can be retained at low cost and configured when needed on a case basis. This reduces the size of the inventory of spares.

8.6 Power Management

Power is a major consideration in the operation of any server or networking platform. Power supplies and modules are commonly flawed elements. Other than component failures, voltage surges due to lightning strikes or problematic local transformers cannot only disrupt platform service, but can damage a platform and destroy the embedded investment. The growing cost of power consumption is also a lurking concern with many enterprises. Advances in power-efficient integrated circuitry are offset by the trend in high-density rack server platforms.

Strengthening a platform's power supply and associated components is the first line of defense against power-related mishaps. The following are some suggested platform-specific measures (other general precautions are discussed later in this book in a chapter on facilities):

- *Redundant power supplies*, once a staple in high-end computing and networking systems, has become more prevalent across a wide range of platforms. Many backplane architectures can accommodate power modules

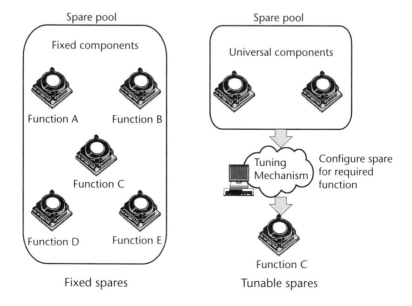

Figure 8.11   Fixed spares versus tunable spares.

directly in the chassis and are hot swappable. Redundant or segmented backplanes will typically each have their own power supply. $N + K$ redundancy can be used, providing an extra power supply than is required to run a platform.

- *Load-sharing* power supplies can be used to spread power delivery among several power supplies, minimizing the chance that one of them will be overstressed. If one of the power supplies fails, the platform would consume all of its power using the remaining supply. Because each supply would run at half the load during normal operation, each must be able to take up the full load if the other supply becomes inactive. As power consumption can vary as much as 25%, having a higher rated power supply may be wise. Load sharing provides an added advantage of producing less heat, extending the life of a power supply, and even that of the overall platform [21].

- *Independent power feeds* for each power supply further eliminates a single point of failure. Each power supply should have its own cord and cabling, as well as power sources. This ultimately includes the transformer and other power plant components. For large-scale mission-critical operations, it may even require knowledge of the power plant infrastructure supporting the facility and the locale. This topic is discussed in the chapter on facilities.

- *Secure power control* requires features that avoid the inadvertent shutting off of power to the platform. A protected on/off switch can avoid accidental or maliciously intended shut off of the system. Secure cabling and connections will also safeguard against power cord disconnects or cuts.

- *Sparing* of replacement components can facilitate hot swapping and improve availability. Many of the sparing practices discussed in the previous section can be applied to power components as well.

- *Power line conditioning* protects against a surge or drop in power, which can be more debilitating to certain equipment than complete power loss. Power conditioning is discussed in the chapter on facilities.

8.7 Summary and Conclusions

This chapter reviewed capabilities that are desired of mission-critical platforms. The trend toward server- and appliance-based architectures has given rise to both server and networking platforms with many integrated COTS components. FT/FR/HA capabilities are the product of vendor integration of these components. Regardless of the type of platform, simplicity, cost, serviceability, redundancy, failover, certifications, and quality are common characteristics desirable of mission-critical platforms.

FT is achieved through hardware and software by incorporating redundant components and building in mechanisms to rapidly detect, isolate, and correct faults. All of this creates extra cost, making FT platforms the most expensive to use. This is why they are often found for specialized applications such as telecommunications, air-traffic control, and process control. FR and HA are lower cost alternatives that may resemble FT platforms on the surface, but they cannot guarantee the low levels of transaction loss found in FT platforms.

Server platforms have evolved into backplane systems supporting several bus standards, including ISA, PCI, cPCI, CPSB, VME, and Infiniband. The use of a bus-based architecture enhances both survivability and performance, as it enables the connection of many redundant components within the same housing. Multiple CPU systems, which are preferred in mission-critical environments, should have the appropriate failover and fault-management mechanisms to ensure a platform's required tolerance level. Because power supplies and modules are commonly flawed elements, extra precautions are required to ensure continuous power. This includes using redundant/load-sharing power supplies, independent power feeds, and power line conditioning.

Many of the qualities desired of mission-critical servers hold true for networking platforms, as they contain many of the same elements as servers. But networking platforms typically run a more focused application—the switching or routing of network traffic. For this purpose, they require interworking with a switch/routing fabric comprised of many versatile network ports. Modularity, controller or switch processor redundancy, reliable timing, and multiple interface modules are common characteristics of a mission-critical networking platform. The platform must also be able to retain all protocol and state information during a failover.

Stability is the most important characteristic one should look for in a mission-critical platform. The ability to predict platform behavior is the key to mission-critical platform success. Manageability and serviceability are also vital qualities. The use of FT/FR/HA platforms must be accompanied with good network design to achieve tolerance at the network level. In the end, the overall efficacy of a mission-critical platform transcends its hardware and software capabilities. The finishing touch lies in an organization's operating environment, encompassing everything from network architecture and management, applications, data storage, and even business processes.

References

[1]   Desmond, P., "Reliability Checklist," *Network World*, August 30, 1999, pp. 53–54.

[2]   Grigonis, R., "Faultless Computing," *Computer Telephony*, May, 1998, pp. 48–50, 71, 92–96.

[3]   Grigonis, R., and J. Jainschigg, "Platforms and Resources," *Interactive Voice Response—Supplement to Miller Freeman*, 1998.

[4]   Wallner, P., "Bringing Carrier-Class Reliability to IP Telephony," *Telecommunications*, April 1999, pp. 54–55.

[5]   Ruber, P., "Server Fault Tolerance," *Network Magazine*, November 1998, pp. 30–37.

[6]   Grigonis, R., "Bullet-Proof Software," *Convergence Magazine*, September 2001, pp. 70–79.

[7]   Sullivan, J., "High Availability RTOSes: A Buyer's Guide," *Communications Systems Design*, April 2001, pp. 44–50.

[8]   Grigonis, R., "Fault Resilience Takes New Forms," *Computer Telephony*, February 2000, pp. 112–116.

[9]   Grigonis, R., "Fault Resilient Failover," *Convergence Magazine*, July 2001, pp. 36–46.

[10] Grigonis, R., "Fault-Resilient PCs: Zippy's Mega-Update (cPCI, Too!)," *Computer Telephony*, May 1999, pp 79–82.

[11] Lelii, S., "Right Technology, Wrong Economy," *VAR Business*, September 30, 2002, pp. 56–58.

[12] Grigonis, R., "Fault Resilience for Communications Convergence," *Supplement to Computer Telephony Magazine*, Spring 2001, pp. 5–16.

[13] Hill, C., "High Availability Systems Made Easy: Part 2," *Communications Systems Design*, December 2000, pp. 45–52.

[14] Lesniak, N., "Challenges of High Availability Software Development," *Integrated Communications Design Magazine*, April 16, 2001, pp. 48–50.

[15] Lawrence, J., "Make It Run Forever," *Computer Telephony Integration Buyer's Guide*, 1999/2000, pp. 44–47.

[16] Caldera, S., T. Manning, and J. Quinlivan, "Guidelines for a Fault Tolerant Network," *Network Reliability—Supplement to America's Network*, December 2000, pp. 26S–28S.

[17] Katan, A., and J. Wright, "High Availability: A Perspective," *Tech Republic*, June 29, 2000, *www.techrepublic.com*.

[18] Telikepalli, A., "Tackling the Make-Versus-buy Decision," *Integrated Communications Design Magazine*, February 2002, p. 20.

[19] Denton, C., "Modular Subsystems Will Play a Key Role in Future Network Architecture," *WDM Solutions*, October 2002, pp. 5–8.

[20] Hirschl, J., "Split Mode is Critical for Upgrading Fault Tolerant Systems," *Internet Telephony*, September 1999, pp. 84–87.

[21] Grigonis, R., "The Rise of the Fault Resilient Computer," *Uptime—Supplement to Teleconnect Magazine*, February 1998, pp. 6–10.

CHAPTER 9

Software Application Continuity

Application software is fundamental to network continuity because it drives most network services. Equal emphasis should be placed on performance and survivability of application as well as network infrastructure and facilities. Compounding this challenge is the fact that today's application software is developed with off-the-shelf component software so that applications can be developed on standards-based hardware platforms. But with this flexibility comes problems with interoperability and product quality, which can escape control of an information technology (IT) organization and ultimately impact network performance.

For this reason, importance must be placed on viewing and measuring application status and performance across a network. Lately, many application performance measurement (APM) tools have become available to supply metrics on the productivity, efficiency, and quality of distributed applications. In spite of the sophistication of these tools, what matters most is the end-user perspective. The best way to gauge how an application is performing is to see what end users are currently experiencing.

The topic of application software is quite broad and beyond the immediate nature of this book. For the purposes of this book, we focus on those aspects of applications that are most pertinent to survivability and that align with the many topics discussed in this book.

9.1 Classifying Applications

Software applications have become more diverse, distributed, and complex and more specific to operational functions—to the point where functions and applications are nearly indistinguishable. Web browsing, email, storage and retrieval, system control, database management, and network management are examples of types of standalone functional applications that must interact with other elements over a network.

A mission-critical network should be an enabler for service applications, be they for revenue generation or support functions. A network should recognize the most critical applications and provide them with the right resources to perform satisfactorily. This means that some form of criteria should be applied to each application to convey its importance to the overall network mission.

We classify applications by applying two criteria: importance to the business (or mission) and how soon they should be recovered. Criteria for business importance lie in the context of use and the type of organization. Applications such as enterprise resource planning (ERP) systems, business-to-consumer (B2C) applications, business-to-business (B2B) applications, customer-relationship management (CRM) applications, and even support applications such as e-mail are considered mission critical. The following are some general categories that can be used to classify applications [1]:

- *Mission critical*: applications that are necessary for business to operate;
- *Mission necessary*: applications that are required for business, but can be temporarily substituted with an alternate service or procedure;
- *Mission useful*: applications whose loss would inconvenience but not necessarily disrupt service;
- *Mission irrelevant*: applications whose loss would not disrupt or degrade service.

The second criterion, application recoverability, is evaluated in terms of recovery time objectives (RTOs). Classes representing an RTO range should be assigned to each application. For example, class 0 applications would have an RTO of 1 hour, class 1 an RTO of 2 hours and so on (there is no real industry standard for classification). Although it is not incorrect to assume that RTO is directly related to criticality, mission-necessary applications can have higher RTOs if they rely on a contingency mechanism. More discussion on this topic is in the section on application recovery, Section 9.6.

9.2 Application Development

Application development is the blending of process, database, and hardware requirements to produce reliable software. Application reliability begins with a well-defined and structured software development process. Table 9.1 shows the steps that make up the traditional development process [2]. The process will vary somewhat depending on the type of application, system, or organizational context. Although the process shown is presented in the context of software development, it is adaptable to almost any system implementation. Shown also are continuity milestones that should be achieved at different stages of development. The purpose of this process is to assure that applications are designed to meet their functional goals, are free of errors, and perform satisfactorily during operation [3].

Any organization pondering a mission-critical system implementation consisting of a single or multiple applications should employ this process and adapt it as needed. These phases can apply regardless of whether software is developed inhouse or purchased. Vendors should certify that their software has undergone a formal and rigorous development process and should be able to produce evidence to the effect. In addition, *secure code review*s should be practiced by organizations and software vendors. This entails testing software against the likelihood that poor programming (e.g., failure to check memory/buffer boundaries or inadequate consideration of failure conditions) will result in an application or OS vulnerability that could be exploited at the expense of an organization.

Surprisingly, the analysis and definition phases are the most valuable phases of the process. It is in these phases where systems engineering is used to define the

Table 9.1   Typical Software Development Process

Phase	Activity	Outputs	Continuity Milestone
Analysis	Customer/mission requirements Systems modeling/simulation Systems analysis Systems engineering	Customer/mission specifications Service/product prospectus Prototype Feasibility study	Desired service-level goals Characterize service behavior Identify service-level objectives Define service-level metrics
Definition	Systems engineering System requirements Project planning Acceptance test preparation	System specifications Project plan and documentation Acceptance criteria	Achievable service metrics Performance and survivability envelope Reporting design Target service levels
Design	System architecture High-level design Integration test preparation System test preparation Acceptance test preparation Project planning	Integration test specifications System test specifications Acceptance test specifications Revised project plan and documentation	Critical resources Contingency design Backup design Failover design
Programming	Low-level design Code and module test Integration test System test preparation Acceptance test preparation OA&M requirements Production and distribution Requirements	Revised design specifications System documentation Revised system test specifications Revised acceptance test specifications OA&M specifications Production and distribution specifications Project planningrevised project plan and documentation	Detection design Recovery design Containment design Montoring design Restoration procedures Resumption procedures Service and repair procedures
System test	System test Site test preparation Migration/deployment Requirements Project planning	Revised system documentation Site test specifications Deployment plan Revised production and distribution specifications Revised project plan and documentation	Committed service levels Backout procedures
Acceptance	Acceptance test User training Project planning	Acceptance checklist Revised system documentation Training materials Revised project plan and documentation Revised deployment plan	Problem resolution procedures User expectations solidified
Deployment	Installation Configuration Site test Change management Project planning	Revised deployment plan Revised project plan and documentation	Service monitoring

context of an application's proposed operation. In today's fast moving, rapid time-to-market environment, these upfront phases are often overlooked. Instead, more emphasis is placed on development and testing activities, as they typically carry the brunt of development resources.

For expediency, organizations will often go directly from user or mission specifications to the software development and design stages. The problem with this approach is that software development is limited solely to satisfying the specifications, without looking beyond the scope of the application or service. Today's applications, especially those that are Web-related, increasingly use *dynamic content* such as active server pages, standard query language (SQL) databases, and user input forms, which are frequently not checked against malicious input. When an application is thrown a "curve ball" by an unexpected event and behaves erratically, it is often because development did not anticipate and design for the event because it was not called for in the specifications.

For mission-critical applications, a sound a systems engineering function is required. The function must absorb user or mission requirements and marry it with the big picture. It must define the events and environment affecting application operation so that development can respond accordingly with the proper design. Upon the completion of the first two phases, a model of an application's or service behavior should exist somewhere. It can either be in computerized form, a written or verbal description, or just a mental picture of how the application should behave under certain conditions. The ability to model and predict behavior is a prerequisite for mission-critical operation.

A software design must produce *well-behaved* performance and enable the software to have its own components monitored and recovered from internal faults, platform faults, or those of other applications. Well-designed software should also leverage the resources of the platform for optimal performance. Application processing deficiencies should not necessarily be compensated by platform upgrades.

This is why proprietary system designs usually prove the most reliable—vendors can design and integrate software to fit well within the platform at hand. Furthermore, proprietary systems only prove more reliable in the respect that the vendor controls the software specification and does not have to comply with external standards or satisfy interoperability requirements. Unfortunately, such designs involve relatively higher cost, lengthy development time, and inflexibility when it comes to migration. In spite of the industry drive towards standards-based building blocks and open systems that are intended to address some of these issues, application instability due to interoperability problems is fast becoming a norm.

Standards compliance does not guarantee interoperability or stable application behavior for several reasons. First, there is neither a standard "standard" nor single point in time when all software or equipment must satisfy a standard. Furthermore, vendors will often interpret the same standards differently. In recent years, vendors have been voluntarily forming consortiums to address these issues by establishing centralized testing, integration, and certification. They often issue certification guidelines so that software can reliably interact with other software and hardware components. Such examples include device driver guidelines and the use of object management group (OMG) Common Object Request Broker Architecture (CORBA)-compliant objects.

Complicating this issue is the incorporation or continued support of legacy applications. Many organizations employ a great deal of legacy systems because of the embedded investment. Many systems are considered irreplaceable. Many legacy application software designs start with a small custom prototype or interim solution, a user interface objective (e.g., a Web browser), and a "make do" platform which in some cases begins as a personal computer (PC)–based system. Many designs are devoid of sophisticated application monitoring, database capabilities, and backup capabilities required for continuity. Furthermore, the original operating requirements, source code, and documentation of many legacy applications is unavailable to those who continue to use them. Despite all of this, legacy applications, having tried the true tests of time, are often found to be more stable than newer front-end systems. In e-commerce environments, many e-business applications serve as front ends to legacy applications, some owned and operated by other entities. The following are some general guiding principles to apply when designing, procuring, or integrating software for mission-critical operation:

- *Requirements*. Continuity requirements should be established early on in a formal software development and integration process. In general, proper requirements should be traceable throughout the entire development process—from design to testing. There are a slew of requirements tools and techniques available on the market today that can be used throughout the entire development process for these purposes. Old-fashioned structured programming can work well too. More discussion on requirements is presented in the chapter on testing.

- *Modularization*. The second chapter of this book includes discussion on the benefits of modularization. This concept lends itself to software design as well. Dividing an application into a set of logically partitioned components avoids the use of "spaghetti code," supports scalability and recovery, and enables a building-block design approach. Such partitioning should start at the requirements level. Traditional software development shops may scoff at this idea, stating that requirements should never dictate design. However, experience has shown that software development and systems engineering are two distinct disciplines. A sound systems-engineering function should develop system specifications that take business process or mission requirements and translate these into blocks of development assignments.

- *Portability*. Application designs should be simple and portable so that a recovering application does not require replacement in entirety. Surprisingly, placing dynamic information such as platform configuration and addresses into parameters is still not common practice, especially among novice developers. Today, there is a multitude of fourth generation languages (4GL) or object-oriented development environments that eliminate the need for addressing platform characteristics within code.

- *Certification*. There is a huge market for prefabricated software components and open-source solutions. Although these products individually can be clean and debugged, issues arise when they try to operate in concert with other products. Use of turnkey applications, although more expensive and restrictive, can often provide a more stable alternative and can be adapted to individual situations. If a component-wise approach is chosen, precertification of purchased components by the vendor is a must.

- *Middleware.* A middleware should interface with platform operating system (OS) and hardware functions. A portable and modular software component will likely look to the availability of middleware to manage redundancy and support continuity. Middleware should have availability capabilities to conduct failover to redundant components. It should also provide application-programming interfaces (APIs) for applications to use without necessarily incorporating platform specifics into the software code.

- *APIs.* The APIs should support *checkpointing*, which is the process of copying state data between the memory areas of different software components. In addition, *heart beating*, sometimes referred to as *keep alive*, should be available to periodically send signals between components. These signals contain information conveying a component's health and operational status and whether it is active or has failed. The industry is fostering availability management through standardized specifications of APIs between middleware, applications, and hardware. Nevertheless, applications must still be designed to leverage use of the APIs.

- *Operations and maintenance.* A well-designed application is pointless without proper operation and maintenance practices. Although an application may run flawlessly, surrounding operating components can be problematic. Application and database servers, load balancers, firewalls, and networking devices, for instance, can all contribute to application degradation. Integration is thus required to ensure flawless operations, maintenance, and recovery across all components.

9.3 Application Architecture

Today's applications have taken on a distributed appearance (Figure 9.1). Service architectures spread applications across numerous network platforms that collectively provide a service [4]. They are referred to as tiered applications because they are layered to provide instances of presentation, logical processing, and data for a given service (Figure 9.2) [5]. The processing logic layers provide input/output (I/O) to the other layers as needed.

When applications execute, they spawn different processes that the host platform must collectively manage to sustain operational performance. A platform may not necessarily fully understand the source or origin of an executing process, unless that information is communicated to the platform in some way. In a mission-critical operation, it is incumbent upon the application to manage itself and/or provide information to the platform for performance and reliability. The fact that different services may rely on common applications further underscores this need.

9.4 Application Deployment

Exposure of mission-critical application software to external end users, by way of the Internet or commercially off the shelf, is fast becoming more a burden than a

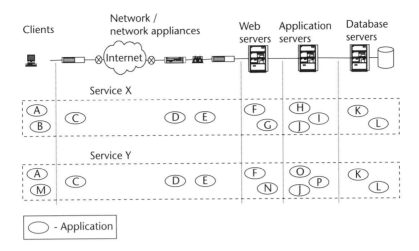

Figure 9.1   Distributed applications architecture example.

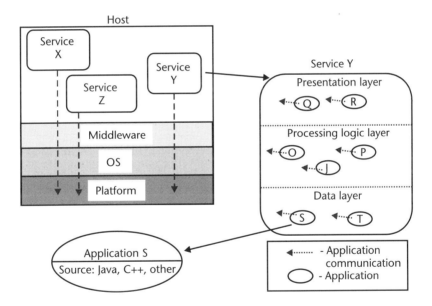

Figure 9.2   Tiered applications.

blessing when it comes to reliability. Applications are still viewed as the weak link to network continuity. In many cases, software still needs to be shut down and reset. Although self-diagnostic, self-healing software coupled with the practice of continuous vendor-supplied upgrades and patches have had a positive effect, they have made life more complicated in deploying and managing software throughout an enterprise. Organizations must ensure they are continuously listed on update distribution lists and must constantly poll vendors for upgrades and patches. Furthermore, managing the volume and diversity of upgrades for large feature-rich networks can be become quite cumbersome.

For reliability, applications are often spread across different network locations on independently run systems. A more preferred approach is to create a single virtual image, or *domain*, of an application that is shared across different locations. Creating application domains across multiple sites can help distribute processing load and aid manageability and survivability. It enables organizations to separate out services so that they can be operated distinctly. It also facilitates the possibility of outsourcing operation of a service by another party, such as an application service provider (ASP).

A major factor affecting software distribution in organizations is software licensing [6]. Traditionally, licensing programs have been designed to maximize vendor return on volume purchases. When multiple images of an application need to be purchased to aid redundancy, vendors have usually required separate software licenses for each, adding to the purchaser's overall cost. As of this writing, many key industry software vendors are reforming their licensing programs to help ease some of this burden. It is still too early to tell whether the effects of this reform are positive.

9.5 Application Performance Management

As an application becomes more business critical, so does its performance and availability. Application performance management (APM) has become a necessity for many organizations in order to ensure service to customers and enforce supplier service-level agreements (SLAs). Use of an APM solution is often mistakenly substituted with other mechanisms:

- *Reliance on external entities.* A typical application interacts with other systems, including application and database servers, load balancers, security, and networking systems, which all may affect application performance. Unless there is a way for an application to provide internal state information vital to its operation to other software or hardware components, it is a mistake to assume that the other components will have any means to manage the application's performance.

- *Reliance on platform hardware.* Quite often too much emphasis is placed on platform and hardware management as a vehicle for APM. Although a poorly running platform can adversely affect application performance, the converse is not necessarily true. Hardware and platform management features are designed to provide optimal resources in terms of central processing unit (CPU), memory, storage, and power for applications to use. Sole reliance on these features for application performance is insufficient—the rest is up to the application. Moreover, application performance is not entirely dependent on any one resource: memory, CPU, disk I/O, switching, encryption processing, network bandwidth, concurrent connections, and more must be tuned to work cooperatively to yield good performance. Even if an application is not running efficiently, one cannot solely rely on what an APM sees to isolate and remedy the bottleneck or vulnerability point.

- *Reliance on architecture.* Although a tiered architecture is a valuable aid to reliability and performance, it is not the only factor. APM capabilities are required that integrate performance across all tiers, and with those of other components that interact with the application.

- *Reliance on vendor warranties.* Software, whether off-the-shelf or custom made, is usually sold as is. Many software licenses contain disclaimers and implied warranties, some offering a limited performance warranty. In such cases, the vendor is only obligated to use their best efforts to fix any reproducible errors. Other than this, there is no assurance whether an application will perform as it is supposed to. Even if such assurances are made contractually, they will not stop degradation or outage when it occurs. APM mechanisms are still necessary to identify such instances so that contingency arrangements can be made.

9.5.1   Application Availability and Response

A mission-critical application must have a predictable response and availability relative to different load conditions. Application availability depends on many factors and is realized through an *availability stack*. Beginning with the application and working down the stack, an application's availability will depend on that of middleware and utilities, database software, networking software, OS, hardware systems and peripherals, and networking infrastructure.

Although the availability of each item in the stack can be measured in its own technically sophisticated fashion, it is best to start with how an end user or receiving system will perceive the application's performance. There are two primary APM measures for this:

1. *Availability.* As already discussed in this book, availability is the percentage of the time that an application can provide or complete service, based on predefined service criteria. For instance, a service criterion for a Web application might be the percentage of time it can complete successful transactions.

2. *Response.* As also already discussed in this book, response is the speed in which an application replies to a request. A commonly desired criterion for most applications is short and predictable response time. Different kinds of applications might require different variants of this metric. For example, a Web page response-time metric would be in terms of the time to process a user request, while a file transfer response metric entails the overall transfer time.

Applying these metrics from the top of the stack rolls up the combined effect of the stack elements into an end-user's view. Such measures should not be used in a vacuum and should be accompanied with additional information. Furthermore, averaging values over a period of time or types of operations can mask important characteristics about an application. For instance, if an application's downtime is primarily during periods of peak demand, there are serious problems worth addressing. Or, response measures that do not reflect end-to-end response or response for different types of operations could also be misleading.

The relevancy of availability and response measures depends on the particular needs of a given application. A communications application might be measured in terms of data or connection availability, while a real-time reservation application may be characterized by user response. To make these measures more relevant to both system administrators and users, quantified values of application availability and response should be used to comprise and convey levels of service. For instance, the percentage of completed transactions could be used in conjunction with response levels to form different categories of service for a given application.

9.5.2 APM Software

APM software tools have become widely available in recent years [7]. As a result, various approaches to APM have emerged. To understand them, one must first understand the concept of a *transaction*, which is the centerpiece of most approaches and is the fundamental item addressed by most APM software. A transaction is an action initiated by a user that starts and completes a processing function. A user in this sense can be a person, system, application, or a network element that initiates a request for service. The transaction ends when the work is completed. In real-time transaction processing, this is often conveyed through the delivery of some type of acknowledgment or confirmation.

Most APM software implements management protocols for three basic types of transactions. *Request-response* transactions are characterized by high volumes of transactions with constant amounts of data. *Transport* transactions involve the transfer of large and varying amounts of data, such as file transfers. *Streaming* transactions transfer data at a constant rate, such as in video or voice streaming.

The use of availability or response metrics to characterize these types of transactions can vary. Request-response transactions are typically characterized by a response time metric in terms of the elapsed time between a request for service and its completion. Transport transactions are usually characterized using response as a metric indicative of the data rate. Streaming transactions are known to use availability metrics that reflect the percentage of time service is degraded or interrupted. An application can also involve all three types of transactions and thus require several types of metrics to convey performance.

The following sections describe several basic techniques used by APM software for application performance measurement. They are not mutually exclusive and can be integrated together in a common reporting infrastructure:

- *Simulated transactions.* Simulated transactions, sometimes referred to as *synthetic* transactions, are dummy transactions generated to gauge the behavior of an application, such as in the case of heartbeat or polling operations. They are typically submitted to an application as a way to sample and measure its response. They are often employed in various types of measurement and control scenarios where use of the observed behavior of live traffic is infeasible or impractical.

- *Checkpoints.* Probably the oldest technique used, *checkpointing,* involves the insertion of markers or break points within application code to either write data or communicate with another entity [8]. They are usually placed at critical points in the application-processing stream, providing precise knowledge of the application processing state. This technique is often used in custom programming, where measurement granularity can be designed as needed. It is also used for specialized applications where generic or common performance measurements are not directly applicable or sufficient.

• *Application agents. Agents* are programs that are placed alongside applications, in the same platform or network, to track their behavior [9]. Some agents can recognize faults or failures, even before they happen, and initiate an appropriate response. They can monitor OS-level events as well as obtain information from the application itself. Agents must be told beforehand what to measure or how to infer what to measure from a prevailing set of circumstances. They can incorporate and retain administrative responses to events in order to recommend corrective actions based on history.

Agents learn to recognize patterns on the cause, response, and corresponding results of prior events to recommend future preventive actions. They can recognize abnormal memory or CPU usage patterns that can signal an impending problem. They must be able to distinguish normal patterns as well so as not to unnecessarily initiate corrective actions. There are two general types of agents. *Active agents* typically simulate transactions to different entities and record measurements. They can be placed in desktop or server systems, network equipment, or probe devices. *Observational agents* monitor actual application activity and record transaction performance.

Agents are known to pose installation and compatibility problems with various OSs, hardware, and applications, making their upgrade process somewhat difficult and costly. Agents can also further tax platform processing and may require platform upgrades for acceptable performance. Finally, general-purpose agents may not always provide the desired type and level of information that can be otherwise achieved through checkpointing or a customized approach.

• *Traffic inference.* Traffic inference involves gathering all or selected measurements from a network environment, correlating the information together, and making assumptions or a "best guess" regarding application performance. It could involve collecting information from a variety of sources involving a combination of the prior approaches, or perhaps include specialized custom methods. For this reason, the technique is less exact, is limited to common performance measures, and may not be able to detect measures specific to certain applications. It can also overlook performance bottlenecks.

Recently, the Internet Engineering Task Force (IETF) has undertaken the development of a standard for the collection of APM information. Called the APM management information base (MIB), it is to be used in conjunction with simple network management protocol (SNMP) traps and messages to provide information and control of an application process at various stages [10]. It requires that applications become SNMP aware, either at the time of their development or converted at a later time using special tools. Applications may also be required to be registered so that they can be uniquely identified by the MIB. Using the MIB, an APM system can define the measurement *collection buckets* for specific applications under different sets of circumstances. Through SNMP, real-time performance reports and alerts for an application can be generated.

9.5.3   Data Collection and Reporting

APM tools statistically summarize measurements into reports. An administrator can typically specify the desired frequency and level of detail of reports. Different APM systems use different approaches to accumulate measurements. Some create a record for each or selected groups of transactions. Others collect *peg counts* of different measures from specific memory registers. Having records written and stored somewhere can facilitate post real-time aggregation and analysis, which can aid in root-cause analysis and managerial reporting.

To generate reports, the administrator can specify, among other things, the application to be monitored, the types of transactions, the minimum time interval for tallying counts, the desired buckets of aggregation, the maximum size of each report, the number of reports to save, and the reporting interval. The reporting interval is the time reports are to be issued. Although real-time or near-real-time reporting is desirable, small reporting intervals of a minute or less are known to consume more resources and should be used when transaction volume is small [11].

For a selected types or groups of transactions, typical reports should contain at minimum information that includes:

* The total number of attempted and successful transactions during an elapsed period;
* The percentage of successfully completed transactions;
* The average, minimum, and maximum response for successful transactions;
* Counts of successful transactions whose availability or response metrics fall into the predefined collection buckets, each tied to specific criteria. This in essence produces a histogram indicating which transactions meet or fail a criterion, something that a computed average can obscure.

A transaction may not necessarily be classified as a completed transaction and added to a report until the transaction is officially concluded. In order for a transaction to be deemed complete, it must have time to finish. In the case of transport and streaming transactions, which have the potential to live a long time, capturing near-real-time measures of completion may prove inconsequential. This is why it is important that reports distinguish counts by application and the type of transaction. Some APM tools will measure and convey information on transactions in progress.

APM agents do not necessarily perform count aggregations, as they are charged mainly with the task of data collection. Also, they may not individually have all of the required information to create intelligible reports. A management station will usually collect information from the various agents and compile reports. However, the IETF APM MIB does identify several standard levels of data aggregation that APM agents can report on [10]:

* *Application-based aggregation* is the highest level of aggregation and is used to compare the relative performance of applications.
* *Server-based aggregation* accumulates transactions at the server level so that the relative performance of servers can be compared.
* *Client-based aggregation* accumulates transactions at the end-user client level. It is used not only to compare the relative performance of clients, but it can be used to view performance as seen by the end user.

- *Flow-based aggregation* is the lowest level of granularity, and consequently results in larger volumes of information. Transactions are accumulated for each distinct combination of application, client, and server over a given reporting interval.
- *Geographic-based aggregation* accumulates client-based aggregations by geographic location. Locations can be defined in any desired way, such as office location, city, or some other administratively defined business grouping. Its purpose is to identify locations or groupings that may be experiencing performance degradation.
- *Media-based aggregation* accumulates client-based aggregations by network access technology, such as local area network (LAN) or dial up. This information is used to identify network access problems for particular users.

Although standard and packaged applications on can be easily accommodated by APM software that use this common MIB information, custom or proprietary applications may require special handling. Instead of an agent-based approach, some APM systems will employ probes, which are hardware or software devices that passively collect packets across a network interface. They examine packets for signatures of custom applications to infer performance.

Overall, APM reporting is mostly a near-real-time capability. It is not a replacement for instant notification and should not be used an alarm mechanism. If transactions are dropped or some other problem occurs, an alert mechanism is required to notify administrators of the problem. Many APM software tools offer exception mechanisms that let administrators define performance thresholds for individual applications. An APM agent then uses the criteria to immediately notify administrators if the thresholds are violated.

9.6   Application Recovery

Many of the principles related to system recovery discussed thus far in this book apply to application recovery as well. As already stated, application interruptions can come from a variety of sources. The actual cause of an application interruption may not be easily distinguishable at first. Programming errors, incompatibilities with other applications, platform problems, resource limitations, peripheral problems, security breaches, and vulnerability exploits are often not readily recognizable.

APM software is only a partial solution to the problem of application recovery. Although APM software may have the general capability to alert managers of performance degradation or malfunction, error handling by the APM software or the application may likely be nonexistent, unless they both have been designed to interact with one another in some way. APM software alone is usually no match for an application gone wild.

Application recovery should encompass both manual and automated recovery mechanisms. To this end, application recovery plans should be defined for each individual application or application group. At minimum, they should include the following:

- The classification of the application, as discussed previously in this chapter, should be identified. As recovery of mission-critical applications takes precedence over noncritical applications, its place in the recovery chain should be designated.

- The alternative processes required to back up the application should be defined. These can be other applications, services, or manual procedures.

- The failover procedures for the application should be developed. Many of the failover concepts discussed in this book can apply.

- The RTO for both the application and the supported service should be specified. An application-level RTO should reflect the time to restore a failed application to a state in which it can provide active service. It may not necessarily be directly tied to the service-level RTO, which is the time from when a service can no longer perform its function until it is resumed. A service-level RTO may reflect immediate failover to another application.

- The application resumption sequence relative to other pertinent applications should be defined. Conflicts could occur when multiple applications are brought up simultaneously. These conflicts often deal with equitable sharing of resources, such as memory, data, and CPU.

- The recovery or restoration procedure for the application should be developed. Whether self restoration or manual recovery is involved, the application must return to a state consistent with that prior to the fault or consistent for operation with backup applications that may already be active.

- The resources required for recovery should be identified in the plan. This may include any other elements such as front-end, mid-level, and back-end systems, database files, and other applications. The recovery point objective (RPO) discussed in the chapter on storage should be specified for any data required for application operation.

- Data synchronization is a critical aspect of application recovery. It should be based on data mirroring or backup frequency, coupled with any good transactional data that was preserved up to the time of failure.

An application failure could induce failures in other applications. External unexpected events can also spread failures among numerous applications. As many applications try to simultaneously recover and regain hold of common resources such as data files or connectivity, they may not be readily available to all, further exacerbating problems. Thus, an application recovery plan should address the effect of the unavailability of key resources and specify the workarounds.

9.7   Application/Platform Interaction

As application software becomes more complex, it is more prone to failure and vulnerability exploitation. Much attention is placed on instilling fault management in system platforms to compensate for application-software inadequacies. As application software is modified and upgraded more frequently than a platform, the impact on platform stability becomes more uncertain. Many software vendors are often challenged in developing applications that can leverage platform fault management. This is why tools are available to adapt applications to fault-tolerant (FT) and fault-resilient (FR) platforms, as well as instilling fault-detection and fault-management capabilities within the applications. For this reason, it is critical to obtain certifications from software and platform vendors that an application will run in a well-behaved fashion on particular platforms.

A distributed processing environment, whether deployed on a single or multiple platforms, will require some type of communication of application status. Checkpointing and heartbeating services must be provided to facilitate active/standby operation. This may involve platform middleware and/or OS capabilities. A high-level application restart model should be supported that makes use of checkpoint and heartbeat services. Such a model need not require the application to be cognizant of the actual platform configuration and topology, as this can be presented abstractly to the application.

At minimum, a platform should provide an event manager to isolate application-level events from platform-level events and respond appropriately to events at either level. In the case of clusters, applications will likely require modification or upgrade to become cluster aware, enabling them to run on multiple cluster nodes in a virtual domain. A domain identifies all the programs and data required to run an application among several nodes. Cluster management software is charged with event management and automatically restarting or switching over to another node in the case of an application fault within the domain.

9.7.1   Operating System Interaction

Mission-critical applications must be designed to run in concert with the prevailing OS. An OS is fundamentally a complex piece of software that coordinates and abstracts platform resources such as memory, CPU, and peripherals from applications. It is also designed to provide a standard operating environment for all applications. To this end, all software applications typically have equal access to a system's memory address space. Because there is usually no memory protection, any malfunctioning application can overwrite memory used by another. This can also pose a security issue as well if written over by malicious input.

Mission-critical processing, however, places more stringent memory requirements on an OS. An OS should provide mechanisms to distinguish and protect memory for mission-critical applications. While critical applications can have memory protection, lower priority applications such as device drivers, protocols, and file can have separate memory allocated and protected as well. This type of partitioning avoids having a poorly written application corrupting other applications or resources. It facilitates starting and restarting applications without causing others to fail and allows software modules to be upgraded dynamically while a platform continues to run. Loosely coupled resource architectures partition resources so that applications can request them from a resource pool. If a resource fails or becomes corrupted, the platform assigns a new resource.

Some OSs facilitate *task isolation*, in which an application is assigned CPU and memory for critical tasks so that a malfunction in one task does not affect another task or system resource. This avoids having a problematic application affect other applications using different address spaces. Task isolation can also be enhanced to allocate memory resources dynamically, versus a priori, to maximize overall plat-form memory utilization.

Any capabilities that support OS self recovery to any extent are all the more bet-ter. FT and FR systems, although expensive, will have self-correcting mechanisms and redundancies to avoid having to reboot an OS. As stated earlier, an application should at minimum return to a consistent state prior to a fault. If an OS undergoes a severe fault or crashes, the entire system and all applications must return to a consis-tent state. This also applies to communication interfaces as well. For mission-critical services, an OS reboot can thus be catastrophic and disrupt service, unless a backup system is available.

A mission-critical OS should also support a range of utility programs and APIs. Utility programs should enable scripts to invoke platform functions. The APIs should enable integration of applications into the platform software environment so that application failover can be managed. The APIs should provide interfaces to the event-manager and network-management functions, which communicate platform events to a network manager. The end result should be the ability to cohesively por-tray the state of all platform components and allow network management or system administrators to control the states as needed.

As mentioned in the chapter on platforms, some multiprocessor platforms can perform split-mode operation, enabling an upgrade to be tested while the platform continues service operation. To support a split-mode operation, applications need to be *split-mode aware*. From the standpoint of an application, several approaches can be used. Through APIs, some applications can be notified of split-mode operation. In this case, the application must have the ability to manage itself accordingly, requiring built-in capabilities for handling split mode. Another approach is to make one application split-mode aware and assign it the responsibility of managing other unaware applications. Yet another approach is to have no applications with split-mode awareness and rely on external means, such as a manual or platform-based mechanism, to transition them through the split-mode process.

9.8 Application Performance Checklist

Finally, we present a checklist that the reader can refer to when developing or pro-curing a mission-critical application (or system, for that matter). As one examines the potential efficacy of a software application, the lines of questioning can lead the reader down deeper paths and trigger additional questions in those areas of interest. Expectations are that they will be rearranged, modified, and restated depending on the situation at hand.

1. Software function:
   - What mission goals and objectives does the application achieve?
   - What does the application do?
   - What external or internal data resources does the application rely upon for proper operation?
2. Development process:
   - What software development methodologies, if any, were used?
   - Was a secure software methodology and review applied?

- Was the software produced using a formal development process?
- To what extent was systems engineering involved in the process?
- To what extent does the software fully satisfy system specifications?

3. Standards and certifications:
   - What standards were followed in the software development?
   - With which standards does the software comply?
   - What platforms, software, and utilities are certified to work with the software?
   - With which platforms, software, and utilities is the software certified to work?

4. Software design:
   - How is performance and survivability designed into the application?
   - Does a performance model exist for the software architecture? If so, what model was used?
   - To what extent has the software design been compartmentalized?
   - What source language(s) were used in development?
   - Were any new technology or techniques used in development? Have they been proven?
   - Does the software rely on any other software (e.g., legacy, common software, software packages, utility) for operation?
   - What other applications depend on successful operation of the application?
   - Are there any known errors, defects, or weaknesses in the design?
   - What are the defined modes of error?
   - What can cause the application to hang, freeze, or lock up?
   - What are the failure modes and corresponding failure group sizes?
   - What internal measurement mechanisms have been incorporated in the software design?
   - How many processes comprise an instance of the application under different operational modes?
   - Are there any single processing threads?
   - What application in the processing chain is most likely to be taxed first or the most?
   - Does the application rely on static or dynamic memory allocation? What are the maximum memory limitations in each case?
   - How does the application handle buffers during processing?

5. Application operations:
   - What are the platform requirements for operation?
   - What platform changes or conditions violate the integrity of the application?
   - What other entities (e.g., applications, interfaces, or drivers) must be relied upon for successful operation?
   - How much overhead is the OS expected to contribute in supporting the application?

- What type of transactions does the application perform? What are the processing/performance budgets for each type?
- How can the application be made to run on multiple processors?
- How can the application be made cluster aware?
- In either case, can the application run in a primary/secondary relationship? How are the roles determined and when?
- What type of auditing is done by the application?
- How does the application communicate with network or device interfaces?
- What level of control does the application have over use of these interfaces?
- How does the application communicate with other processes or resources (e.g., shared memory, domain, or transactions)?

6. Performance and reliability:
   - What performance criterion was the application designed to meet?
   - What level of availability has the application been designed for? What are the mean time between failure (MTBF) and mean time to recover (MTTR)?
   - What is the application's reliability over time? What is required to sustain the reliability?
   - What has been done for it to meet this criterion?
   - What levels of performance are guaranteed?
   - How much headroom has been designed into the application in meeting the criterion?
   - How is the application's usage or workload characterized?
   - How are performance envelope budgets modified during adverse situations to sustain full or partial operation?
   - What are the levels of response when the performance envelope is violated?
   - In these cases, what activities are given priority, and which ones are shed?
   - How sensitive is the application's performance to variations in workload, operating environment, and resources?
   - How much does the application's performance rely on that of the host platform?
   - How does the application support performance monitoring and management?
   - To what extent will the application monitor and recover itself?
   - Can the application be made SNMP (or equivalent) aware?
   - How is the application health and status conveyed to middleware, OS, platform, or other components?
   - Does the application employ heartbeating or checkpointing with other processes?
   - How does the application communicate faults/errors to other entities?
   - Does the application support any platform alarming?
   - With which APM packages is the application compatible?
   - What resources are taxed by the application's performance-monitoring functions?

- How does the activation of different features affect application performance?

7. Application recovery:
   - Can the application meet the target RTO objectives?
   - To which unexpected external/internal events has the application been designed to respond?
   - How can errors and faults be isolated from other processes to minimize corruption?
   - What capabilities does the application have in response to hardware errors?
   - What are the defined recovery modes?
   - What type of failover can the application employ?
   - How much transactional data loss is expected during each recovery mode?
   - What is the recovery sequence in each case?
   - After a reboot, what state does the application return to? What information is saved or lost?
   - How is error control handled? What type of cleanup is done after errors?
   - What error/recovery codes are used and what are their meaning?
   - Can the application state prior to failure be preserved in some way for resumption?
   - Can the application support tracing for troubleshooting purposes?
   - Can a sparing strategy be used in conjunction with the application?
   - What manual overrides can be used in recovery?
   - Is there a manual command to shut down the application if needed?

8. Data:
   - What minimal data, if any, is required for reliable operation?
   - What is the required RPO of the data for successful resumption?
   - To what level is data abstracted or masked to the application?
   - What happens if data is corrupted?
   - How are other dependent systems' data affected if the application changes?
   - Can data be mirrored or backed up while the application is running?
   - Is the data the application employs static or dynamic?
   - What measures are taken to assure the integrity of and authorized access to the data used by the application?
   - What kind of user input is expected?
   - Are measures taken to assure that only appropriate data is input to the application?

9. Operations, administration, and maintenance (OA&M):
   - Does the application need to be taken off-line for test or upgrade?
   - Can the application be made to support split-mode operation?
   - How often are upgrades and fixes released?
   - How are upgrades in subservient software modules, packages, or utilities handled?

- What is the migration plan for the application?
- How are diagnostics performed?
- How is the software configured for site operation?
- How is access to software features administered or configured?
- What is the sequence of functions for initialization and shutdown?
- Can a selective installation be done?
- How is backout achieved? What rollbacks must be applied to data to facilitate backout?
- How is site testing accomplished? Can it be done with real data in a nondestructive way?

9.9   Summary and Conclusions

A mission-critical network is an enabler for applications. Performance and survivability of critical applications is just as important as network and facility infrastructure. Critical applications should be identified based on their mission or business importance and recoverability requirements. Recoverability should be stated in terms of an RTO range.

Procuring or developing applications for a mission-critical operation mandates that they have undergone a well-defined and structured software systems-development process, more or less reflecting the traditional systems-development cycle. It is also important to ensure that the application has had substantial systems engineering so that conformance to performance and survivability requirements is assured. It should also ensure that the application performs in a predictable, well-behaved fashion.

Mission-critical software applications should have modular design and should enable monitoring of their performance, either internally or externally, through the hosting platform. An application must somehow interact with other entities in order to be managed. APM software tools offer a viable solution to manage and control mission-critical applications. Several monitoring approaches are used; the most popular is the use of software agents to monitor and report on application status. The recent introduction of an APM MIB fosters standardization of the information handled by software agents.

External components or a system platform should not be solely relied upon to compensate for poor application performance. A platform can only facilitate the running of an application. When hosting mission-critical applications, a platform OS should have the ability to segment and protect application memory. It should also support middleware with APIs that support failover and recovery.

References

[1]   Jost, S., "Year 2000 Conversion Services at John Deere," *Enterprise Systems Journal*, March 1998, pp. 22–27.

[2]   Londeix, B., *Cost Estimation for Software Development*, Cornwall, UK: Addison-Wesley Co., 1987, pp. 3–7.

[3]   Metzger, P. W., *Managing a Programming Project*, Englewood Cliffs, NJ: Prentice-Hall, 1981.

[4]   Steinke, S., "The Well-Mannered Application," *Information Week*, January 29, 2001, pp. 69–78.

[5]   Boar, B. H., *Practical Steps for Aligning Information Technology with Business Strategies*, New York: John Wiley & Sons, 1994, pp. 114–160.

[6]   Savage, D., "Buggy Code Blues," *CIO Insight*, January, 2002, pp. 77–78.

[7]   Margulius, D. L., "Reeling in the Tiers," *Infoworld*, May 20, 2002, pp. 42–43.

[8]   Steinke, S., "Don't Let Applications Get out of Hand!" *Network Magazine*, January 2001, pp. 124–131.

[9]   Shafer, S. T., "Diagnosis Dilemma," *Infoworld*, May 20, 2002, pp. 40–41.

[10] Waldbusser, S., "Application Performance Measurement Grows Up," *Network Computing*, May 14, 2001, pp. 123–130.

[11] Hollows, P., "Essentials of a Smooth-Running Network," *Communications News*, July 2002, p. 16.

CHAPTER 10

Storage Continuity

For many organizations, information is a valuable asset. Information is also the "fuel" that drives the mission-critical network engine. Business continuity planning must include measures to preserve and protect information so that operations can continue in the event of any network mishap. In the past, this amounted to backing up data on to magnetic tapes and then shipping them to an off-site storage company for safekeeping. However, given the real-time nature of today's business environment and the need to support transactions with up-to-the-minute information, daily data backups are only a partial solution. Furthermore, the need for real-time processing efficiency has placed a whole new set of requirements on data storage. In many instances, there is a trade-off between storage system cost, performance, and data integrity. Integrity is based on the data currency, consistency, and accuracy required to restart and continue operation after an outage. In this chapter, we will review data storage techniques and strategies that address today's mission-critical requirements.

10.1 Mission-Critical Storage Requirements

To determine the mission-critical storage requirements of a firm, the first question to ask is how information intensive is a company, and, more importantly, what information is needed to continue hourly operations. For example, a retail chain operation will complete business transactions in the physical environment of a store. Access to pricing and physical inventory databases will be required to complete store transactions. On the other hand, retail businesses that are exclusively on-line require additional data to complete transactions, such as customer shipping and credit data. "Click and mortar" companies conduct both off-line and on-line commerce via Web sites. Their storage requirements might straddle those of the previous two types of firms. Then there are firms whose business is managing and aggregating information, such as an on-line auction house or search engine. Their data needs might be completely different than the other types of firms.

The level of a company's information intensity is a key factor in driving mission-critical storage requirements. One must identify what portions of a business operation need to stay alive amidst outages. The databases required to support business functions must also be identified. Because different business functions and services drive different kinds of mission-critical storage needs, there is typically no uniform solution. For example, data warehousing involves the use of data mining queries that can potentially slow on-line transaction processing. Data warehouse transactions also usually involve more reads than writes. These characteristics imply that separating data warehousing from other real-time data transactions could provide on-line processing benefits. A possible data strategy would be to store the real-time transaction data separately and load it into the data warehouse at a later time.

Mission-critical storage requirements can be achieved through a process of reviewing data components of a business. This process involves asking the following questions for each data component:

- *What business functions and related applications are involved?* For example, e-commerce, e-mail, and data warehousing all drive different kinds of storage requirements. E-mail requires a relatively small amount of network storage. Video data streaming, on the other hand, requires continuous reads of large blocks of data from storage. Relational databases are usually viewed as s single large file. Application data often is categorized by their importance, with transactional application data earning a higher priority than archived data.

- *What operations will be applied to data?* Examples of operations include database lookups, read/writes, backups, content caching, file distribution, and data assembling.

- *What will be the nature of data access?* This question addresses *when* access to the data is required (e.g., continuous or intermittent) for the applications to properly function. It also addresses the question of what is the acceptable length of time the data can be unavailable.

- *What is the interdependency on other data?* It is important to recognize that when applications depend on data from other applications, the data associated with both may require the same level of protection.

- *What distribution of reads and writes is anticipated?* Different types of applications will involve different splits between the number of reads and writes. The quantity of data in each read/write should be estimated. This information ultimately drives requirements for port densities, distance, bandwidth requirements, disk throughput, and database size.

- *How will the data change over time?* How does the data grow over time? Is the content static or can it be modified in real time? Many Web sites, for example, will read static content containing images and text descriptions. Effective planning should address how a data volume size will vary over time.

- *How will the data be archived?* Business needs, laws, policies, and regulations may dictate how long data has to be stored and how it is to be accessed.

- *What user groups or organizations can access the data and from where?* If different users are sharing the same files from different locations, then modifications should be made to a single copy of the files rather than having multiple copies with differing changes. Many database-management systems resolve the problem of sharing data among multiple users.

- *In the event of an application failure, what users should have access?* This question addresses the usage priority of data in light of an outage. It would identify what users should be reconnected to the database and how soon. For example, a catalog operation might require that data supporting their customer-relationship management (CRM) operation will get top priority because it is vital to processing sales orders.

- *What data should be replicated and where?* A list of critical data requiring protection should be prepared. When determining what data should be protected, it is important to be aware of transaction consistency. A single transaction can update several databases. They must be protected as a set, with control data that describes the relationship between the data sets.

- *How current should the data be?* An application that requires data for instant decision making would need the most recent data, or data with high *currency*, to restart operation after an outage. This implies preserving data with sufficient quality as close as possible to the point in time (PIT) a failure begins. Ideally, one would like to freeze data just prior to a failure because of the possibility of data corruption or even loss of the entire data set. However, because one cannot predict the future, a recovery strategy is required that can quickly produce the best and most recent data image prior to the corruption. For many applications such as e-commerce and electronic data interchange (EDI), where there is a continuous volume of on-line data transactions, a data image that is an hour old might still be inadequate to restart operations after an outage. Records of the transactions occurring since the time of that data image could be lost. Heavy consideration must be given to developing a recovery point objective (RPO) that aligns with business needs.

- *What should be the level of consistency between the replicated data image?* Data consistency means that, from an application's perspective, the replicated data contains all updates until a specific PIT, without any missing updates beyond that specific PIT. All updates (read and writes) and their sequence are preserved in the data replica. If the replicated data image is consistent with respect to data, then applications can be restarted without a lengthy and time-consuming data-recovery process.

- *What level of data integrity is needed?* Data integrity is dependent on a number of things; one of them is the sequence of updates made to a database. Operating systems (OSs), file access methods, or applications track the update sequence using both data indices (pointers to data) and the actual data so that updates can be resolved in the event of a sudden failure. Updates are logically and sequentially dependent on each other. An outage can cause multiple write errors, which can corrupt a database and make it useless. If the data was replicated after these errors occurred, then the corruption could spread to the replicated image as well.

- *How should the replicated data be used?* Replicated data images can be used in many ways. For example, they can serve as a backup copy of a production database to be used solely for restoration. If continuous availability is required, they can be used as a standby database, where they can automatically take over processing from the primary production database. If there is a need to temporarily offload query processing from the production database, then a replicated database could serve as a read-only database for running back office applications such as data mining, without placing additional overhead on the production data.

- *How can seemingly duplicate but "stale" data be identified and removed?* Early revisions of documents, reports, and data that contain sensitive information, especially information that's contrary to the latest version, must be managed in some way.

- *What file structures are involved?* Data can be stored in a variety of formats and structures. Relational database-management systems (RDBMSs), used typically to support high-throughput applications, store data in alphanumeric form. It is usually easier to automate processes using RDBMSs rather than a file system. On the other hand, much mainframe data is still stored in virtual storage access method (VSAM) structures, which are nonrelational in nature [1]. Then there are also Web site content databases, which typically require reading large blocks of text and graphic files and may involve different kinds of file structures.

Software, in addition to data, is another information asset that requires storage protection. This includes application software, OSs, and middleware. Commercially available software, as is often the case, can be made available and replaceable by a software vendor. A vendor-provided copy of an OS can help restart a system following an outage, if needed. OS storage should be kept distinct from database storage for quick retrieval. Many firms, particularly software vendors, have in-house software that may require the same level of protection as data.

Like data, software copies should be updated regularly to reflect any upgrades or modifications. E-businesses have exacerbated this problem somewhat by embedding data directly within application code to facilitate quick changes. For example, it is not surprising to see product description and price information written inside hypertext markup language (HTML) application code. As the coupling between applications and data grows tighter, equal consideration must be given to protecting both software and data.

Configuration data is very often overlooked, especially security system configuration data. As months can be spent on implementing a firewall or virtual private network (VPN) policy, loss of such data might require significant effort to reconstruct it. Event data is also critical—keeping log and audit data for analysis and forensic post-incident activity is critical to today's businesses.

10.2   Data Replication

Simply put, the term *replica* means a copy. Data replication involves copying all or some portion of a data source to other locations. The data source can be any type of digital data, including databases and software. Locating the replicated data in a different geographic area removes a single point of failure, as only one replica would be affected if it encountered corruption. A local replica of data might also be created for high availability or throughput. Replicating data along key points in a network can enhance system performance. Content delivery networks (CDNs), for example, locate Web site data closer to users to improve response time and reduce network traffic. Load balancing, another performance-enhancing technique described in a previous chapter, can often require use of data replicas.

The first question to address when replicating data is what kind of replica will be needed. Ideally, an identical replica of the source can be made. To an application, this copy will look exactly like the source, further ensuring that the application can run properly. Many mission-critical applications will likely require an identical replica of the data source. On the other hand, some transformation of the source data might be required to accommodate specialized applications, a different database management system, or to save storage space [2]. The original transformed source data must still be retained, however, as some transformations can lose or add fields to data records or lose timestamps and other critical fields that can render transformed data less meaningful and authentic than the original.

Another issue that requires consideration is the physical structure of the replica. Data images can be divided into separate replicas and stored on multiple platforms to handle different types of transaction categories. The data is divided using unique identifier. For example, telecom networks often geographically separate databases based on subscriber account numbers, so that queries can be routed to different portions of the same database, each located in a different geographical area (often referred to as a *mated pair* configuration).

10.2.1   Software and Hardware Replication

Deciding how replicas are to be made requires careful consideration. There are two basic approaches to making a copy of data. Software-based replication, as the name implies, uses some form of software to create a copy of data. Hardware-based replication uses a special device external to the server to perform the replication. The following sections describe these approaches in more detail [3].

10.2.1.1   Software-Based Replication

Software-based replication is done by software residing on a server. The software can be an application, a utility an OS, or OS driver. Software-based replication provides flexibility in using different types of storage platforms, enabling platform scalability and migration. Software-based replication is usually conducive to less complex and less numerous applications and databases. Software replication can become more difficult to implement as the number of servers, applications, data sizes, or replicas grows. In these cases, special utilities are often used to help automate and manage the process.

Another factor to consider is the server's OS. As the number and types of servers grow, so do the number and types of OSs and OS releases. Because not all server-based replication methods are available to all OSs, variations in OSs can impact the complexity of the replication process.

Server platform constraints play another critical role in server-based replication. Replication will use server central processing unit (CPU) time and memory resources, thereby impacting system performance. Server size and CPU power can be a limiting factor. Furthermore, the speed of a server's input/output (I/O) bus is often a bottleneck in transferring data.

Replication performed by an application typically uses the server's OS to gain access to the server platform facilities. The data being replicated is usually specific to the application. The replication could be part of the application or in the form of a utility that operates between the application and the disk. Examples include backup utilities that can efficiently copy data from one location to another.

Many database applications provide some form of data replication function to a local or remote location. Some replicate data while doing other types of tasks, which can sometimes adversely affect performance. Data integrity concerns can also arise if the database is open during the replication process. Most practices require the database to be closed in order to replicate.

Database applications track database changes using transaction logs. The change information for committed transactions can then be issued and applied to the local primary database and another database at a remote location. The changes to the remote database can then be made instantaneously or at a later time. Database management systems usually create logs detailing a sequence of database changes over time.

File systems offer replication capabilities as well, often transparently to the storage device. But they can sometimes consume extra platform resources to provide the replication services. Customized applications are often developed to address some of these concerns. Customized applications often have to be modified as hardware and software changes are made to a host system. As many OSs offer some means of backing up and restoring data, some applications will use OS-level commands. These applications may have to be modified as new versions of the OS are issued.

10.2.1.2 Hardware-Based Replication

Hardware-based replication duplicates data in its physical form. Usually considered the fastest and easiest approach, it is less dependent on the server and OS. There is reduced impact on system overhead or application performance because the replication functions are not on the server. Because the replication occurs at a lower level, usually the at the disk block level, hardware-based replication solutions often have little, if any, knowledge of the application level or the OS and usually respond to disk drive commands.

Two approaches are used: controller based and appliance based. The controller-based approach provides the replication functionality in the disk driver and controller. Various implementations are used, depending on the OS. In many cases, additional firmware on the controller is needed. The OS views the driver as a distinct device. Appliance-based approaches, on the other hand, use a separate hardware device containing all replication functionality.

Hardware-based solutions can be problematic in the event a write failure occurs at the primary data site, as disk subsystems typically have little, if any, knowledge of the activity of other disk subsystems. Because data flows directly from the primary disk subsystem to the secondary subsystem, a write failure occurring at the primary subsystem does not necessarily prevent an update on the secondary subsystem. A capability on the primary host might be required to monitor the failure activity within both disk subsystems. This capability, sometimes referred to as *geoplex*, is used to ensure data consistency and freezes the data if an error takes place [4, 5].

Another problem with hardware replication is that only data existing on a drive is replicated to another drive. If a host processor fails, there is a possibility of incomplete data writes to the primary drive at the time of failure. Cache-resident data could be lost at the time of failure and never replicated. Software-based solutions avoid these problems, as the data flows through a component in a host processor having the ability to manage activity among multiple disk subsystems.

Although hardware-based approaches are usually easier to implement, they do present some operational issues. They usually require data to have the same format on the source and target devices in use. Consequently, this requires having the same device or copy mechanism at the primary and secondary system. Hardware-based replications can also consume more bandwidth, as data is being transferred in large block quantities. For these reasons, hardware-based replication is often confined to proprietary implementations, requiring the use of the same vendor devices at the primary and secondary sites.

10.3   Replication Strategies

Developing the right storage strategy depends upon a firm's business needs and application requirements. There are many different methods and technologies that can be used to create and maintain storage. The goal for any replication strategy is to ensure that all replicas are identical at any point, particularly at the time of an outage. Redundancy is only a partial solution, as corrupted data can be written to both the source and replicated images. One must consider what mechanisms will be used to create the data, such as those previously discussed. Also to be considered is how the data will be transferred to the remote site. If a network is involved, network redundancy to the remote site will be required. Geographical separation is another important element. The replica must be an adequate distance away from the location of the primary data image so that the same or a collateral outage affecting the primary location cannot affect it. Quite often natural disasters such as hurricanes can potentially damage a company's facilities in one city and take out power grids in another city, where the replicas are maintained.

10.3.1   Shared Disk

Shared disk involves sharing the same storage device between different servers [6]. Several servers share a common disk array, such that one of the servers can take over the disk resources in the event another server fails. Failover is conducted effectively by preserving the server and data states after failover. Because such systems can read or update a data set while other systems are simultaneously reading or updating the same data, these systems must maintain data integrity. In situations where many servers access a shared disk system, the system may require some degree of fault tolerance if data is not replicated somewhere else.

Shared disk requires servers to have some disk-sharing capabilities. Shared-disk configurations use less resources for data protection and enable the use of large amounts of data. An example of a shared-disk application is the use of VSAM data sets, which are often shared among different applications either in a single system or on different systems (VSAM is a file access method used on IBM mainframes whereby records in a file can be created and accessed in the sequential order they were entered).

10.3.2   File Replication

File replication involves making copies of files and placing them on either the same or different device. Often the files have to be closed before replication is done;

otherwise, the replica can be out of synch with the original file. This can provide challenges when replicating database files used by 7 × 24 applications. In this situation, software may be required to manage the open files, and records may have to be locked prior to replication. Many applications sharing the same files can add complexity and time. Knowledge must be maintained regarding which applications use the files so that their operations are not adversely affected after a recovery. As file replication works at the file level, additional system overhead may be required to replicate a file.

10.3.3 Mirroring

Mirroring involves simultaneously writing identical blocks of data to multiple drives. The drives are synchronized with each other to assure that all data written to the second drive is always up to date. In the event of a drive failure or outage at the primary site, operation can instantly switch to the other drives without data or transaction loss. Mirroring essentially creates an identical failover data image, which can improve meeting the specified RPO and recovery time objective (RTO).

Figure 10.1 illustrates how mirroring is achieved. The storage devices can be collocated or geographically separated (usually preferred) and can be connected to different servers. Mirroring can be done over a local network, typically in conjunction with a clustering solution. Geographic mirroring, sometimes referred to as *remote mirroring*, entails using a network to provide connectivity between the primary and remote storage devices.

Mirroring can be accomplished using either software-based or hardware-based replication solutions. Software-based mirroring enables time stamping of the data by the server-based software, assuring that data is written in the correct sequence. Hardware-based solutions are done at the controller level and do not usually offer

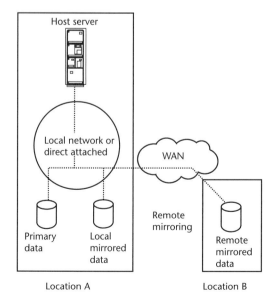

Figure 10.1 Data mirroring example.

this capability, as disk subsystems may not necessarily use the same clock. Controller-based mirroring, being a hardware-based replication solution, usually requires use of the same proprietary I/O process at the remote site. As with most proprietary solutions, operation and management are greatly simplified, but scalability and evolution can be limited. Software-based mirroring, on the other hand, often allows the ability to use server hardware made by different vendors. Using a software-based solution from the same vendor at the primary and remote sites can ensure better interoperability and smoother recovery.

Although mirroring provides excellent recovery from server outage, storage device failure, or site disasters, it is still prone to data corruption because erroneous data can be written to a primary drive can also be written to a mirrored drive. It also mandates use of a backup storage mechanism, either on-line or off site, so that backup copies of critical files can be used to return the data to an accurate state and rerun an application should the data be corrupted. An alternative is to use fault-resilient or fault-tolerant servers with mechanisms to reinitialize an application to avoid data corruption.

Preserving the update sequence (i.e., the order of which the updates occur) is a critical factor in avoiding data corruption at a mirrored site. The time across these sites should be synchronized as well to preserve the chronology and sequence of events. The sequence of the updates at the mirrored site must be in the same order as they occurred at the primary site; otherwise, the mirrored image will be inconsistent. This means that that any time delay encountered between updates as they are received at the mirrored site can result in their being out of sequence if proper mechanisms are used. For these reasons, mirroring is often considered an expensive option.

Only critical data should be mirrored. Buildup of mirrored noncritical data could drive up storage requirements. Mirroring also consumes additional system resources in order to write to multiple drives and to ensure that all writes are time-stamped. Although having a mirrored database can drastically improve recovery time after a failure, the amount of time to remirror drives following a failure could be lengthened when large data stores are involved. Server failover is often disabled until the drives are remirrored to avoid further data corruption.

Although mirroring can be considered a form of file replication, the two differ in several ways. Mirroring involves replication at the disk level. Every write is mirrored to a remote copy. The write is considered complete when both the primary copy and the mirror are updated. Unlike file replication, mirroring is a block-based procedure that works below the file level.

Mirroring results in the creation of an identical data image on another storage device. In an approach called *striping*, segments of a block of data are written or read across several disk drives at the same time. This is done to improve performance and throughput by having multiple drives simultaneously share the work in performing the read or write operation. Mirroring can be used in conjunction with disk arrays containing striped data, as the array is viewed as a single disk system.

Using a mirrored data image for remote site testing is not recommended. If a problem occurs during testing, the mirrored data can be possibly corrupted. Running tests against a PIT image of the mirrored data, in lieu of a production image, does not place the production mirror at risk. It preserves the production image and avoids disrupting operations.

There are two general approaches to transferring data to a mirrored image: *synchronous* and *asynchronous*. The following describes each approach in further detail [7–9].

10.3.3.1 Synchronous Data Transfer

Synchronous data transfers involve writing data to multiple data sets at the same time. A transaction causes both the primary data set and replica to be updated simultaneously. This keeps the replica completely synchronized with the primary copy, ensuring that both disk systems are identical at all times. Synchronous replication is often used for applications where RPOs are close to zero, as no transactions can afford to be lost in case of an outage (RPOs are discussed later in this chapter and in the chapter on metrics). Synchronous systems tend to be more expensive than other data transfer systems and can double storage requirements. Hardware-based synchronous transfer has been a popular approach until the recent introduction of software-based synchronous systems.

Synchronous transfers are sometimes referred to as *two-stage commit* transfers because no disk update is committed at the primary system until it has also been committed at the mirrored site [10]. This process is illustrated in Figure 10.2 for both the software-based and hardware-based scenarios. When the primary server or disk subsystem writes to its own data drive, the software can send messages to a mirrored server's disk subsystem. The primary system first sends a Read/Write request to the mirrored disk system and waits. Once the mirrored site acknowledges the request, it returns confirmation to the primary system. When the primary system receives the confirmation, it commits the write to its own disk system. In addition to keeping the replica in lockstep with the primary, this process also preserves the sequence of updates, as no update can be initiated until a previous update is successfully completed at both locations. The two-stage commit process, however, can increase I/O response times and introduce additional processing overhead. Furthermore, synchronous transfers do not necessarily mean that all data is in sync with all other data at the same PIT. In fact, it is likely that different data stores will have different transfer frequencies.

Systems performing synchronous data transfers are usually connected on the same network segment. In the case of geographic mirroring over a wide area

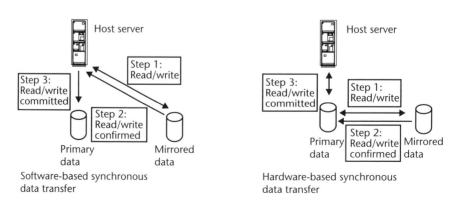

Figure 10.2 Synchronous data transfer scenarios.

network (WAN), the possibility of delay when transmitting the commit messages can further increase I/O response times and affect the performance of software that relies on prompt I/O response. Thus, high-quality networking with high bandwidth, low loss, and little delay is required. Some synchronous solutions can throttle the rate of updates according to network load. Furthermore, the frequency and size of data updates sent over the WAN can consume network bandwidth [11].

10.3.3.2 Asynchronous Data Transfer

Asynchronous replication involves performing updates to data at a primary storage site and then sending and applying the same updates at a later time to other mirrored sites. The updates are sent and applied to the mirrored sites at various time intervals. This is why asynchronous data transfer is sometimes referred to as *shadowing*. With asynchronous data transfer, updates committed at the primary site since the last PIT of issue to the mirrored site could be lost if there is a failure at the primary site. Thus, the mirrored site will not see these transactions unless they are somehow recovered at the primary site. Thus, some consideration is required in specifying the right interval to meet the RPO. The smaller the RPO, the more frequent are the update intervals to the mirrored database.

Asynchronous replication is illustrated in Figure 10.3. A read/write request is immediately committed at the primary disk system. A copy of that same request is sent at a later time to the mirrored system. The primary system does not have to wait for the mirrored system to commit the update, thereby improving I/O response times at the primary system. The local disk subsystem at the primary site immediately informs the application that the update has been made, usually within a few milliseconds. As the remote subsystem is updated at a later time, the application performance at the primary site is unaffected.

As stated earlier, a caveat with asynchronous replication is the potential for some transactions to be lost during failover. However, from a practical standpoint, this may be a small amount depending upon the time interval used to issue updates to the mirrored site. Some vendors have developed mechanisms to work around this

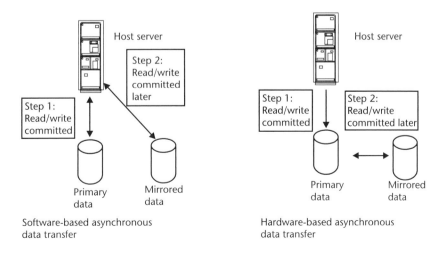

Figure 10.3 Asynchronous data transfer scenarios.

caveat with procedures to preserve the order in which transactions are written. Some solutions even allow sending updates to multiple mirrored servers. As different vendors offer various approaches, adhering to a single vendor's solution at every location is often preferred.

Although asynchronous overall is a more cost-effective mirroring approach, it creates a delayed mirrored image, unlike synchronous, which produces a more up-to-the-minute image. However, the synchronous approach can result in greater I/O response times, due in part to the extra server processing delay and network communication latency. Synchronous also has greater dependency on network resources because updates must be sent as soon as possible to the mirrored site(s). Asynchronous systems are less affected by networking resources and have mechanisms to resend updates if not received in adequate time. However, a network outage could adversely impact both synchronous and asynchronous transfers. Updates, if not resent across a network, would ultimately have to be either stopped or maintained in some way until the outage or blockage is removed.

10.3.4 Journaling

Remote transaction journaling involves intercepting the write operations to be performed on a database, recording them to a log or *journal* at a primary site, and then transmitting the writes off site in real-time [12]. The journal information is immediately saved to disk. Figure 10.4 illustrates the journaling process. This process in essence records transactions as they occur. An entire backup of the primary database is still required, but this does not have to be updated in real or near real time, as in the case of mirroring. If an outage occurs, only the most recent transactions, those posted since the most recent journal transmission, need to be recovered. Remote journaling is cost effective and quite popular. It can also be used in combination with mirroring techniques.

10.4 Backup Strategies

Backing up data involves making a copy of data on a storage medium that is highly reliable and long lived and then putting it away for safekeeping and quick retrieval.

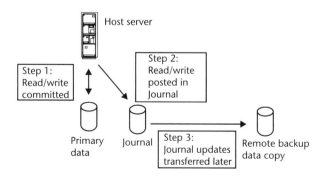

Figure 10.4 Remote journaling example.

The most common form is the daily backup of data to tapes and storing the tapes in a secure, temperature controlled off-site storage facility; however, other reliable media, including various types of removable disk–based media, are used as well. The tapes can be sent to an alternate location, such as a recovery data center, so that they can be loaded onto a computer and ready for use. Unlike mirroring, which replicates data simultaneously so that continuous access is assured upon failure, data backup is used to safeguard and recover data by creating another copy at a secure site. This copy can be used to restore data at any location for any reason, including if both the primary and mirrored databases were to be corrupted. Loading and restoring backup data, particularly when stored on tape, can be lengthy but is typically done within 24 to 48 hours following an outage.

The tapes, disks, or other media that contain the backup set should always be stored completely separate from those that contain the operational database files, as well as primary on-line logs and control files. This means maintaining them on separate volumes, disk devices, or tape media. They should be stored in a location physically distant from either the primary or mirrored sites, far enough away so that the facility would be unaffected by the same disaster occurring at either of those sites. Data backup should be performed routinely, on an automated basis, and with minimal or no disruption to the operation of the primary site. Software and hardware systems and services are available that can back up data while databases are in use and can create the backup in a format that is device independent.

Several basic steps are involved when a system is creating a data backup. Usually, a backup software agent residing on the system manages the backup. First, the data is copied from disk to memory. Then it is copied from one block of memory into a file. At this point, if the backup occurs over a network, the backup agent divides the data into packets and sends them across a network where a server receives and unpackages them. Otherwise, the backup agent reads the metadata about the file and converts the file into tape format with headers and trailers (or a format appropriate for the intended backup media). The data is divided up into blocks and written to the media, block by block. Many backup software agents do not necessarily have to open files during a backup, so files can be backed up while in use by an application.

The type of backup plan to use depends on an organization's key objectives. Optimal plans will most likely use a combined backup and mirror approach customized to specific business requirements and recovery objectives. For example, critical data can be mirrored for instantaneous recovery, while less critical data is backed up, providing a cost-effective way to satisfy RTO and RPO objectives. The volume of data and how often the data can change is of prime importance. The *backup window*, or the length of time to create the backups, must also be defined. Backup windows should be designed to accommodate variability in the data sizes, without having to redesign backup processes to meet the expected window. The smaller the RPO, more frequent backups may be required, leading to a tighter backup window. The length of time to recover files, sometimes referred to as the *restore horizon*, must also be defined.

The data files to be backed up should at minimum be the most critical files. Backing up everything is usually the most common approach but can be more expensive and resource intensive. Database files, archive logs, on-line log files, control files, and configuration files should be likely candidates for backup. There are also many instances where migrated data sets will also be required in a backup. These are data sets originating from sources other than the primary site but are necessary for proper recovery. They could have been created at an earlier time, but must be kept available and up to date for recovery. Application and system software can be backed up as well, depending on how tightly the software is integrated with the data. Software file copies usually can be obtained from the software vendor, unless the software has been customized or created in house. An OS software copy should be made available if it must be recovered in the event of an entire disk crash. Furthermore, a maintenance and security strategy for off site data storage should be developed.

Backups should be performed on a regular basis and automated if possible. A common practice is to keep the latest backups on site as well as off site. Critical or important data should be backed up daily. A full backup of every server should be made least every 30 days. Backups should be retained for at least 30 days. Regulated industries may have even more extreme archival requirements.

A separate set of full backups should be made weekly and retained for 90 days or even longer if business requirements dictate. Data that is to be retained for extended periods for legal reasons, such as tax records, or for business reasons should be archived separately on two different types of media (e.g., disk and tape).

There are three basic types of data backups: normal, incremental, and differential. The following sections describe these various backup types [13].

10.4.1 Full Backup

A full backup, also referred to as a normal *backup*, involves copying all files, including system files, application files, and user files to a backup media, usually magnetic tape, on a routine basis. Depending on the amount of data, it takes a significant amount of time. To meet the backup window, multiple tape drives could be used to back up specific portions of the server's disk drive. Most full backups are done on a daily basis at night during nonbusiness hours when data access activity is minimal.

Backups involving data that does not need to be taken off line for a full backup are often referred to as *fuzzy* backups [14]. Fuzzy backups may present an inaccurate view of a database, as some transactions can still be open or incomplete at the time of backup. This can be avoided by making sure all transactions are closed and all database records are locked prior to backup. Some backup and recovery software include capabilities to help restore fuzzy backups using transaction logs.

Figure 10.5 illustrates a full backup process. A popular backup scheme is to make a full backup to a new tape every night, then reusing tapes made on Monday through Thursday on a weekly basis. Tapes made on Friday are maintained for about a year. The problem with this scheme is that a file created on Monday and deleted on Thursday cannot be recovered beyond that week. Use of four sets of weekday tapes can get around this problem [15].

Knowing the amount and type of data requiring backup is fundamental to devising an appropriate backup strategy. If data is highly compressible, such as user documents or Web sites, it will require less backup resources than incompressible data, such as that associated with database files [16]. If the amount of data to back up is

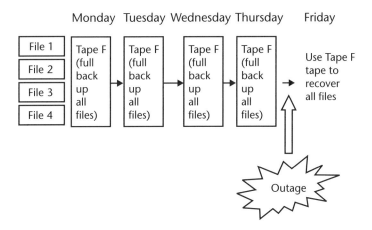

Figure 10.5 Full backup example.

voluminous, issues could result with respect to fitting the data on a single or set of media (e.g., tapes) as well as completing the backup within the backup window. In these situations, incremental or differential backups (discussed in Section 10.4.2 and Section 10.4.3 respectively) may be a more attractive option.

As was earlier discussed, mirroring can be used to create a high-currency replica of a database. Making frequent full backup copies of data is both costly and resource intensive if it has to be done at the frequency typically used for mirroring. Relying only on a daily backup replica for recovery implies the possibility of losing up to a day's worth of transactions upon restoration, which might be unacceptable for many real-time intensive applications. For situations involving high-frequency backup, mirroring provides a much better option [17].

There do exist variations and extensions of the full-backup approach. One method called *copy backup* will perform full backups several times during a day at specified time intervals. Another approach called *daily backups* will perform full backup of files based on specific file descriptor information. Full backups can also be used in conjunction with mirroring in several ways. A full-backup copy can be used to initially create a replica of a database. Once established, frequent updates using one of the previously discussed mirroring techniques can be applied. Even when mirroring is being used, full backups of the primary and mirrored sites are required to recover the data if it gets corrupted.

10.4.2 Incremental Backup

For large data volumes, incremental backups can be used. Incremental data backups involve backing up only the files that have changed or been added since the last full backup. Unless a file has changed since the last full backup, it will be not be copied. Incremental backup is implemented on a volume basis (i.e., only new or changed files in a particular volume are backed up). Incremental backups involve weekly full backups followed by daily incremental backups [18].

Most files are backed up during a weekly full backup, but not every file is necessarily updated on a daily basis. In an incremental backup, if a file is backed up on Monday at the time a full backup is routinely performed, then it will not be backed up in subsequent incremental backups if it does not change. If the file is lost on Friday, then the backup tape from Monday is used to restore the file. On the other hand, if the file is updated on Wednesday, an incremental backup will back up the file. If the file were lost on Friday, the Wednesday tape is used to restore the file [13]. These strategies may not only apply to user data, but also configuration files and system software as well.

Figure 10.6 illustrates the use of an incremental backup with four files that must be recovered on Friday. Because files A, B, and C are modified on Wednesday, Tuesday, and Thursday, respectively, incremental backup tapes are created. A recovery required on Friday would necessitate the use of the Monday's full backup tape plus all of the incremental tapes made since then. The incremental tapes must be restored in the order they were created.

To completely reload software and data onto a server, sometimes called a *dead server rebuild*, first the OS is installed on the server, followed by the last available full data backup of the server. This is then followed by loading, in order, each incremental backup that was made since the last full backup. This requires keeping track of when updates are made to files. Some software can simultaneously load both full and incremental backups, as well as keeping track of deleted files. Some packages also allow users to choose the specific data sets to backup from a group of volumes. Migrated data sets usually are not included in incremental backups and have to be copied using other means.

As a dead server rebuild can be a painful process, involving reconfiguration, patching, and applying hot fixes, some firms use a disk/volume imaging approach, sometimes referred to as *ghosting*. This involves creating an image to a media when a system is initially installed to a desired baseline configuration. This image is then used to rebuild the server, along with the appropriate data backups.

There do exist variations and extensions to the process of incremental backup. Some applications can perform *continuous backups*, sometimes referred to as *progressive* backups. They perform incremental backups on a regular basis following a full backup and maintain databases indicating what files exist on all systems and their locations on backup storage media. They use various file descriptors including file size and time stamp to determine whether files should be backed up.

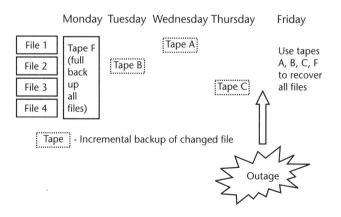

Figure 10.6 Incremental backup example.

10.4.3 Differential Backup

Like an incremental backup, a differential backup is used in conjunction with a full backup and backs up only those files that change since the last full backup. However, it differs in that all changed files appear on one differential backup. Figure 10.7 illustrates the use of a differential backup using the previous example. All updated files since the last full backup are copied and recopied on every subsequent differential backup [13]. If a server has to be restored using tape backups, only the last full backup tape and last differential backup tape are required. Unlike the incremental backup, the need to know when files were last updated is eliminated. Regardless of when the file was last updated, only the most recent tape is required. This makes server rebuilds much easier to do than using incremental backups.

Because of the cumulative nature of differential backups, the amount of media space required for differential backups could be greater than that required for incremental backups. Like incremental backups, differential backups are usually performed on a daily basis after a full backup, which is typically made on a Monday.

10.5 Storage Systems

A mission-critical storage strategy will utilize different types of storage media. Disk systems are normally used for instant reading and writing between a host and a data set. Magnetic tapes are widely used for data backups. In each case, a disk media can present a single point of failure in and of itself. It was earlier stated that multiple copies of the same backup disk or tape is usually wise in the event a disk drive fails or a tape wears out. The following sections describe in further detail the main categories of storage media.

10.5.1 Disk Systems

Disk drives, the most critical storage component found, are a prime candidate for redundancy. Often referred to as *fixed disks* or *hard disk drives*, they use moving arm devices called *actuators* to read or write to an area on disk media. Some designs employ dual actuators that read or write to the same or different areas of disk media. Because the read/write process is somewhat mechanical, it introduces a

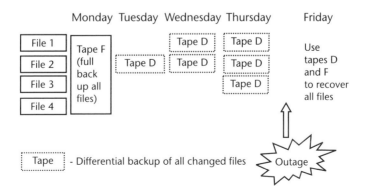

Figure 10.7 Differential backup example.

potential performance bottleneck as well as a point of failure. It is also the reason why disk drive systems are the more cost-intensive ways to store data and why magnetic tape is used for data backup.

Portable disks, such as compact discs (CDs) and digital video disks (DVDs), have become quite popular in supplanting magnetic floppy disks and tapes as a backup and transport media. However, they are unsuitable for large data backups due to space limitations. Optical disks last longer and present a more attractive archive medium that magnetic disk or tape. Unlike magnetic media, they do not have to be rerecorded periodically. These disks are also very useful to inexpensively distribute data or software to users at remote sites.

10.5.2   RAID

RAID is an acronym for *redundant array of independent disks*. The concept behind RAID is straightforward: store data redundantly across a collection of disks combined into a single logical array. Using several inexpensive disks operating in conjunction with each other offers a cost-effective way to obtain performance and reliability in disk storage.

RAID uses software or firmware embedded in hardware to distribute data across several drives, thereby enabling quick recovery in the event a disk drive or disk controller fails. In the firmware implementation, the controller contains the processing to read and write to the multiple drives, removing the burden from the host server CPU. If the controller fails, all drives and data are lost. The software solution, on the other hand, has the ability to copy data to a disk on a different controller to protect against controller failure. Because of the greater number of heads and arms that can move around searching for data, the use of multiple drives provides better performance for high volume I/O of many individual reads/writes versus using one large single drive.

There are two types of hardware-based RAID array controllers: host based and small computer systems interface (SCSI)-to-SCSI based. A host-based controller sits inside the server and can transfer data at bus speeds, providing good performance. The controller connects directly to the drives. Multichannel SCSI cabling is often used to provide the connectivity. As each drive occupies a SCSI ID on the controller, up to15 drives for each controller can be used, which can limit scalability. A special driver for the server OS is required. Many vendors do provide drivers for many of the most widely used OSs. Using the driver, the server performs all of the disk array functions for every read and write request. Many controllers have their own CPU and RAM for caching, providing performance advantages.

The SCSI-to-SCSI array controller is located in the external disk subsystem, which connects to existing SCSI adapters in the server. Any SCSI adapter recognized by the OS can thus be used to attach multiple subsystems to one controller. The SCSI controller uses a single SCSI ID for each subsystem, rather than an ID for each drive, as in the case of the host-based controller.

RAID employs several key concepts. *Duplexing* is the simultaneous writing of data over *two* RAID controllers to two separate disks. This redundancy protects against failure of a hard disk or a RAID controller. *Striping* breaks data into bits, bytes, or blocks and distributes it across several disks to improve performance. *Parity* is the use of logical information about striped data so that it can be re-created in the event of a drive failure, assuming the other drives remain operational. This avoids the need to duplicate all disks in the array. Typically, parity is used with three disks. The parity information can be stored on one drive, called the *parity drive*, or it can be stored across multiple drives. In an array with five drives, for example, one drive can serve as a parity drive. If one drive fails, the array controller will recreate the data on that drive using information on the parity drive and the other drives. Figure 10.8 illustrates a RAID system using striped parity. Some higher order RAID systems use error correction code (ECC) that can restore missing bits. Unlike parity, ECC can recognize multibit errors. This involves appending bits to the stored data indicating whether the data is correct or not. ECC can identify which bits have changed and immediately correct the incorrect bits.

RAID can be implemented in different configurations, called levels, that each use different techniques to determine how the drives connect and how data is organized across the drives. Originally there were five RAID levels, and the standard was later expanded to include additional levels. Although all RAID vendors follow the RAID-level standards, their implementations may differ according to how the drives connect and how they work with servers. The following is a description of the original five levels, along with the popular RAID 0 and RAID 10 levels [19, 20]:

- *RAID level 0 (RAID 0)*. RAID 0 employs disk striping without parity. Data is spread among all of the available drives. RAID 0 does not provide any data redundancy—data loss would result upon disk failure. If one drive in the array fails, the entire system fails. However, RAID 0 systems offer good performance and extra capacity. RAID 0 requires at least two drives and can be accomplished with as many drives as a SCSI bus allows. RAID 0 is often used for high-performance versus mission-critical applications.

- *RAID level 1 (RAID 1)*. RAID 1 involves writing data to two or more disks in a mirroring fashion—but always an even number of disks. This provides

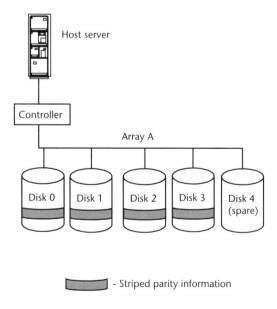

- Striped parity information

Figure 10.8   RAID example.

redundancy and allows data recovery upon disk or controller failure. If one drive fails, another will take over with little if any degradation in system performance. (Systems that employ *hot-spare* drives can even withstand failure of additional drives.) The simplest configuration would involve one or more disk controllers simultaneously writing data to two drives, each an exact duplicate of the other. Simultaneous writes to multiple disks can slow down the disk write process, but can often speed up the read process.

RAID 1 provides complete redundant data protection. However, because RAID 1 requires paying the cost of an extra drive for each drive that is used, it can double the system cost. RAID 1 is best used for mission-critical data.

- *RAID level 2 (RAID 2)*. RAID 2 stripes data bit by bit across multiple drives. This level is use with disks that don't have ECC error detection. To provide error detection, ECC drives must be used to record ECC data, which can create some inefficiency.

- *RAID level 3 (RAID 3)*. RAID 3 stripes data into identically sized blocks across multiple drives. Parity is stored on a separate drive to quickly rebuild a failed drive. Data is striped across the remaining drives. If a drive fails, the parity drive can be used to rebuild the drive without any data loss. RAID 3 is effective if the application can survive short outages, while the disk drive and its contents are recovered. If multiple drives fail, then data integrity across the entire array is affected.

  One dedicated drive in the array is required to hold the parity information. The parity drive can be duplicated for extra protection. RAID 3 can be expensive, as it requires an entire drive to be used for parity. It requires paying the cost of an extra drive for every four or five drives. It is used for applications that read or write large quantities of data.

- *RAID level 4 (RAID 4)*. RAID 4 stripes data files block by block across multiple drives. A block of data is written to disk prior to writing another block to the next disk. Parity is again stored on a separate drive. As data is stored in blocks, it provides better performance than RAID 2 or RAID 3.

- *RAID level 5 (RAID 5)*. RAID 5 involves striping data in the same fashion as RAID 4 and spreading parity information across all of the available drives in an array. Consequently, RAID 5 is one of the most fault-tolerant levels available. RAID 5 arrays typically consist of four or five drives. It is similar to RAID 4, except that parity is distributed among the disks, versus a single parity disk. RAID 5 distributes data among disks similar to RAID 0, but it also includes the parity information along with the striped data.

  A RAID 5 array can withstand a single drive failure and continue to function using the parity information. RAID 5 is effective if the application can survive short outages, while a disk drive and its contents are recovered. If a hot spare or a replacement drive is used, the system can run while the data rebuild is in progress. However, if another drive fails during this process, the entire RAID 5 array is inoperable. RAID 5 will require paying the cost of an extra drive for every four or five drives.

- *RAID level 10 (RAID 10)*. RAID 10 is also referred to as *RAID 0 + 1* because it is essentially a combination of RAID 1 and RAID 0. Data is striped across multiple disks, and then all of those disks are duplicated. Although it provides excellent fault tolerance and access speed, it is one of the most expensive levels to implement.

RAID 1, RAID 3, and RAID 5 are commonly used for high-availability implementations. RAID 5 can be used in conjunction with mirroring techniques to provide higher levels of continuity. For example, a pair of RAID 5 systems can each be connected to a server that supports software-based mirroring [11]. If a RAID array has multihosting capabilities, the array can be connected to a second server in either a load share or standby configuration.

In general, RAID systems provide enormous advantages over single-disk systems. Outages due to write failures are extremely rare. If a single drive fails, the RAID subsystem can determine the missing bits using parity information and reconstruct the data, providing the opportunity to repair the failed drive.

Parity errors might occur for various reasons—usually faulty hardware such as connectors, terminators, or cables. Many RAID controllers come with diagnostics to identify such problems. Host systems can cause parity errors if data or parity is lost prior to a system crash.

A drawback to RAID is its vulnerability to multiple simultaneous disk failures, although this event is extremely rare. Furthermore, to prevent single points of hardware failure, RAID systems should be configured with redundant hardware resources, such as dual cooling fans and dual power feeds. Most RAID systems offer hot swappable drives allowing replacement of a failed drive without disrupting the system.

10.5.3   Tape Systems

Magnetic tape is a widely deployed technology. It is relatively inexpensive and is highly reliable. Tape has been found to be an extremely secure media for protecting data from damage or loss. Many firms invariably back up data regularly to external tapes. Tape technology is often used on data that does not need to be immediately backed up or restored in minutes. The reason for this is that data backup and restoration using tape is typically a tedious process that requires longer time frames. Tapes usually must be wound forward or backward to a file's starting point in order to read and write individual files.

When compared to tape, disk systems usually provide a faster option. To improve recovery time using tapes, a common practice is to keep a backup tape copy on-line in a tape changer or drive. This practice is okay as long as there is another backup copy of the tape located in a remote site. Fortunately, some vendors are beginning to offer tape drives that can be recognized as a bootable disk device by the host system. This can speed up the recovery process, reducing it to a one-step process. In this instance, both the restoring application and backup data would be stored on the recovery tape.

Tape is a less expensive storage media than disk. Uncompressed data of up to 40 GB can fit into a digital linear tape (DLT) cartridge and up to 100 GB can fit in a linear tape open (LTO) cartridge (these technologies are discussed later in this section). Choosing the right tape systems requires comparing cost as well as capacity, speed, and reliability of various tape backup technologies [21].

Tape condition is subject to wear and often requires annual replacement. Data on tape should be recopied every 2 years at the latest. Tape management programs should include a tape maintenance strategy to make sure tapes and drives are kept in good working order. Tapes traditionally require manual inspection to check for wear. This can literally involve visually inspecting the tape—a process that can be quite labor intensive and inexact if large quantities of tapes are involved. Tape drives require periodic inspection and cleaning to remove flakes from the tape media that occur. Some tape drives, such as DLT drives, have lights on the front panel of the drive to indicate that cleaning is required.

The number of tape volumes typically found in data center tape libraries can be staggering—on the order of tens, if not thousands. Management of these libraries involves keeping track of age, usage, and performance for each and every tape. Fortunately, tape library software packages and services are available that simplify the management. Some packages are quite comprehensive and perform such tasks as tape library preventive maintenance and problem solving. These packages automatically collect and collate tape hardware and media performance data. They then use the data to automatically compute and track performance indicators and recommend tape library preventive maintenance and enhancement functions. They can compare drive utilization data to determine the nature of a tape library problem. These packages allow customization to the needs of individual data centers in areas such as management reporting.

10.5.3.1 Tape Performance

The use of sense bytes on tapes provides the foundation for obtaining tape performance information. For example, 3480/90 tape drives contain 32 sense bytes. Sense bytes are used to generate error indicators that are returned from a drive when there is an error. Sense information can identify the specific error in terms of an error fault symptom code, which can be issued to the OS of the host computer. When a tape drive error occurs, it is flagged to the OS. The operator can use the error code information to determine the severity of the error. Sense bytes are based on industry standards and are capable of reporting a data check, a hardware unit check, and a channel check. Error fault symptom codes generated from sense bytes can vary depending on the drive manufacturer. Many manufacturers use sense information in some kind of error reporting system in their equipment [22].

By counting the number of errors encountered and the volume of data a drive or control unit has processed, key performance indicators can be compiled that can be used for preventive maintenance. The number of megabytes of data a drive or control unit has written can be used to quantify the volume of data. Key error indicators are temporary write errors (TWE), correctable errors (ECC), erase gaps, and transient errors, which are caused by drive speed variations and block count errors [22].

Megabytes per temporary write error (MB/TWE) is a popular indicator to track media and hardware problems. Industry standards for the media technology involved can be used to establish a performance benchmark. The MB/TWE of a device should be evaluated periodically to verify the behavior of the tape hardware and tape management systems.

Many indicators have been developed to track tape library efficiency. These include single volume dataset capacities, the number of multivolume datasets, block size, density, the number of opens, expiration date, and last date used. Such indica-tors are used in eliminating small datasets, inefficient block sizes, or densities, plan-ning for scratch tapes, and planning storage requirements [22].

Segmenting a tape library enables the comparison of performance and efficiency data on library segments containing smaller volume ranges, rather that the overall library. This allows one to isolate the poorly performing or inefficiently used vol-umes. Monthly reports should be generated to spot trends in performance and effi-ciency over time. The reports should include summaries of year-to-date averages and monthly comparisons indicating whether a library is meeting a set of perform-ance and efficiency benchmarks. The analysis should develop a pull list of poor and/or inefficiently used volumes. The volumes in the list should be replaced or removed from the library when they reach scratch status.

10.5.3.2   Tape Technology

There are various tape technologies and formats available as of this writing. As data requirements and tape technologies evolve, there comes a time when conversion to a different technology or format is required. When changing over to a new technol-ogy, it is important to retain at least one or two tape drives that can read tape back-ups made with the older technology. Backup libraries should be reviewed to determine what backups should be initially stored using the new technology. Invariably, this will most likely include the most recent backups and those that con-tain critical data. The review should also include identifying what backups should be retained or discarded.

DLT is a magnetic tape technology that has been predominantly used for many years. DLT cartridges provide storage capacities from 10 to 35 GB. DLT drives have offered very high sustained data rates. Linear tapes use linear scan tech-nology. Data tracks are written to tape from front to back in a linear serpentine fashion. That is, when the head reaches the edge of the tape, it drops down a row and switches direction. LTO is a new linear technology jointly developed by HP, IBM, and Seagate. It uses the latest in ECC, data distribution, data compression, and head technologies. With data compression, capacities of 2.4 TB per cartridge are expected [21].

Helical scan is a tape recording method that uses fast spinning read/write heads. Helical scan writes data tracks diagonally, or at an angle, across the entire tape media. The angled position enables writing more information in the same amount of tape space. The spinning head writes high-density magnetic images without stretching the tape. The heads usually must be kept clean for proper operation. Heli-cal scan technology is widely used for video cassette recorder (VCR) tapes [23].

When compared to linear scan technology, helical scan offers higher density and performance, while linear scan provides increased reliability. Helical-scan drives pull the tape from the cartridge and wrap the tape around a rotating drum that con-tains the heads. Linear scan uses stationary heads and a less complex threading method, providing better reliability [21, 23].

An example of a popular helical scan technology is the advanced intelligent tape (AIT) specification. AIT was developed by SONY and uses 8-mm cassettes. It can be used only in those drives that support the format. It can store up to 25 GB of uncom-pressed data on a single cartridge. To increase capacity, AIT devices use a technique called adaptive lossless data compression (ALDC), which allows a single cartridge to hold up to 65 GB of data.

Digital audio tapes (DAT) are a helical tape technology that has been in use for quite some time. DAT is designed for high-quality audio recording and data backup. DAT cartridges resemble audiocassettes. DAT cartridges provide over 2 GB of storage and until now have found popular use in medium-to-small work groups. However, newer systems are expected to have much larger capacities. The DAT format has been subdivided into specifications called digital data storage (DDS). There are currently three DDS specifications: DDS-1, DDS-2, and DDS-3. DDS drives use a small 4-mm cartridge and differ by the length of the tape media. With compression, a single DDS-3 tape can carry up to 24 GB of data. DDS drives are quite versatile because they are SCSI devices and thus can work with many available backup software programs [23].

New generations of tape technologies are under development. For example, intelligent tape storage systems are being devised that will place additional management intelligence in the controller. Combined with intelligent disk subsystems, they will enable data movement between disks and tapes with minimal host server intervention. This would further improve backup and recovery times. Intelligent tape cartridges use built-in flash memory chips and software to provide file and record navigation capabilities directly inside the cartridge.

10.6 Storage Sites and Services

Various alternatives to maintaining data replicas and backups have arisen in the past decade. These options offer companies a cost-effective way of maintaining mirrored and backup images of data in a secure environment.

10.6.1 Storage Vault Services

Storage vault services have been in use for some time. These services typically involve using a contracted service provider to pick up data backups from a customer's site and transporting them to a secure facility on a regular basis. This is known as *physical vaulting*. The vaults usually consist of fireproof, environmentally safe, and intrusion-secure facilities. Physical vaulting is quite popular because it is inexpensive and is less resource intensive with respect to a firm's IT department. Physical vaults should be located off site in a location away from primary location so that it cannot be affected by the same physical disaster, yet close enough so that it does not result in slower recovery. Depending on the nature of the crisis at hand, successful delivery of backup data to the primary site can be unreliable. The longer delivery time creates a potential for greater loss of data during an outage.

Electronic vaulting involves mirroring or backing up data over a network to a remote facility. Data can be transmitted in bulk at the volume, file, or transaction level using mirroring data transfer methods. Electronic vaulting moves critical data off site faster and more frequently than physical vaulting and enables faster retrieval of the data for recovery. It also makes it easier to keep off-site mirrored or backup data images up to date.

By reducing or eliminating tape recall times and shortening recovery windows, electronic vaulting can ultimately reduce the cost of downtime. By improving or even eliminating the handling of tapes as well as the need to physically transport, electronic vaulting can reduce on-site labor and operation costs. It can also be scaled to achieve desired RPOs and RTOs.

Electronic vaulting is seeing more widespread use. *Recovery service providers* use electronic vaulting as a means of keeping data at recovery sites or vaults as updated as possible. Many vendors allow transferring data across a WAN or the Internet. Electronic vaulting may prove more expensive if not used in a cost-effective fashion. Fees and bandwidth requirements for electronic vaulting for full backups can be extensive. It will often require constant network bandwidth. Vaulting to closer facilities can minimize bandwidth costs, as long as they are distant enough not to be affected by the same physical disaster. Moreover, networking from a reliable third party communications service provider is required. Proving in the use of electronic vaulting requires weighing its costs with the expected improvement in downtime costs. A cost-effective alternative is using electronic vaulting in combination with physical vaulting by electronic journaling updates as they occur between regular physical backups [24, 25].

10.6.2 Storage Services

Storage service providers (SSPs) manage data storage for customers in a variety of ways. They can manage storage for a customer at the customer's location, sending support staff to perform tasks such as backup and library management. Companies that don't want their most critical data residing off site often favor this approach. The customer's storage can also be kept and managed at the SSP's own secure, physical vault site, where it is usually collocated with storage of other clients, allowing the SSP to operate more cost effectively. The SSP will pick up storage daily, weekly, or monthly and deliver it to the customer premise when requested. Many SSPs maintain physical vaults for storing backups and also offer electronic vaulting along with their services. Some employ electronic vaulting to centralize storage for multiple customers, using broadband networks connected from their site to the customer sites.

SSPs charge monthly fees based on capacity and levels of service. Fees are usually based on the amount of storage specified in cost per gigabyte per month for basic service. Fees may also include items such as a storage assessment, installation and configuration of hardware and software, and remote performance management. Additional optional services are usually available. SSPs enable firms to avoid the costs of building and operating their own physical or electronic vaulting facilities. Customers pay based on the service levels they require. This is done using a service-level agreement (SLA), which specifies the required levels for capacity, availability, and performance for the subscribed services [26]. (SLAs are discussed in further depth in Chapter 12.)

Important features should be considered when selecting or using an SSP. As the SSP is charged with overall responsibility for data availability, they should be bonded and have an operation with built-in business continuity. Off-site storage should be at a secure site. It is important to know what other clients are served by the SSP and what preference they will be given to retrieve backups in the event of a regionwide disaster. As concurrent tape-restore operations for multiple customers can be taxing during a regional crisis, electronic vaulting could provide a safer solution. The SSP should also provide a way for clients to monitor activity with respect to their storage and confirm SLA conformance. Some SSPs provide on-line reporting to the customer, going so far as allowing customers to remotely manage their storage at the SSP's site.

10.7   Networked Storage

Enterprises are gradually shifting away from the direct attached storage device (DASD). This is the traditional approach where a storage device is directly attached to a host server via the server's chassis or through a SCSI bus extension. DASD is still a very good option in situations where high-speed direct access between a storage device and a server is required, without the use of an intervening network. However, as geographically dispersed companies have been creating networks to support distributed computing, storage networks are now being devised to network together distributed storage. In these cases, a networked storage approach can improve the mission-critical characteristics of storage. Not only can it improve storage scalability, but it also leverages traditional network planning techniques aimed at improving reliability and performance.

A storage network is a network whose sole responsibility is to enable storage devices to communicate with one another. A storage network is usually separate from the computing communication network. Figure 10.9 illustrates the concept of networked storage, which can take on decentralized and centralized network topologies. In a decentralized topology, each storage device is closely connected to a host server, such that processing and storage is distributed. In a centralized topology, storage is centralized and accessed from multiple systems. These two topologies can also be used in combination as well.

Host servers have traditionally transferred data to storage devices directly through an I/O bus or across a local area network (LAN) to a storage device or host server directly connected to a device. Data transfers, especially those associated with a backup operation, could impose heavy traffic across the LAN and degrade LAN performance for other users as well as increase backup times. Eliminating the LAN from the backup path, or *LAN-free* backup, avoids these issues completely.

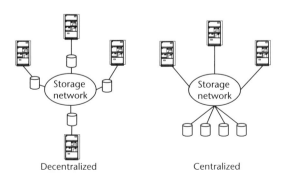

Decentralized   Centralized

Figure 10.9   Networked storage topologies.

In this approach, data transfers are moved across a dedicated storage network, separate from the LAN. This approach reduces network traffic, providing better network performance, and can reduce backup time. As software on the server still has to spawn and maintain the backup process, CPU overhead is not necessarily improved.

An approach that can further improve CPU overhead performance is called *serverless* backup. The method moves the processing from the server to a separate device, freeing up the server to perform other tasks. These devices talk to each other and can move data in one copy directly to one another. The backup application runs on the device and queries the server to identify the data that must be transferred. The application then performs the copy block by block from one device to the other over the storage network. In a typical backup operation, the origination device would be a disk drive and the destination device would be a tape drive. This approach is sometimes referred to as *extended copy* or *third-party copy* [27].

10.7.1   Storage Area Networks

The Storage Networking Industry Association (SNIA), a voluntary group of storage vendors working to standardize storage methods, defines a storage area network (SAN) as "A network whose primary purpose is the transfer of data between computer systems and storage elements and among storage elements. A SAN consists of a communication infrastructure, which provides physical connections, and a management layer, which organizes the connections, storage elements and computer systems so that data transfer is secure and robust."

A SAN is a network that is separate from the primary production LAN. The SAN can provide connectivity between storage devices and between application servers and storage devices. Direct high-speed channels between storage devices and servers are established via special hubs or switches. The SAN's physical network can connect directly to the server buses. The primary purpose of a SAN is to separate storage functions from host computing functions. A SAN can logically connect a set of shared storage devices so that they can be accessed from multiple servers without affecting server or LAN performance.

In a SAN, the server is still considered as the *gateway* to the storage devices. A storage device that directly connects to a SAN is called *SAN-attached storage* (SAS). These devices provide the traditional data access services in the form of files, databases, or blocks to the storage subsystems.

Like any network, a SAN consists of a set of interconnected nodes. These may include storage devices, servers, routers, switches, and workstations. SAN connectivity consists of an interface, such as Fibre Channel or SCSI, which interconnects the nodes. (These are described further in Section 10.7.1.1 and Section 10.7.1.2 respectively.) A protocol such as Internet protocol (IP) or SCSI controls traffic over the access paths between the nodes. The servers and workstations have special SAN adapter boards, called host bus adapters (HBAs), which are network adapters fitting in the server that allow communication between the server bus and the SAN.

The storage devices in a SAN may include disk subsystems, tape backup devices, and libraries. Special hub devices are used to interconnect devices, similar to a LAN. Devices called *directors* are similar in nature to switches and often have fault-tolerant capabilities.

Applications are typically unaware that a SAN is in use. Applications usually send storage requests to the OS, which then communicates with the storage device through the SAN. Communication is done typically using SCSI or Fibre Channel commands. Once the commands are delivered, communication between the controller and the physical drives is typically SCSI or integrated drive electronics (IDEs).

SANs use locking mechanisms to resolve conflicts that can arise between multiple users trying to share the same file. When users share files across a SAN, file transfers are not involved, unlike traditional networks. A copy of the file's content is not created on local storage, nor is the file attached to the local file system. Instead, the file appears as if it resides on their local system, even though it is stored somewhere else [28].

Nodes on a SAN are identified by their worldwide names (WWNs). These are 64-bit identifiers that uniquely identify ports on the SAN during initialization or discovery. WWNs are useful for reestablishing network connections following a site outage. Logical disks are represented by logical unit numbers (LUNs). In the case of a RAID controller, all of its logical disks are presented to the SAN. SANs employ *mapping tables* to associate LUNs with WWNs, providing node-to-logical disk relationships. Mapping tables can be implemented at the server, switch, or storage array levels [29].

When designing a SAN for a mission-critical environment, the desired level of tolerance to be provided should be considered. SANs are not intrinsically fault tolerant, and therefore require careful network planning and design. Ideally, single points of failure must be eliminated wherever feasible. Because a storage device can be shared among different users, a failure in this system can be detrimental. Redundancy in disk systems and connections to each server should thus be incorporated into the design. Use of fault-tolerant or high-availability platforms can provide additional reliability. SANs can be designed in various ways to provide extra bandwidth capacity to handle periods of high demand. SAN growth should be carefully planned.

Figure 10.10 illustrates two SAN topologies. A tree topology is considered a basic configuration, where switches cascade off one another in a single-tier fashion. This topology introduces latency through the single-port interface, which limits bandwidth and is a single point of failure. Storage devices could be multihomed and redundant links could be established between switches, but this may not be cost effective. An alternative topology involves introducing another switch so that each

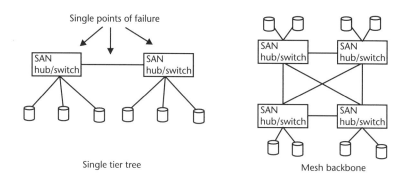

Figure 10.10   SAN topologies.

switch connects to at least two other switches, creating a mesh network. This eliminates a single point of failure by providing each switch with redundant data paths
through the SAN. This rule can be used if more switches are added and can result in
a mesh backbone if each switch connects to every other switch [30]. Additional reliability can be incorporated into the topology by using switches and storage devices
nodes that support dual HBAs, allowing storage devices to be multihomed. The following are some examples of situations where SANs can be most effective:

- Organizations spread among multiple locations seeking to recentralize server
  operations and consolidate storage operations;
- Large systems where many users require access to the same bandwidth-
  intensive content, such as news environments and architecture/engineering
  design firms;
- Video streaming broadcast applications, where the video content can be
  played from multiple servers to multiple channels;
- Applications using large databases such as data warehousing, where a high
  degree of storage capacity and efficiency may be required;
- Environments where changing business requirements foster continuous
  changes in application and storage needs. SANs provide the opportunity to
  adapt to changes in software applications, data capacity needs, and storage
  technology.

The following are some ways in which SANs can be used to facilitate backup,
mirroring, and clustering tasks for mission-critical networks:

- Backup operations can be performed without using the production LAN or
  WAN. Backup disk and tape units are situated on the SAN. During a backup
  process, data is simply transferred across the SAN from the production storage disk to the controller of the backup media. This allows little if any disruption to the operation of the production server and storage disk. Devices
  known as SAN *routers* can back up directly from disk to tape without the use
  of a server. These devices use embedded commands that can automatically
  trigger a backup [27].
- A SAN architecture can simplify data mirroring. Data sets can be mirrored
  across a SAN to another storage device, or to another remote SAN, such as
  one operated by an SSP. Using mirroring software embedded in the SAN network hardware, I/Os can be sent from the primary storage subsystem to the
  mirrored storage subsystems, using either synchronous or asynchronous data
  transfer. This offloads processing from the server and offloads traffic from the
  production LAN [31].
- SANs are ideal for clustering because they provide the ability for many servers
  and processors to share common storage. A SAN can support a single cluster
  as well as geographically dispersed clusters. For example, a cluster can be split
  so that it can operate in separate data center locations. One whole cluster mir-
  rors another whole cluster. The nodes of one cluster and SAN operate in one
  location while nodes of the other cluster and SAN operate in another location.
  The two locations are connected over a WAN or equivalent network.

In all, SANs can provide a multitude of benefits when they are effectively used. These benefits include:

- *Increased availability*. Because the storage is external to the server and operates independent of the application software, processing and bandwidth requirements for mirroring and backup can be offloaded. They also improve availability by making it easier for applications to share data with legacy applications and databases. SANs also make it possible to perform maintenance on storage without having to disrupt servers.
- *Greater scalability*. SANs offer economies of scale gained by pooling storage software and hardware across locations. When adding storage devices to a SAN, fewer, higher capacity systems can be added versus numerous smaller capacity devices.
- *Efficient management*. SANs allow centralized management of data volumes, reducing time and costs. More amounts of data can be managed using a SAN versus a decentralized system.
- *Improved flexibility*. SAN devices can be spread over a wide area to allow integration of different storage media. Storage can be added seamlessly on an as-needed basis with minimal service interruption.
- *Better protection*. SANs create new possibilities for business continuity. They offer cost-effective implementations for data mirroring, backup, and migration. With multiple storage devices attached to multiple servers, redundant paths can be established. Physical data migration can be simplified.
- *Improved bandwidth*. SANs have broadband capability. As devices are added, bandwidth can be allocated based on the speed of the storage device, providing better bandwidth utilization.
- *Greater accessibility*. SANs provide universal access to data from all types of server platforms almost simultaneously, thereby improving workflow efficiency.

Finally, SANs enable companies to integrate older equipment with newer and more efficient devices. This allows them to preserve their investment in the legacy equipment and save money in the long run by extending equipment life cycle. In spite of all the aforementioned benefits, there are two major caveats with SANs:

- *High entry costs*. The number of switches and other components to accommodate large storage requirements could be cost prohibitive in the beginning. SANs may not necessarily be a cost-effective solution when a small number of I/O channels and large amounts of storage are needed. A stand-alone shared disk system might be a better choice.
- *Limited interoperability*. There is still limited SAN interoperability among server platforms and between SAN vendors. Unlike a LAN, SANs cannot be built using a wide range of different vendor equipment. This is mainly attributed to the lack of available standards for SANs. As a result, vendors will differ in their communication protocols, file interchange, and locking methods. For these reasons, SANs will remain mostly as single-vendor solutions until progress is made in developing standards.

10.7.1.1   Fibre Channel

Connectivity within a SAN is linked using a network transmission standard called Fibre Channel. The Fibre Channel standard specifies signaling and data transfer methods for different types of connection media, including coaxial and fiber optic cable. As of this writing, data can be transmitted at speeds of up to 1 Gbps.

Fibre Channel was developed by the American National Standards Institute (ANSI) in the early 1990s as a way to transfer large amounts of data at high speeds over copper or fiber optic cable. Fibre Channel can support SCSI, IP, IEEE 802.2, and asynchronous transfer mode (ATM) over the same medium using the same hardware. When Fibre Channel was originally introduced, it ran over fiber-optic cable. Fibre Channel can be supported over single-mode and multimode fiber optic cabling. Copper Fibre Channel implementations are supported mainly using coaxial cabling [32].

Fibre Channel is a physical layer serial interconnect transport standard similar to Ethernet. Fibre Channel has been used as an alternative to SCSI for high-speed connectivity between network and storage devices. It can be used as either a direct storage interface or to create network topologies. Fibre Channel works as a shared SCSI extender, allowing local systems to treat remotely located storage as a local SCSI device. It uses SCSI-like bus arbitration. Almost all Fibre Channel SCSI devices are dual port. Fibre Channel can accommodate multiple protocols. It is capable of carrying both SCSI and IP traffic simultaneously. This allows existing products that are either IP or SCSI based to easily migrate to Fibre Channel. Fibre Channel supports speeds from 132 Mbps all the way up to 1.0625 Gbps, which translates into a theoretical maximum of roughly 100 Mbps (200 Mbps full duplex) [33].

At one time, wide area connectivity was somewhat limited and was only achievable with channel extension. Initially, Sysplex SANs (S-SANs) were developed for mainframes and used bus and tag technology. This technology supported a limited number of devices with distances not exceeding about 400 ft. Enterprise system connection (ESCON), a fiber-optic I/O connection method used by S/390 computers, was later introduced. It accommodated a greater number of devices, improved overall throughput, and increased the distance between devices to 20 km [34]. Fibre Channel technology currently supports distances of up to 47 miles over coax, but can be extended to longer distances when used in conjunction with high-speed WANs and fiber optics. This enables storage devices to be located on different floors in a building or even in different cities.

Fibre Channel transfers files in large blocks without a lot of overhead information. Data is sent as payload contained in Fibre Channel frames. Figure 10.11 illustrates the Fibre Channel frame [35]. The payload is surrounded by a header and footer, which help direct the frame through the network and correct errors that may have occurred in transmission. The header contains information about where the frame came from, where it is going, and how it is sequenced. After the payload, there is a four-byte cyclic redundancy check (CRC), followed by an end-of-frame marker. Although an individual packet payload size is 2,048 bytes, a large number of payload packets can be strung together into sequences as large as 4 GB.

The faster performance in Fibre Channel is attributed to a number of features. A key feature is the use of packet-parsing functions integrated into hardware using direct memory access interfaces. Because processing is done at the hardware level,

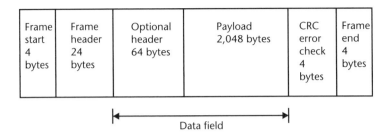

Frame start 4 bytes	Frame header 24 bytes	Optional header 64 bytes	Payload 2,048 bytes	CRC error check 4 bytes	Frame end 4 bytes

Data field

Figure 10.11   Fibre Channel frame.

large sequences of frames can be transferred with little processing power needed to read and interpret header information. Furthermore, when packet errors occur, packets are sent again without the added delay of status checks. Additional performance enhancing features include the use of absolute addressing, simplified media access rules, and a SCSI command protocol that is optimized for bulk data transfer [36, 37].

SANs can be constructed from servers and storage devices connected through Fibre Channel HBAs to a network. An HBA is a network adapter fitting in the server that acts a port, enabling communication between the server bus and the Fibre Channel network. The HBA hardware houses much of the Fibre Channel communication processing capability. HBAs can perform error detection and correction and flow control. As a Fibre Channel SAN grows, other devices such as hubs and switches can be incorporated. Three network topologies are possible, depending on operating requirements and desired performance: point to point, arbitrated loop, and switched. The following is an explanation of each topology:

- *Point to point.* Point to point is the easiest topology to implement and is illustrated in Figure 10.12. A single direct connection is made between two Fibre Channel devices. The devices can be, for example, two servers, or a server and its storage device, such as a RAID system. No hubs or other devices are involved. Only one device can transmit at a time. Because of its simplicity, it is the lowest cost approach. Point-to-point connections are established on a per-channel basis. For each channel, the two end communicating ports define the port communication criteria. Depending on the type of connection, different

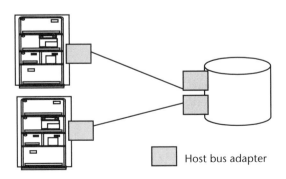

Host bus adapter

Figure 10.12   Fibre Channel point-to-point topology.

criteria can be defined for different channels. Because each device sends and receives data in block form, no file translation is done [36].

- *Arbitrated loop.* An *arbitrated loop* involves sharing bandwidth and forwarding data in a round-robin fashion among hubs. A Fibre Channel arbitrated loop (FC-AL) has the ability to deliver any-to-any connectivity among storage devices and servers. FC-AL is a shared bandwidth loop where each system *arbitrates* for permission to place information on the loop. Once a connection has been established between two devices, the devices own the loop and all associated bandwidth. When one device is ready to transmit data, it first arbitrates and gains control of the loop. It sets up a point-to-point connection to the receiving device. Any device in between will simply repeat data frames being sent between the two devices as long as the destination address of the frame does not match its own. Although there is no limit on how long a device can control the loop, an optional access fairness algorithm can be used to prohibit a device from arbitrating again until all other devices have had a chance to communicate. Once a device relinquishes control of the loop, other devices can gain access [30, 32, 33, 35, 36].

FC-AL networks are wired in a star fashion using a hub, switch, or router. Figure 10.13 illustrates an example of an FC-AL. A hub is a device used to connect the servers and workstations to the storage devices. Multiple devices can connect to a hub, switch, or router, with each sharing available bandwidth. The hub device manages the bandwidth sharing and arbitrates signals from any computer to any storage device, disallowing simultaneous conversations across ports. Unlike the 15-device limit of a SCSI chain, FC-AL can support up to 126 devices, or nodes, per loop. Like the point-to-point topology, small FC-AL configurations can be achieved at low cost.

In the FC-AL topology, the hub device offers either no redundancy or partially redundant paths. For reliability, they are equipped with dual ports, enabling the creation of a redundant path to each storage device, or a dual loop, in the event one of the loops fails or is busy. Hubs rely on port bypass circuits (PBCs) to allow a failed node to be bypassed, providing an instant *patch* to any breaks in the loop. This concept is illustrated in Figure 10.14. A dual-loop FC-AL also allows simultaneous communications between devices, greatly

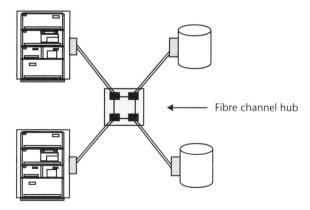

Figure 10.13 Fibre Channel arbitrated loop topology.

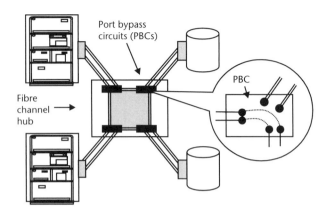

Figure 10.14   Fibre Channel dual arbitrated loop topology.

increasing available bandwidth and reliability. The failure of one loop will not bring down the other loops.

FC-AL operating issues can arise when an HBA is initially connected or reconnected to a Fibre Channel hub. In this situation, the HBA must perform a loop initialization process (LIP), which essentially resets the loop. During this process, the loop is in a somewhat unstable state until devices are able to renegotiate.

Another component found in a FC-AL is a Fibre Channel–to–SCSI (FC-SCSI) bridge (sometimes called a Fibre Channel router). This device enables SCSI-based devices to connect into the SAN. The bridge multiplexes several SCSI devices through one HBA. This allows legacy SCSI devices, such as tape libraries or RAID systems that do not have Fibre Channel controllers, to be attached to an FC-AL.

• *Switched fabric.* The fiber channel switched (FC-SW) fabric uses Fibre Channel switches to provide connection to the SAN. A fabric is a set of point-to-point connections. Similar to switches in a LAN, the fabric connects logical devices so they can communicate, moving data from an incoming port directly to the proper outgoing port. Unlike hubs where media is shared, the switched fabric protocol allows devices the full connection bandwidth from port to port.

Fibre switches are generally nonblocking and can operate multiple concurrent transmissions at full duplex. Unlike hubs that require an LIP, nodes connecting to Fibre Channel switch ports perform a fabric logon process that is seen only by the switch, without interrupting the other devices on the network. It is also possible to connect an FC-AL to a switch port. In order to make the devices on the FC-AL compatible with other devices on the fabric, some vendors have integrated fabric loop ports into their switches. This allows the switch to accept LIPs from the FC-AL and allows devices to communicate across the loop and fabric [30].

FC-SW configurations typically involve servers and storage systems connecting point to point through a central switch. Figure 10.15 illustrates an example of an FC-SW architecture. By connecting switches to other switches, large storage networks can be developed. An FC-SW SAN can support up to

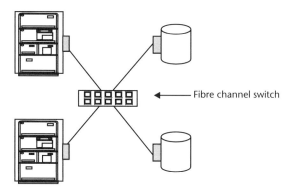

Figure 10.15   Fibre Channel switched fabric topology.

6 million devices. Fibre switches are commonly found with port densities in the range of 8 to 64 ports.

Doubling switches for redundancy, provisioning multiple paths for data traffic between the server and storage device, and implementing failover between paths can further achieve reliability. This failover uses a routing protocol recently accepted by ANSI called fabric shortest path first (FSPF). When a failure occurs in a SAN, FSPF detects the failed paths and reroutes data through another path [38].

Fibre Channel fixed-port switches offer failure resiliency but not stand-alone fault tolerance. For higher port counts, scalability, and reliability, chassis-based Fibre Channel *director* switches are used. Director class switches can sustain bandwidth up to 128 ports, and it is expected that manufacturers will offer products with even greater port densities over time. Director-class products offer fully redundant paths, interconnects, and backbone interconnects and typically have redundant power supplies, fans, and hot-swappable components. Fibre Channel switches are more expensive than hubs, while director-class products tend to be the most expensive.

Fibre Channel is designed in a five-layer model [35]. It defines the physical media and transmission rates, encoding scheme, framing protocol and flow control, common services, and the upper level protocol interfaces:

- *FC-0 (physical layer)*. This defines the characteristics and electrical parameters of the physical link to be used to connect the interface and media (including cables, connectors, transmitters, and receivers).
- *FC-1 (transmission layer)*. This defines the way data is encoded and decoded, and error handling.
- *FC-2 (signaling protocol layer)*. This defines the framing and sequencing of data transfer between ports. It also defines the mechanisms for managing three classes of service (CoS):
  1.   Class 1 is a dedicated connection for point-to-point operations.
  2.   Class 2 provides a connectionless operation that requires a confirmation from the receiving node.

3. Class 3 is a connectionless service that requires no confirmation and is typically used for storage subsystems.

- *FC-3 (common services layer)*. This layer defines common services required for advanced features, such as aggregating multiple ports and multicasting.
- *FC-4 (application interface layer)*. This layer defines the application interfaces that can be used over Fibre Channel; the predominant ones are SCSI and IP.

When compared to SCSI, Fibre Channel connectivity can further increase the distance between the main production network and remote storage devices. This feature is quite useful for mission-critical networks, as enterprisewide backup, mirroring, and recovery can be more easily accomplished than distributed storage. Because Fibre Channel block sizes are quite large, it can also be used for applications requiring large file transfers, such as video.

Fibre Channel removes the distance restrictions and scalability constraints of SCSI. As discussed earlier, a compelling feature of Fibre Channel is the extended distance it allows between storage and the computer. Extensions are being developed to enable multiple Fibre Channel SANs to be interconnected over longer distances using available telecommunication capabilities. Gateway devices can be used to encapsulate Fibre Channel for transport over IP, ATM, or synchronous optical network (SONET). SONET can provide low latency, high-bandwidth transport of Fibre Channel traffic over long distances. Fibre Channel over SONET gateways can introduce latency of less than 100 ms, making it desirable for mirroring operations. Additional latency introduced by the fiber-optic network corresponds to the distance between SANs across the network [39].

Overall, Fibre Channel SANs may be more cost effective for enterprises that value high scalability and reliability. A final caveat with Fibre Channel is the heavy reliance on hardware-based, versus software-based, functionality. This creates uncertainty with respect to which devices perform what functions. This uncertainty also makes the engineering and managing the network more challenging.

To migrate from an existing storage architecture to a Fibre Channel SAN, the following must be considered:

- Determine what storage tasks can be offloaded from the server and what storage resources should be allocated to what servers.
- Identify which existing SCSI devices, including RAID systems, should be included in the SAN. SAN functionality can be added to devices through the use of HBAs and bridges to attach older storage systems to the SAN.
- Develop and choose the appropriate architecture (i.e., FC-AL or FC-SW) for the SAN. Determine how the SAN will be accessed from devices at a redundant location.

10.7.1.2   SCSI

For quite some time, servers were typically connected to disk drives and other peripherals using the IDE or SCSI bus. SCSI was developed more than 15 years ago, evolving from the old Shugart Associates standard interface (SASI) bus. SCSI is still used as the most basic way of connecting disk arrays and other storage devices to servers and transferring data between them at relatively high speeds. It is a high-speed bus that allows computers to communicate with peripherals such as disk and tape drives, as well as printers, scanners, and other external devices. The use of SCSI requires a bus termination on the host machine. SCSI interfaces can move data at rates ranging from 5 to 40 Mbps. SCSI supports a maximum of one master host machine and a number of peripherals [32, 37].

SCSI has slowly evolved over the years with respect to speed and the number of supported devices. Distance limitations have kept SCSI from evolving very rapidly. Over the years, SCSI has seen increases in the size of the data path from 8 to 32 bits, the ability to handle more peripherals, and better error. The ultra2 SCSI standard enables data transmission at 80 Mbps. The ultra3 SCSI standard promises a throughput rate of 160 Mbps. The ultra160 SCSI specification is a subset of ultra3 that is gaining popularity [32, 37].

SCSI's limitations are cable distance and the number of drives that can be supported. SCSI is currently limited to 13 drives per SCSI channel, and a maximum limit of 15 SCSI identifiers (IDs), with one ID used per device controller. A maximum of two channels, or *chains*, can be supported without performance degradation. With ultra SCSI, distance is limited to a range of 1 to 25m between the host and device.

Although the distance limitation is not a problem when using SCSI to connect a storage device to a local server, it does limit the applicability of SCSI in building mission-critical SANs. Unless a medium such as Fibre Channel is introduced, storage devices cannot be connected to other devices at remote locations. The device limitations hinder its ability to scale. Furthermore, the SCSI bus termination itself introduces a point of failure such that if a device fails, connectivity through the bus's termination point is lost.

SCSI is best used for direct connectivity between a host and local device. For small reads and writes of 8K blocks or less, ultra SCSI is actually faster than Fibre Channel. It is possible to construct local SANs comprised solely of SCSI if faster performance is desired, or if Fibre Channel is unavailable. SCSI SANs would consist of parallel chains of devices, unlike a SAN constructed with FC-AL involving a series of serial links [32, 37]. However, as was earlier indicated, SCSI protocol can be transported over Fibre Channel, providing the opportunity to extend a local SCSI SAN into a Fibre Channel SAN.

10.7.1.3 InfiniBand

It is well known that servers are typically seen as bottlenecks in enterprise networks. Microprocessor speeds have been unable to keep up with network speeds, thanks largely in part to advancements in fiber-optic technology. To address this issue, the InfiniBand Trade Association (ITA), a consortium of vendors, is devising a new bus technology. The technology, called InfiniBand, is designed to speed up servers with respect to server clustering, internal memory, and bus I/O transfers. InfiniBand is a new switched bus architecture that connects to a computer's motherboard, replacing the common PCI bus. InfiniBand products, yet to be released, promise to provide wire connections in the range of 2.5 to 30 Gbps (current bus technology supports speeds up to 1 Gbps) [40].

InfiniBand is a channel-based system area bus network made from a point-to-point switch fabric. Servers and storage devices can be connected to the network without requiring the use of network interface cards (NICs), which are typically traffic bottlenecks. The fabric design eliminates the possibility of I/O being a single point of failure. Furthermore, RAM or a memory cache can be added by plugging the components into the bus. These features enable direct memory-to-memory connectivity between clustered servers. This offloads the overhead associated with transaction control protocol (TCP)/IP processing from the server CPUs, freeing them up for application processing. This also enhances network performance by offloading I/O traffic from the network to the InfiniBand bus.

The InfiniBand bus will also enable more versatile server architectures. It enables CPU and I/O resources to be scaled independently. Servers can be constructed from blades that can run clustering software, allowing several blades to act as a single server. This feature will enhance processing related to high-volume data-intensive applications and enable systems to be partitioned or reconfigured for different tasks.

SANs made entirely from InfiniBand components are not foreseen in the near term. In the mean time, routers and bridge products are being devised to translate traffic from InfiniBand protocol to TCP/IP, SCSI, and Fibre Channel. Adapters will be required on the host and storage devices to interconnect them through an InfiniBand fabric. Special software and I/O protocols will also be required to use and manage the fabric.

A goal of InfiniBand is to enable preservation of investment in legacy infrastructure and interoperability with existing, incompatible equipment. Like any new technology, investment in InfiniBand equipment is likely to be significant at first, in terms of cost, management complexity, and limited compatibility with other products.

10.7.2   Network Attached Storage

An alternative approach to shifting away from DASD storage is the use of network-attached storage (NAS). The SNIA defines NAS as a "term used to refer to storage elements that connect to a network and provide file access services to computer systems." A NAS is a storage device that attaches directly to a network. Figure 10.16 illustrates an architecture using NAS devices. The NAS device is a scaled down

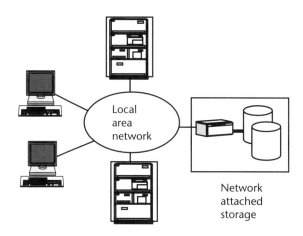

Figure 10.16   Network-attached storage.

server, similar to a thin server, which can connect directly into a LAN, instead of attaching to a host server connected to a LAN.

A NAS device works at the file level, providing file access services. It consists of an engine, which operates the file services, and one or more storage devices. Because file services reside on the NAS, the OS of the application servers is freed up. Storage arrays are attached to the NAS using a high-speed bus, typically SCSI. External storage devices can also be attached. Instead of containing a full network OS, NAS devices use a microkernal, which is a specialized OS hosted within embedded processors. The microkernal is designed to handle specific I/O functions, allowing heterogeneous data sharing over various types of networks. NAS devices also use a highly optimized file system which, when combined with the microkernal, enables them to read and write files with very high performance. Capacity on low-end NAS devices ranges from 6 to 12 GB, while higher end devices range upward to 4 TB [41–43].

A key benefit to using NAS is the scalability that it provides with respect to storage growth. The ability to connect to almost any type of network enables users to access data from all types of OSs and platforms. NAS does have several potential drawbacks. First, a NAS device must have an interface that can keep up with the speed of the connecting network; otherwise, the device will pose a bottleneck. Moreover, network traffic is increased, as I/O traffic must be transmitted over the production network. This could create problems with networks having large numbers of hosts requiring a high degree of data access or use of large file sizes. Unless the network switch can effectively segment traffic and reduce contention, a switch upgrade may be necessary.

10.7.2.1   NAS and SANs

NAS and SANs differ in several ways:

- In a NAS, the storage device is accessed through the production LAN. In a SAN, a dedicated network between the server and the storage arrays is used. The NAS device can impose a bottleneck because NAS software processes operate between clients and their data.

- In a NAS, the data moves as packets and as files at LAN speeds. In a SAN, the data moves as blocks at Fibre Channel or SCSI speeds. However, some NAS systems are on the horizon that will provide block-level access to data.

- NAS provides high-performance shared access to file system data from a number of different clients for read-intensive applications, such as Web browsing and content caching. SANs, on the other hand, are optimized for applications involving bulk data transport between devices, such as backup, transport of stored data, mirroring, or data replication.

- Fibre Channel SANs are channel based and offer at least three times the performance of a NAS solution. Unlike a NAS, data throughput is unaffected by congestion on the production network.

- Fibre Channel SANs have inherent distance limitations, which can be overcome with special gateway devices. NAS, on the other hand, relies on standard networking technology and IP protocols found in nearly all LAN and WAN devices.

NAS and SAN devices can work with each other in a given environment. NAS elements can be attached to SANs just as other storage devices, allowing SANs to serve as back-end storage infrastructures for NAS systems. Figure 10.17 illustrates this concept. NAS devices allow the attachment of external storage and have Fibre Channel ports that enable the NAS file system to connect to a SAN device. For example, several NAS devices storing databases can be networked together through a SAN, while the NAS devices allow users to access the SAN over a LAN. The SAN can be used to backup or mirror of the data, or can even allow application servers direct access to the same storage devices [44].

10.7.3 Enterprise SANs

Enterprise SANs (E-SANs) is a storage concept that integrates various storage components, including different SANs, across an enterprise. E-SANs are designed to link individual SANs, DASDs, and NAS devices such that a centralized data pool for the entire enterprise results. Figure 10.18 illustrates this concept.

The pool can be accessed by enterprise LANs or WANs from all corporate and branch locations. E-SANs are created with a high degree of fault tolerance. E-SANs are designed with multiple objectives that typically include [34, 45, 46]:

- Support storage that is centralized and accessed by many heterogeneous servers from different locations;
- Allow the addition of newer storage and servers on line without service interruption;
- Connect storage devices using wide area connectivity;
- Have intelligent, automated, and centralized storage management;
- Enable choice of connectivity (IP, Fibre Channel, ESCON, and SCSI);
- Allow high levels of data protection, security, and availability;
- Provide scalability;

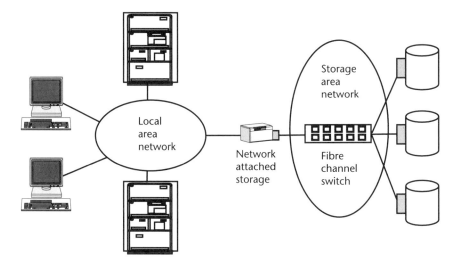

Figure 10.17   Interworking NAS with SAN.

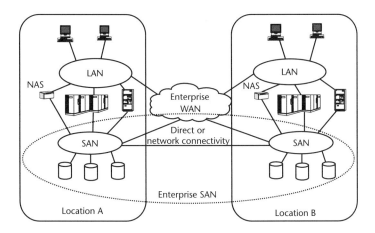

Figure 10.18   Enterprise SAN.

- Use open interfaces wherever possible;
- Consist of well-integrated systems and components.

10.7.4   IP Storage

From the previous section, it is evident that NAS can be used as a cost-effective way to provide data access over a traditional IP network. Several approaches to transmitting storage data over IP networks are now being developed. Moving stored data over an IP network requires setting up virtual pipes between the storage devices across the network. The devices would have to manage connections for data transfer and convert data frames to and from IP packets. Although IP, as a layer 3 protocol, can efficiently route packets across networks, TCP, as a layer 4 protocol, is not designed to ensure efficient transmission of large blocks of data typically seen in volume data transfers. A layer 4 protocol that can replace or enhance TCP to efficiently handle large-volume data transmissions it still required.

Devices accessing data through a SAN often assume that the storage device is local. If the data is shipped using IP packets traveling over wide areas, any latency between the local and distant storage systems can affect data coherency between the two, as discussed earlier with respect to mirroring. An outage at either site can thus lead to lost transactions.

A class of solutions referred to as *on-line storage* uses standalone network applications and Internet-based services to enable storage transfer over IP [47]. The most popular application for on-line storage is allowing remote or traveling workers to back up their local storage to a corporate backup server over the Internet. To cut down on the amount of data to be transferred over the IP network without taxing the abilities of layer 4, most on-line storage products perform single-instance storage. This replicates only the changes made to a user's files, similar to incremental and differential backups. Changes are typically at the block or byte level.

This technique, however, requires that a full backup image of the user's files be first created on the on-line backup storage device. This can be quite time consuming if a large quantity of data is involved and only slow communication links are available. On-line storage vendors do provide compression methods to get around this.

They also keep track of how much data has been stored in the event a backup is not completed because of a dropped connection, and they can resume the backup when the user reconnects. Some use agents that begin sending data whenever an Internet connection takes place. Most products encrypt data for added security.

If on-line storage is conducted over the Internet, then the transfer is subjected to the Internet's instability and unreliability. On-line storage can also be expensive if used to back up all users on a network. Because most users find it easier to back up all of their files, backup server disk space can be heavily consumed and inefficiently used.

Another problem that can arise is the treatment of open files during backup. If the on-line storage application uses a single volume or database to store all of the files, it may be frequently open and in use if large numbers of remote users are making backups.

Another storage application involving IP is Fibre Channel over IP (FCIP), sometimes referred to as *storage tunneling*. This involves encapsulating Fibre Channel control codes and data within IP packets. Encapsulating packets can be inefficient. FCIP can help overcome interoperability issues between different vendor SANs. FCIP can be an effective way of connecting geographically distributed Fibre Channel SANs using IP over a WAN or the Internet. An added advantage is that the I/O and SAN management functions are done over a single network. An alternative implementation to FCIP is Internet Fibre Channel protocol (iFCP). iFCP maps Fibre Channel commands and addresses over IP, rather than using the encapsulation approach, as in FCIP [38]. This enables leveraging the inherent capabilities of IP.

Another evolving IP storage approach is using SCSI in conjunction with IP networks. IP SCSI (iSCSI) encapsulates SCSI commands for transport over an IP network. In systems using iSCSI, storage commands and data to and from a SCSI storage device are encapsulated in IP packets that are then passed to software or to an iSCSI NIC card or HBA. An HBA appears to a host system as a SCSI storage device and to network as a NIC with an IP address. iSCSI requires the use of iSCSI adapters installed in both the server and the storage devices. Figure 10.19 illustrates the use of iSCSI.

Alternate pathing or load balancing usually requires two adapters in a server to balance the network traffic and provide failover capabilities if one adapters fails. The iSCSI adapter cannot be used to perform alternate pathing or load balancing with adapters residing in the same server. In some products, they cannot be assigned to the same domain as the network adapters, creating the potential for domain management conflicts.

iSCSI adapters use standard IP addressing and commercial OS stacks. For this reason, iSCSI installs, manages, and troubleshoots in the same manner as Ethernet and leverages use of the LAN. iSCSI allows different types of storage devices to be connected to create a smaller localized SAN using common Ethernet infrastructure. Compatibility with IP enables the creation of geographically dispersed SANs that can communicate over IP networks.

A key issue with respect to networking multiple Fibre Channel SANs across an IP network is with respect to disruptions in the IP network. In cases where connections between Fibre Channel switches are broken, the individual SANs can become separated. They then must rely on their own autonomous network addressing and

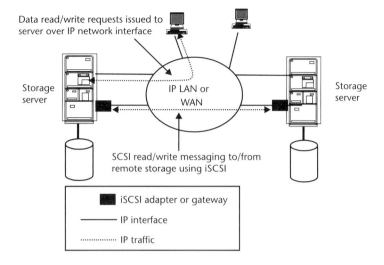

Figure 10.19 iSCSI example.

thus might have to reestablish their own address space. This can be problematic if address conflicts result between the different SANs [48].

10.8 Storage Operations and Management

A successful mission-critical storage operation entails the use of a cost-effective management approach. Consolidation and centralization of data and storage is a popular industry trend. Enterprise data consolidation can potentially increase the amount of data that can be managed and lower the total cost of storage. Although this may seem rational, in some cases it can be impractical or extremely challenging to implement. Many enterprise operations are decentralized and need to remain that way for business purposes. Furthermore, an underlying mission-critical tenet is to maintain multiple replicas of data images used throughout the enterprise in separate locations so that they can be restored when needed.

Storage operations and management entails many activities, including disk backups, database reorganizations, disk defragmentations, and database planning. Many vendors offer automated software tools and systems to assist in these activities, enabling isolated storage elements and networks to be connected and managed. The following are some general characteristics of a well-developed storage-management system [37]:

- *Multiple data replica management.* A storage management system should be able to effectively manage multiple replicas of the same data image. In a mission-critical environment, at least two replicas of the same data image are required: one on a server (or storage device attached to the production server) and one backup copy. If mirroring is used, then additional replicas of the image is required, particularly if vaulting or SSP services are used for archiving.

- *Desktop/mobile data management.* A storage management system should be able to manage mobile and desktop images. Quite often, a significant portion of a company's critical data is stored on the hard drives of laptops and desktops. Automating on-line backup solutions that were earlier described can be an effective way for organizations to protect critical data maintained by mobile users. These solutions initiate backups automatically and transparently to the user. These solutions could also be effective for small organizations that do not have the sufficient IT resources to perform rigorous backups.

- *Cross-platform support.* Several backup approaches have been discussed earlier in the chapter. Regardless of the specific approach, using a single cross-platform backup system could be very cost effective. The system should be compatible with a wide range of storage platforms and be able to evolve independently.

- *Data location.* Storage-management systems must be able to easily and quickly locate critical data whether on site or off site. In mission-critical environments, identifying the location of data anywhere in the enterprise, both physically and logically, is mandatory.

- *Maintain data integrity.* A storage-management system should be able to inspect and validate the integrity of mirrored or backup data anywhere in the enterprise. It should also provide access protection and encryption mechanisms for security purposes.

- *Optimize storage capacity.* A storage-management system should be able to cope with massive increases in data growth. It should also support archiving and be able to remove inactive records from files and databases so that they are kept to a manageable size. It should allow capacity reallocation, monitoring, and tuning to optimize performance. Storage could be reallocated from one device to another, without physically moving storage, depending on need.

- *Reporting:* A storage-management system should be able to provide information used to monitor usage and estimate growth rates, so that storage upgrade decisions can be made. It should also report the status of all storage network components.

10.8.1 Hierarchical Storage Management

Hierarchical storage management (HSM) systems provide a means to deal with many of the aforementioned management issues. HSMs are software and hardware systems that manage the migration of data across many devices [49]. HSM software can compress and move files from an operational disk drive to slower, less expensive media for longer term storage. It bases such transfers on rules and observations of file activity and data usage. For example, the newest data, or the data with the most chance of being recalled, will reside on faster devices, while older data is transferred to slower and cheaper media. A key HSM function is the ability to locate and retrieve archived data when needed. An HSM should perform management functions in a fashion that is entirely transparent to users.

An HSM system is comprised of hardware and software components. In addition to a computing platform, other hardware components might be included, depending on the vendor. These may include different types of automated and robotic devices to retrieve media. The supported media could include magnetic disks, tapes, cartridges, CDs, and optical disks. An HSM is most effective in environments where large volumes of data are involved, such as mainframe environments. It keeps large volumes of data accessible to users while optimizing use of storage capacity.

Deciding which files to move can involve different criteria, depending on need. For example, if the amount of stored transaction data reaches a specified threshold, the HSM can transfer the least active files (or records) to an HSM system. From there, the data is transferred to a secondary device, such as an optical library. If storage space on the secondary device is approaching capacity, then the HSM could transfer files to less expensive tertiary media, such as tape. Once data is transferred, the HSM software generates a placeholder file for each file that is migrated. The placeholder works as a shortcut to the stored file. When the placeholder is accessed, the HSM application retrieves the actual stored file. Depending on the file size and location, some delay may be encountered during the process of file retrieval.

10.8.2 SAN Management

Despite the many merits of SANs that were described earlier in this chapter, the storage hardware comprising a SAN is only half of the equation. The flexibility offered by SANs fosters added complexity with respect to management. It remains difficult to mix the best and most cost-effective SAN devices and still have them function as a single, easily managed pool. SAN management software is required to effectively integrate the operation and management of these devices and provide a centralized view of all of the storage pools involved. A sound SAN management system should address both the physical and logical elements of a SAN.

We previously noted some key desirable storage management system characteristics that were general in nature. In addition to these general management functions, the following are some specific capabilities that should be required of a SAN management system [37]:

- *Sharing control.* SAN management should manage the physical sharing and partitioning of disk or tape subsystem hardware for simultaneous access by multiple servers. It should also address file-level sharing and locking between different types of platforms so that multiple servers can access the same data simultaneously. This eliminates the need for having unnecessary replicas of data to manage.
- *Device zoning.* Zoning is an approach used to manage the hardware in a SAN. A zone is the logical isolation of a port or a SAN-attached device. This allows devices to communicate with each other only if they are assigned to the same zone. Zones can be defined at the port level and LUN level. Devices are assigned to a zone based on their LUN. A port zone segments devices based on the ports used to connect to a hub or switch. Only nodes attached through ports sharing a common zone can communicate [30].
- *Device pooling.* SAN managers can group logical devices together using LUNs to create virtual storage pools [30]. Pools can be comprised of volumes, disk and tape drives, and even RAID. For example, devices from an array can be

pooled with drives in another location and appear to the end user as a single drive. Multiple drives can be configured within a tape library to appear as a single drive, making it possible to allocate them to different departments. Dynamic changes can be made to the size and configuration of these storage pools without having to reboot servers.

- *Security*. The SAN management system should ensure that devices communicate only with those that are in their assigned zone. This reduces the potential for data corruption or deletions caused by accidental changes or replications. Reporting should be provided to alert personnel about any such errors. As SANs incorporate IP addressing, storage devices and networks will be able to incorporate security measures sufficient to defend against a myriad of threats. Strong authentication, data origin verification, and data integrity and confidentiality are issues now under consideration by the standards community.

- *Cross-platform integration*. SANs are comprised of a variety of platforms from different vendors, each having a their own element management systems. The SAN management system should allow interoperability among these disparate systems. It should also allow using existing IT infrastructure and other network management systems in use within the enterprise. The software should support the discovery and monitoring of HBAs, hubs, switches, routers, disk arrays, and other hardware components.

- *Application integration support*. A SAN management system should enable the dynamic reconfiguration of storage based on the needs of host applications. It should have the ability to automatically manage how applications use storage so that priority is given to mission-critical applications.

- *Scalability*. The SAN management system should be scalable throughout the SAN and the enterprise. SAN management system elements will likely be distributed among different locations in a large geographically dispersed enterprise. The system architecture must be able to grow cost effectively across all the SAN locations without disruption.

- *Capacity monitoring*. A SAN management system should monitor storage resources and capacity utilization. Given that many such solutions will be customized at first, the storage industry is favoring the use of SNMP to provide compatibility among diverse solutions. This will involve the definition of a standard Fibre Channel management integration (FCMGMT-INT), a simple network management protocol (SNMP) management information base (MIB) that will establish a common base of parameters that characterize the status and performance of network elements [50]. This enables SAN management information transfers to take place outside the SAN through an IP-based network. Use of IP also enables integration with other network management platforms.

The MIB will ultimately enable a number of features. It will permit autodiscovery of components and subcomponents and allow a network management station to retrieve system and environmental information. With this information, network management can compile real time and historical statistics on the health and performance of components and their subcomponents. Many systems will be able to generate graphical displays of SAN topology, statistics, and color-coded alarms. In addition to event monitoring and performance reporting, a SAN management system should also correlate error information and execute corrective actions based on prespecified policies.

Event monitoring is accomplished by sending and receiving information from SNMP agents residing within various SAN devices. These are specialized software processes that report information from the MIB. Agents will send information to the main network management system based on requests and event traps. For example, a disk controller agent will monitor disk space use, based on predefined rules. If space exceeds a predetermined threshold, the agent will record the event in a log for user viewing and reporting. When a receiving agent obtains the message, it determines which rule to invoke in response to that event. Each rule will dictate the corrective action to perform and log. So, for example, in response to a disk-full event, the agent might delete or move unnecessary files to increase disk space. The types of files to be removed are dictated by other predefined rules.

10.8.2.1   *Virtualization*

Storage virtualization is a technique that creates an abstract view of a physical storage environment. It acts as a layer between physical storage and applications. In this layer, storage media can be logically joined together and viewed as a single entity. Software can access the contents on the entity without caring about where the content is physically stored. This logical view hides the physical SAN complexity, making it more cost effective to manage storage. It can be an efficient method to centrally manage storage in large, complex IT environments, where storage is spread across a variety of systems [51].

Virtualization is not a new concept. IBM's virtual machine (VM) concept, developed in the 1970s, allowed users to define *virtual machines*, which included virtual disk devices. Storage virtualization extends this concept. Different physical devices can be pooled together into one virtual device. Storage capacity can be allocated among servers according to business needs, rather than physical capacity limitations. A virtual disk volume can comprise many physical drives. Conversely, large RAID arrays could be made to look like many smaller virtual disks that can be shared by a wide variety of servers, making better use of hardware investment.

Virtualization simplifies SAN manageability by creating a single view of all SAN components. Many disperse storage pools can be tied together in a consolidated view for better management and utilization. Figure 10.20 illustrates several approaches to implementing virtualization: on the host computer or server, on an appliance, or on the storage array. Within these approaches, virtualization processes can be performed in band (within the data path) or out of band (outside the data path). Inband virtualization, also referred to as network-based virtualization, uses a device with virtualization intelligence that sits in the data path between a server and storage device. Although server processing is unaffected, this approach can introduce latency and throughput problems because the inband appliance can affect data transfer and create a bottleneck [51, 52].

In the out-of-band approach, intelligence is housed in a host server or storage device (typically arrays). Many of the early product offerings were out-of-band implementations. Server-based virtualization offers better scalability, as it is

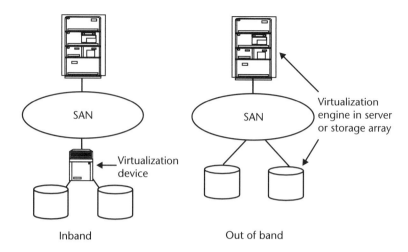

Figure 10.20   Storage virtualization techniques.

divorced from the storage devices, but can add additional processing overhead to a server. Because storage-based virtualization places more of the intelligence on the storage device, it is viewed more favorably.

Virtualization products available as of this writing have yet to live up to the expectation of grouping all storage devices and hosts under a scalable and open framework. Some vendors virtualize only storage devices of their own making, while others virtualize a variety of devices. Other vendors virtualize storage and incorporate mirroring, replication, and backup functions. Some vendors are offering hybrid approaches using out-of-band devices that do not affect the data transfer. Overall, storage virtualization can provide several distinct benefits with respect to mission-critical environments:

- Data redundancy for mission-critical needs can be more easily managed across different storage platforms. The reduction in management complexity also improves the ability to respond to and correct storage network problems. Greater flexibility and simplicity in organizing storage makes it easier to meet application data requirements.
- Administrative efficiencies provided with virtualization increases the amount of storage resources that can be managed by a firm. It also improves the ability to grow a network in a nondisruptive fashion. Devices can be added, upgraded, removed, or reassigned within a virtualized pool without bringing down the rest of the storage system.
- Easier storage provisioning makes more capacity available to users upon request. Storage utilization can be improved by pooling multiple resources such as disk arrays to appear as a single logical resource to users.
- Hardware utilization is improved, as hardware from different vendors can be pooled. This avoids single vendor implementations and the need to purchase extra high-end hardware.

There are, however, several caveats associated with storage virtualization [51]:

- Virtualization can add complexity to the data recovery process. Because the virtualization layer must track data across the disks and partitions that comprise a virtual volume, there is uncertainty as to whether data can be restored correctly and quickly in the event of a device failure. For example, if a physical device supporting multiple pools fails, then all of the applications feeding off of those pools can fail, unless the pools are mirrored or backed up.

- Inband virtualization can pose performance issues because data runs directly through an inband appliance, which can become a bottleneck. Although out-of-band virtualization can be an alternative, a server-based virtualization solution can still pose performance issues on the server.

- There is some controversy regarding application awareness of virtualization in the storage network. One argument for physical device transparency to an application is that it makes application development and migration much easier. On the other hand, there may be instances where an application will need to take advantage of the specific capabilities of a device. For example, a backup application may be able to take advantage of some built-in components of the storage device to increase the efficiency of the operation.

10.8.3   Data Restoration/Recovery

Until now, much discussion has been devoted to creating and managing replicas of data for mission-critical reasons. Backup and mirroring are well-known tactics, but data restoration and recovery can be far more challenging. Data recovery is a process that involves restoring image copies of data and logs to disk and executing forward recovery utilities to apply updates to the image copies. Data restoration is just one component of an overall recovery. Recovery involves the restart of operations from a known PIT where data has guaranteed integrity. Many business applications can tolerate a loss of some transactions, but usually not corrupted data. A proper data recovery strategy should achieve the company's recovery objectives. It should meet the specified RPO and RTO objectives and restore data in a nondisruptive fashion. In the long term, a good data-recovery scheme should reduce service disruptions due to data errors and associated resource expenditures and costs.

There are many types of incidents that may require restoration of data. Common examples include software error, communication network problems, loss of a storage subsystem (e.g., a component fails on a storage device) or media failures. The use of high availability or fault-tolerant storage platforms can help isolate platform problems from data operations. For example, if a platform supports hot swapping, then a failed component can be replaced without having to take the device out of service. Data and media failures are failures that affect the data stored in data files. Although less common, they can be destructive when they occur. These types of failures are not easily recoverable and often require some human intervention.

Many high-availability database products used on RAID devices provide features to automatically recover from a storage device failure, effectively masking the problem. The software places the affected data files off-line until recovery completes. The RAID subsystem then automatically recreates the lost data using its built-in redundancy.

In many restoration scenarios, multiple storage platforms and file images will be involved. Data recovery should be as automated as possible so that it can be invoked following an unwanted event. Keeping backups on-line in a tape changer or library aids restoration when an important file is accidentally deleted or corrupted. However, backups located in the same location as the primary data image could be vulnerable to the same physical disaster. Storing backups off site might be the better solution, as long as the main data center or recovery center has immediate access to the backup image. As was discussed earlier, a hot site can be retained in a standby or mirrored state. If in a mirrored state, it is likely to be continually used to share production load and/or to maintain a mirrored database image. A reciprocating backup/recovery arrangement can be made between the primary and mirrored sites. Mirrored sites should retain a backup of their data, in the event data is corrupted. This backup could also be used to recover the primary site's image as well. The primary site could reciprocate in the same manner.

The reciprocating relationship between the primary site and mirrored site mandates reliable communication between the sites. Even if a mirrored site is not involved, retaining a backup image copy at the primary site and a copy off site and far away from the primary site is suggested. Having a backup image at the primary site can safeguard against problems that might occur in the event of communication or ground transport disruptions. Prestaging and testing of the recovery configuration can reduce recovery time.

10.8.3.1   Point-in-Time Recovery

At the very first sign of a software fault or error, the primary data image and secondary data image (if data is mirrored) should be frozen. Processing might continue beyond this state, resulting in unintended changes to the data files by incorrect operations. For example, an update transaction could be run against the wrong rows, or a table could be dropped or truncated. This could in turn affect other applications and databases, creating a rolling disaster. Rolling the database forward to the current time will reintroduce the errors, which can be stored in log files as well as data files [53].

In this scenario, the most effective approach to recover the primary data image is to use a PIT copy. This is a copy that is a *snapshot* (also referred to as a *checkpoint* copy) of the data at a particular PIT. This allows rolling a backup data file to a predetermined PIT just before the error was introduced (sometimes referred to as a *roll forward* recovery), capturing the transactions that took place from the last backup to the point of failure. PIT recovery can be performed on an entire database or on a portion of it. However, it still might be quite difficult to determine the exact time when bad data began to be created. The gap between the PIT of the error and PIT of the restoration copy will determine the amount of lost or invalid transactions. The time interval between the PIT of the restoration copy and the PIT of the snapshot should be within the required RPO window. Figure 10.21 illustrates this concept.

PIT recovery requires generating a backup copy at specified intervals and recreating updates from the PIT of the last backup to the PIT just before the error occurred. Rolling a database forward from a backup will require the replay of many operations, often recorded in archived redo logs. The older the PIT of the backup, the more work will be required to replay logs to roll the data forward to the current

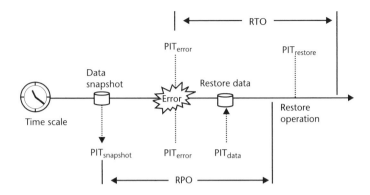

Figure 10.21   Data recovery timeline.

PIT. Also, if many associated databases are involved, they all must be recovered to a consistent PIT [54].

Tools are available on the market to easily create snapshots of a database at periodic intervals. The database must be placed in hot backup mode, so that recovery using the snapshot can be instantaneous. Periodic snapshots can lag production data up to a few hours, depending upon the frequency of snapshots. A mirrored database can be used for recovery, but if data is corrupted, the mirrored data is of no use. An alternative is to use a standby backup of the production database. This is a backup database that is kept in synch with the primary image by periodically applying archived redo log files. The standby continually rolls its data forward as new logs are applied after a log is changed. This can be an easier way to achieve a smaller time lag. If the time lag cannot be tolerated, multiple staggered backups can be maintained, each lagging the production by a different amount of time.

PIT data images can also be used to run recovery tests on either the primary or mirrored data images without risk, as testing is done on a replica of these images. This enables more frequent and thorough recovery testing without jeopardizing operations.

10.8.3.2   Time to Data

The time required to restore data is often referred to as the time to data (TTD). This time encompasses all activity required to bring a database back into operation. It includes the following tasks:

- *The time to retrieve the backup data.* This time is minimized if the backup media is readily available. For example, if the media is a tape stored off site, then this time includes obtaining the tapes from the off-site storage and transporting them to the recovery site. If the tape is housed in a robotic library, the time to select and place it in the drive should be included.
- *Time to transfer the data from the backup media to the primary storage media.* This time will vary depending on the media involved and can be challenging to estimate. For example, in a transfer from tape, the speed at which the tape drive can read data from tape is a significant factor. Another factor to be considered is the speed at which the drive's interface (e.g., SCSI or Fibre

Channel) can transfer the data to another device. Some devices use parallel data streams to help speed the process. If tape data is being sent to a storage device directly, the speed at which that device can record the data stream is important [11].

• *Restore horizon.* This is the time to recover files or records required for operation and verify their integrity. A factor to consider in estimating this time interval is the extent that a server is involved in the restoration process. In many cases, a software application to manage the data restoration is required and is usually executed on a host server. The server can impose additional latency in the TTD through its OS, file system, and logging processes.

10.8.3.3 Data Recovery Process

The prerequisites to a sound data recovery process include many of the things that have been discussed in this chapter to ensure that data can be readily accessed at the time of recovery. Data and applications should be backed up, and archive logs should be maintained. Tape or optical libraries and recovery software should be made available. Recovery software should be able to manage the backup and recovery process, specify and maintain backup policies, and catalog all backup and recovery activities. An automated recovery mechanism is recommended. Whether automated or manual, a data recovery process should involve these basic steps [53]:

1. *Failover/restart to a replica of the primary data image.* If an error is encountered and the primary storage device is affected, failover to another working device should be made. If the nature of the outage is such that no data corruption occurred, then applications can be restarted, beginning where they left off. The application could recover the image, as it existed prior to the failure. If the data is corrupted, it should be frozen. If a mirrored image exists, it should be frozen as well.

2. *Identify the data to be recovered.* If possible, effort should be made to narrow down what data actually must be restored. This could be an entire volume, database, and set of files or portions of a file. The goal is to reduce the extent of the recovery effort.

3. *Obtain the appropriate backup for restoration.* The backup data and media to be used for recovery should be located and placed on the proper device for recovery. The data can include an entire replica of the primary image, differential/incremental backups, transaction journals, redo logs, and any other items needed to restore the primary data image.

4. *Transfer backup data.* Data from the backup replica is transferred to the primary storage device for production. The transfer can take place over a storage network or even a WAN.

5. *Reconfigure application software.* In order to restore operations with the backup data, application programs should be adjusted accordingly, if necessary, to use the new image of the data. This may not necessarily involve code changes but may involve reconfiguration. This process is sometimes referred to as *application revectoring.*

6. *Restore the primary image to the latest possible time.* The primary image of the production data should be restored. This involves recreating or

recovering any lost or damaged data, applying any necessary updates to roll up to the latest time possible. This process may even include manual reconstruction of the latest transactions.

An alternative approach is to take the damaged primary image and undo the inadvertent changes, rather than rolling the database up to a specific PIT. This requires knowledge of what was changed in order to undo the changes. Some database management software products maintain log files to help remove erroneous file changes. These capabilities do not work for all types of errors, but can be useful in many situations.

One final note: file protection through encryption and confidentiality is an increasing priority in many companies. Although encryption can improve security, it can cause problems for backup software. Some programs simply transfer the raw read and write operations of the data in encrypted form, without disturbing the security of encrypted data. However, if portions of a database or file system need restored, decryption may be required [15].

10.9   Summary and Conclusions

Backing up data is no longer sufficient in today's real-time business environment. Other forms of data protection are required in order to recover the most recent data images following an outage. At a minimum, duplicate images must be created at locations distinct from the primary data source. Updating such images typically involves either a software or hardware replication strategy. Data mirroring has become a popular replication strategy for mission-critical operations. Although mirroring provides excellent data recovery following an outage, it may not necessarily protect against data corruption caused by erroneous data writes. Because mirroring can be expensive, it is most cost effective when used to protect the most critical data.

Data replication can be conducted in couple of ways: synchronous data transfers involve writing data to a primary and secondary site at the same time; asynchronous replication involves writing data to a primary site and then applying the same write to a secondary site at a later time. Asynchronous data transfer is less expensive to implement but may not guarantee the level of protection that synchronous transfer offers. A lower cost alternative to these approaches is journaling, which involves intercepting write operations, recording them to a log, and then transmitting the writes to a secondary site.

Backing up data involves making a copy on a storage medium that is long lived and putting it away for safekeeping and quick retrieval. The tapes, disks, or other media that contain the backup set should always be stored at a completely separate location. Backups should be performed on a regular basis and automated if possible. There are several classes of backup strategies that were reviewed in this chapter: full, incremental, and differential backups.

Storage systems have evolved from the simple hard drive to multidisk RAID systems. RAID spreads data across a logical disk array. The use of multiple drives provides better performance, as many individual reads/writes can be done simultaneously versus using one large drive. The multiple drives enable RAID systems to instantaneously recover from read/write and drive failures.

Magnetic tape has long been a staple for data backup for many years. Advancements in tape technology have given rise to tape formats that can store up to 100 GB of data in a single cartridge. However, advancements in disk technology have resulted in disk alternatives that enable faster and less tedious backup and recovery operations.

Storage vault services have been in use for some time. They involve using a service provider to pick up data backups from a customer's site and transport them to a secure facility on a regular basis. Electronic vaulting goes a step further by mirroring or backing up data over a network to a remote facility. Firms can also elect to outsource the management of their entire data operations to SSPs.

A storage network is dedicated to connecting storage devices to one another. A SAN is a network that is separate from a primary production LAN. A SAN provides connectivity among storage devices and servers. SANs are not intrinsically fault tolerant and require careful network planning and design. But they do improve scalability and availability by making it easier for applications to share data. They also offer a cost-effective way to implement data mirroring and backup mechanisms. Fibre Channel technology has found widespread use in SANs. Fibre Channel transfers files in large blocks without a lot of overhead. Point to point, FC-AL, and FC-SW are common Fibre Channel SAN architectures.

Rather than implementing entirely new networking technologies to accommodate storage, alternative strategies are being devised that can leverage more conventional legacy technologies. A NAS is a storage device that attaches directly to a LAN, instead of attaching to a host server. This enables users to access data from all types of platforms over an existing network. NAS can also be used in conjunction with SANs. There are also several approaches to further leverage legacy IP networks for block storage transport. FCIP, iFCP, and iSCSI transfer Fibre Channel commands over IP. iSCSI, in particular, is of great interest, as it leverages the installed base of SCSI storage systems.

SANs enable isolated storage elements to be connected and managed. Many vendors offer automated software tools and systems to assist in this management. HSMs are software and hardware systems that manage data across many devices. They compress and move files from operational disk drives to slower, less expensive media for longer term storage. Virtualization is a technique that abstracts the data view from that of the physical storage device. This can be most appealing to organizations having large, complex IT environments because it offers a way to centrally manage data and make better use of storage resources. Data redundancy for mission-critical needs can be more easily managed across different storage platforms.

Data restoration is just one component of an overall recovery operation. A most effective approach to recover primary data images is to make snapshot copies of the data at specified time intervals and restore from the most recent snapshot. Recovery software and systems can automate the recovery process. Regardless of the approach, there are several basic steps to data recovery that should be followed—these were reviewed earlier in this chapter.

Networked storage and storage management applications provide the ability to efficiently archive and restore information, while improving scalability, redundancy, and diversity. Not only do such capabilities aid survivability, they improve accountability and auditing during an era when organizations are being held liable for information and, depending upon the industry, must comply with regulatory requirements [55].

References

[1]   Kirkpatrick, H. L., "A Business Perspective: Continuous Availability of VSAM Application Data," *Enterprise Systems Journal*, October 1999, pp. 57–61.

[2]   Golick, J., "Distributed Data Replication," *Network Magazine*, December 1999, pp. 60–64.

[3]   Collar, R. A., "Data Replication Is the Key to Business Continuity (Part 2)," *Disaster Recovery Journal*, Summer 2001, pp. 64–65.

[4]   Bruhahn, B. R., "Continuous Availability…A Reflection on Mirroring," *Disaster Recovery Journal*, Summer 2001, pp. 66–70.

[5]   Dhondy, N. R., "Round 'Em Up: Geographically Dispersed Parallel Sysplex," *Enterprise Systems Journal*, September 2000, pp. 25–32.

[6]   Adcock, J., "High Availability for Windows NT," *Enterprise Systems Journal*, July 1999, pp. 46–51.

[7]   Gordon, C., "High Noon—Backup and Recovery: What Works, What Doesn't and Why," *Enterprise Systems Journal*, September 2000, pp. 42, 46–48.

[8]   Flesher, T., "Special Challenges Over Extended Distance," *Disaster Recovery Journal*, Winter 2002, pp. 54–58.

[9]   Mikkelsen, C., "Disaster Recovery Using Real-Time Disk Data Copy," *Enterprise Systems Journal*, November 1998, pp. 34–39.

[10] Talon, M., "Achieve the Level of Disaster Recovery that the Enterprise Requires," *Tech Republic*, January 25, 2002, www.techrepublic.com.

[11] Toigo, J. W., "Storage Disaster: Will You Recover?" *Network Computing*, March 5, 2001, pp. 39–46.

[12] Elstein, C., "Reliance on Technology: Driving the Change to Advanced Recovery," *Enterprise Systems Journal*, July 1999, pp. 38–40.

[13] Rhee, K., "Learn the Different Types of Data Backups," *Tech Republic*, December 7, 2001, www.techrepublic.com.

[14] Chevere, M., "Dow Chemical Implements Highly Available Solution for SAP Environment," *Disaster Recovery Journal*, Spring 2001, pp. 30–34.

[15] Marks, H., "The Hows and Whens of Tape Backups," *Network Computing*, March 5, 2001, pp. 68, 74–76.

[16] Baltazar, H., "Extreme Backup," *eWeek*, Vol. 19, No. 32, April 15, 2002, pp. 39–41, 43, 45.

[17] Wilson, B., "Think Gambling Only Happens in Casinos? E–Businesses Without Business Continuity Processes Take High Risk Chances Daily," *Disaster Recovery Journal*, Winter 2001, pp. 62–64.

[18] Fletcher, D., "The Full Range of Data Availability," *Computing News & Review*, November 1998, p. 26.

[19] "Using RAID Arrays for Data Recovery," *Tech Republic*, November 6, 2001, www.techrepublic.com.

[20] Rigney, S., "Server Storage: Rely on RAID," *ZD Tech Brief*, Spring 1999, pp. S7–S8.

[21] Fetters, D., "The Emerging Tape Backup Market," *Network Computing*, July 24, 2000, pp. 86–92.

[22] Pannhausen, G., "A Package Deal: Performance Packages Deliver Prime Tape Library Performance," *Enterprise Systems Journal*, November 1999, pp. 52–55.

[23] Rigney, S., "Tape Storage: Doomsday Devices," *ZD Tech Brief*, Spring 1999, pp. S9–S12.

[24]   "Vaulting Provides Disaster Relief," *Communications News*, July 2001, pp. 48–49.

[25]   Murtaugh, Jerry, "Electronic Vaulting Service Improves Recovery Economically," *Disaster Recovery Journal*, Winter 2001, pp. 48–50.

[26]   Edwards, M., "Storage Utilities Make Case for Pay-as-You-Go Service," *Communications News*, August 2000, pp. 110–111.

[27]   Connor, D., "How to Take Data Storage Traffic off the Network," *Network World*, April 10, 2000, p. 30.

[28]   Gilmer, B., "Storage Area Networks," *Broadcast Engineering*, December 2000 pp. 42–44.

[29]   Eaton, S., "The Fibre Channel Infrastructure," *Enterprise Systems Journal*, November 1999, pp. 38–40.

[30]   Fetters, D., "Building a Storage Area Network," *Network Computing*, May 15, 2000, pp. 169–180.

[31]   Toigo, J. W., "Mission: Impossible? Disaster Recovery and Distributed Environments," *Enterprise Systems Journal*, June 1998, pp. 48–52.

[32]   Karve, A., "Lesson 136: Storage Area Networks," *Network Magazine*, November 1, 1999, pp. 28–30.

[33]   Massiglia, P., "New I/O System Possibilities with Fibre Channel," *Computer Technology Review*, April 1998, pp. 52–54.

[34]   Hubbard, D., "The Wide Area E-SAN: The Ultimate Business Continuity Insurance," *Enterprise Systems Journal*, November 1999, pp. 42–43.

[35]   Gilmer, B., "Fibre Channel Storage," *Broadcast Engineering*, June 2001, pp. 38–42.

[36]   Gilmer, B., "Fibre Channel Storage," *Broadcast Engineering*, August 2000, pp. 34–38.

[37]   Fetters, D., "Siren Call of Online Commerce Makes SANs Appealing," *Storage Area Networks, CMP Media Supplement*, May/June 1992, pp. 4SS–22SS.

[38]   Clark, T., "Evolving IP Storage Switches," *Lightwave*, April 2002, pp. 56–63.

[39]   Helland, A., "SONET Provides High Performance SAN Extension," *Network World*, January 7, 2002.

[40]   Jacobs, A., "Vendors Rev InfiniBand Engine," *Network World*, March 4, 2002.

[41]   McIntyre, S., "Demystifying SANs and NAS," *Enterprise Systems Journal*, July 2000, pp. 33–37.

[42]   Baltazar, H., "Deciphering NAS, SAN Storage Wars," *eWeek*, April 9, 2001, p. 26.

[43]   Wilkinson, S., "Network Attached Storage: Plug and Save," *ZD Tech Brief*, Spring 1999, pp. S12–S13.

[44]   Clark, E., "Networked Attached Storage Treads New Turf," *Network Magazine*, July 2002, pp. 38–42.

[45]   Lewis, M., "Creating the Storage Utility—The Ultimate in Enterprise Storage," *Enterprise Systems Journal*, November 1999, pp. 67–72.

[46]   Tsihlis, P., "Networks: How to Avoid the 'SAN Trap'," *Enterprise Systems Journal*, July 2000, pp. 38–41.

[47]   Rigney, S., "On-line Storage: Protecting the Masses," *ZD Tech Brief*, Spring 1999, p. S18.

[48]   Connor, D., "IP Storage NICs May Disappoint," *Network World*, February 25, 2002, pp. 1, 65.

[49]   Wilkinson, S., "Hierarchical Storage: Taking it from the Top," *ZD Tech Brief*, Spring 1999, p. S20.

[50]   Swatik, D., "Easing Management of Storage Devices," *Network World*, March 20, 2000, p. 69.

[51]   Toigo, J. W., "Nice, Neat Storage: The Reality," *Network Computing*, May 27, 2002, pp. 36–45.

[52]   Connor, D., "Every Byte into the Pool," *Network World*, March 11, 2002, pp. 60–64.

[53]   Zaffos, S., and P. Sargeant, "Designing to Restore from Disk," Gartner Research, November 14, 2001.

[54]   Cozzens, D. A., "New SAN Architecture Benefit Business Continuity Re-Harvesting Storage Subsystems Investment for Business Continuity," *Disaster Recovery Journal*, Winter 2001, pp. 72–76.

[55]   Piscitello, D., "The Potential of IP Storage," www.corecom.com.

CHAPTER 11

Continuity Facilities

All network infrastructures must be housed in some kind of physical facility. A mission-critical network facility is one that guarantees continued operation, regardless of prevailing conditions. As part of the physical network topology, the facility's design will often be driven by the firm's business, technology, and application architecture. The business architecture will drive the level of continuity, business functions, and processes that must be supported by the facility. The technology architecture will define the physical requirements of facility. The application architecture will often drive location and placement of service operations and connectivity, as well as the physical facility requirements.

11.1 Enterprise Layout

Any individual facility is a single point of failure. By reducing a firm's dependence on a single location and distributing the organization's facilities across several locations, the likelihood of an adverse event affecting the entire organization is reduced. Geographic diversity in facilities buys time to react when an adverse event occurs, such as a physical disaster. Locations unaffected by the same disaster can provide mutual backup and fully or partially pick up service, reducing the recovery time.

Recognizing this, many firms explore geographic diversity as a protection mechanism. Decentralization avoids concentration of assets information processing systems in a single location. It also implies greater reliance on communication networking. However, a decentralized architecture is more complex and usually more costly to maintain. For this reason, many firms have shifted to centralizing their data center operations to at least two data centers or one data center and a recovery site. In the end, a compromise solution is usually the best approach.

The level organizational centricity will often drive logical and physical network architecture. The number and location of branch offices, for instance, will dictate the type of wide area network (WAN) connectivity with headquarters. Quite often, firms locate their information centers, Internet access, and key operations personnel at a headquarters facility, placing greater availability requirements on the WAN architecture. Quite often, Internet access is centralized at the corporate facility such that all branch office traffic must travel over the WAN for access. On the other hand, firms with extranets that rely heavily on Internet access will likely use a more distributed Internet access approach for greater availability.

11.1.1 Network Layout

Convergence has driven the consolidation of voice, data, and video infrastructure. It has also driven the need for this infrastructure to be scalable so that it can be cost effectively planned and managed over the long term. The prerequisite for this is an understanding of a network's logical topology. A layered hierarchical architecture makes planning, design, and management much easier. The layers that comprise an enterprise network architecture usually consist of the following:

- *Access layer*. This is the where user host systems connect to a network. These include elements such as hubs, switches, routers, and patch panels found in wiring or telecom closets (TCs). A failure of an access layer element usually affects local users. Depending on the availability requirements, these devices may need to be protected for survivability. These devices also see much of the network administrative activity in the form of moves, adds, or changes and are prone to human mishaps.
- *Distribution layer*. This layer aggregates access layer traffic and provides connectivity to the core layer. WAN, campus, and virtual local area network (VLAN) traffic originating in the access layer is distributed among access devices or through a network backbone. At this layer, one will see high-end switching and routing systems, as well as security devices such as firewalls. Because these devices carry greater volumes of traffic, they likely require greater levels of protection within a facility than access devices. They will usually require redundancy in terms of multiple components in multiple locations. Redundant and diverse cable routing between access and distribution switches is necessary. Access-layer devices are often grouped together with their connecting distribution layer devices in *switch blocks*. An access-layer switch on will often connect to a pair of distribution-layer switches, forming a switch block.
- *Core layer*. Switch blocks connect to one another through the core layer. Core layer devices are typically high-density switches that process traffic at very high line speeds. Administrative and security processing of the traffic is less likely to occur at this layer, so as not to affect traffic flow. These devices normally require physical protection because of the amount of traffic they carry and their high-embedded capital cost.

11.1.2 Facility Location

Network layout affects facility location. Locations that carry enterprisewide traffic should be dependable locations. As geographically redundant and diverse routing is necessary for mission-critical networks, locations that are less vulnerable to disasters and that have proximity to telecom services and infrastructure are desirable. The location of branch offices and their proximity to carrier point of presence (POP) sites influences the type of network access services and architecture.

Peripheral to a decentralized facility strategy is the concept of *ruralization*. This entails organizations locating their critical network facilities in separate locations but in areas that are local to a main headquarters. Typically, these are rural areas within the same metropolitan region as the headquarters. The location is far enough away from headquarters so that it is still accessible by ground transportation. This is done for the reason that a disaster will less likely simultaneously damage both locations. If one location is damaged, staff can relocate to the other location. However, this strategy may not necessarily protect against sweeping regional disasters such as hurricanes or earthquakes.

11.1.3   Facility Layout

Compartmentalization of a networking facility is fundamental for keeping it operational in light of a catastrophic event [1]. Many facilities are organized into a central equipment room that houses technical equipment. Sometimes, these rooms are subdivided into areas based on their network functionality. For instance, networking, data center, and Web-hosting systems are located in separate areas. Because these areas are the heart of a mission-critical facility, they should be reinforced so that they can function independently from the rest of the building.

Size and location within a facility are key factors. As many of these rooms are not within public view, they are often not spaciously designed. Such rooms should have adequate spacing between rows of racks so that personnel can safely move equipment without damage. Although information technology (IT) equipment is getting smaller, greater numbers of units are being implemented, resulting in higher density environments. *Swing space* should be allocated to allow new technology equipment to be installed or staged prior to the removal of older equipment. A facility should be prewired for growth and provide enough empty rack space for expansion.

11.2   Cable Plant

A mission-critical network's integrity starts with the cable plant—the cables, connectors, and devices used to tie systems together. Cable plant is often taken for granted and viewed as a less important element of an overall network operation. However, many network problems that are often unsolvable at the systems level are a result of lurking cabling oversights. Quite often, cabling problems are still the most difficult to troubleshoot and solve, and sometimes are irreparable. For these reasons, a well-developed structured cabling plan is mandatory for a mission-critical facility [2]. Mission-critical cabling plant should be designed with the following features:

- Survivability and self healing capabilities, typically in the form of redundancy, diversity, and zoning such that a problem in a cable segment will minimally impact network operation;
- Adequately support the transmission and performance characteristics of the prevailing networking and host system technologies;
- Support easy identification, testing, troubleshooting, and repair of cable problems;
- Easily and cost effectively facilitate moves, adds, and changes without service disruption;
- Support scalable long-term growth.

11.2.1   Cabling Practices

Network cabling throughout a facility should connect through centralized locations, so that different configurations can be created through cross connection, or *patching*. But inside an equipment room, a decentralized approach is required. Physically separating logically clustered servers and connecting them with diversely routed cabling avoids problems that could affect a rack or section of racks. Centralization of cable distribution can increase risks, as it becomes a single point of failure.

Redundant routing of cables involves placing multiple cable runs between locations. Quite often, two redundant host systems will operate in parallel for reliability. However, more often than not, the redundant devices connect to wires or fibers in the same cable, defeating the purpose of the redundancy altogether. Placing extra cables in diverse routes between locations provides redundancy and accommodates future growth or additional backup systems that may be required in an emergency.

This notion revisits the concept of logical versus physical link reliability, discussed earlier in this book. If a cable is damaged and a redundant cable path is available, then the systems on each end must continue to either send traffic on the same logical path over the redundant cable path or redirect traffic on a new logical path on the redundant cable path. In WANs, synchronous optical network (SONET) ring networks have inherent disaster avoidance so that the physical path is switched to a second path, keeping the logical path intact. In a local area network (LAN), a redundant physical path can be used between two devices only after reconvergence of the spanning tree. Newer LAN switching devices have capabilities to logically tie together identical physical links so that they can load share as well as provide redundancy.

11.2.1.1   Carrier Entry

A *service entrance* is the location where a service provider's cable enters a building or property [3]. Because a service provider is only required to provide one service entrance, a redundant second entrance is usually at the expense of the property owner or tenant. Thus, a redundant service entrance must be used carefully in order to fully leverage its protective features and costs. The following are some practices to follow:

- A redundant service entrance should be accompanied by diverse cable paths outside and within the property. A cable path from the property to the carrier's central office (CO) or POP should be completely diverse from the other cable path, never converging at any point. Furthermore, the second path should connect to a CO or POP that is different from that which serves the other path. (See Chapter 7 for different access path configurations.)
- The redundant cable should be in a separate cable sheath or conduit, instead of sharing the same sheath or conduit as the primary cable.
- Circuits, channels, or traffic should be logically split across both physical cables if possible. In a ring network access topology, as in the case of SONET, one logical path will enter through one service entrance and exit through the other.

- Having the secondary access connect to an additional carrier can protect against situations where one access provider might fail while another might survive, as in the case of a disaster.

There should be at least one main cross connect (MC) [4]. The MC should be in close proximity to or collocated with the predominant equipment room. For redundancy or security purposes, selected cable can be passed through the MC to specific locations.

11.2.1.2   Multiple Tenant Unit

Multiple tenant units (MTUs) are typical of high-rise buildings. Figure 11.1 illustrates a typical MTU cable architecture. It is clear that many potential single failure points exist. Vertical backbone cables, referred to as *riser cables*, are run from the MC to a TC on each floor in a star topology. For very tall MTUs, intermediate cross connects (ICs) are inserted between the TC and MC for better manageability. Sometimes, they are situated within a TC. Although risers are predominantly fiber cables, many installations include copper twisted pair and coaxial cable. All stations on each floor connect to a horizontal cross connect (HC) located in the TC [5].

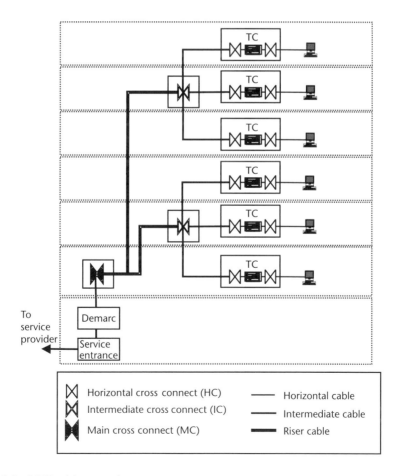

Figure 11.1   MTU cable example.

A mission-critical cable plant (Figure 11.2) should be designed to ensure that a cable cut or equipment damage in any location does not cause loss of connectivity to any TC. Although installing two TCs on a floor, each connecting to a separate riser, can add redundancy, it is most likely that each user station on a floor will be served from one TC [6]. If one of the TCs is damaged, a portion of the floor can remain operational. The TCs on each floor can be tied together through a cable that connects between the HCs [7]. Adding tie cables between TCs produces a mesh-like topology, enabling paths around a damaged IC or TC to be created. If ICs are present, connecting multiple HCs together on a floor can provide a redundant path between an HC and IC.

For further survivability, redundant riser backbones can extend from the two MCs located on different floors to each IC. The two MCs can also connect to each

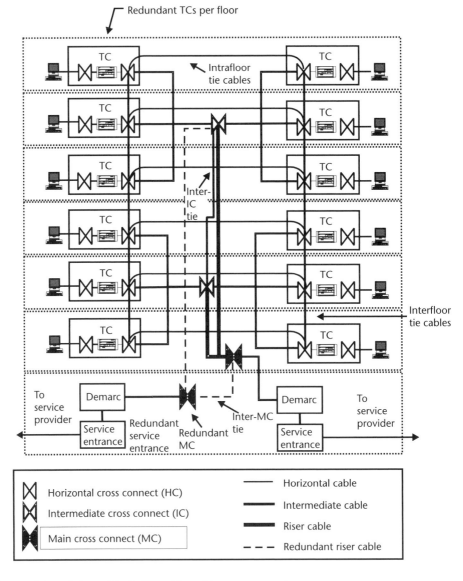

Figure 11.2 Mission-critical MTU cable architecture example.

other for added redundancy. From each MC, a separate riser cable runs to each IC. This creates a ring-like architecture and can thus leverage networking technologies such as SONET or resilient packet ring (RPR). The redundant riser is routed via a separate conduit or shaft for diversity [8]. Although not explicitly shown, a primary and secondary cable run between each cross connect is essential in the event a cable is damaged. Although they aid cable management and reduce riser cable runs and attenuation, particularly in very tall buildings, ICs are highly vulnerable points of failure, because of their collective nature.

The architecture in Figure 11.2 illustrates a highly redundant architecture. Although it may seem like expensive overkill, it is the price for embedding surviv-ability within a mission-critical facility. Variants that are more economical can be used, depending on the level of survivability desired. In MTUs, organizations typi-cally do not operate in all the tenant units. Thus, an organization should include terms in leasing agreements with the landlord or provider to assure that they can implement, verify, and audit the architecture redundancy that is prescribed.

11.2.1.3   Campus

Many of the cabling principles that apply to MTUs also apply to campus networks. Although campus networks can be viewed as flattened versions of the MTU archi-tecture, they must also reliably connect systems across several buildings, as well as systems within each building. A simple campus configuration (Figure 11.3) is typi-cally achieved by running backbone cable from a central MC residing in one build-ing to ICs in every building on campus, in a star-like fashion. Although all endpoints are routed to the MC, logical point-to-point, star, and ring networks can still be supported [9]. For ring topologies, connections are brought to the central MC and

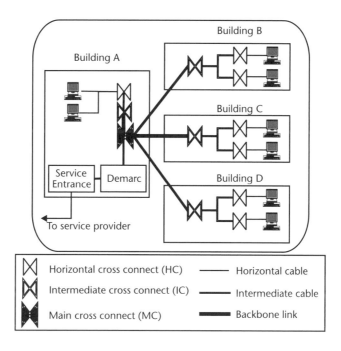

Figure 11.3   Simple campus network design.

routed back to their ICs. The star topology also enables patching a new building into the network from the central MC [10].

The campus network design can be enhanced with features to add survivability, similar to the MTU case (Figure 11.4). Redundant MCs and tie links between ICs add further redundancy in the event a building, cross connect, or cable is damaged. The result is a mesh-like network, which adds versatility for creating point-to-point, multipoint, star, or ring topologies within the campus. Ring networks can be made by routing cable from each IC to the MC (as in the simple campus case), as well as across tie links between ICs.

In campus environments, outdoor cable can be routed underground or on aerial plant, such as poles. Underground cable often resides in conduit that is laid in the ground. The conduit must be sealed and pressurized to keep water out of the pipe and must be sloped so that any water will drain. A major misconception is that underground cable is more reliable than aerial plant. Surprisingly, field experience has shown that underground cable is more vulnerable to damage than aerial plant. Underground cable must often be exhumed for splicing and distribution.

As an option to cable, wireless technologies such as microwave or free-space optics can also be used to establish redundant links between buildings. Because buildings within a campus are usually in close proximity, line of sight is much easier to achieve. For microwave, the closeness also enables higher bandwidth links to be established.

11.2.2 Copper Cable Plant

The ubiquity of copper cable plant still makes it favored for LAN implementations. However, as enterprise networks migrate toward 100-Mbps and 1,000-Mbps speeds,

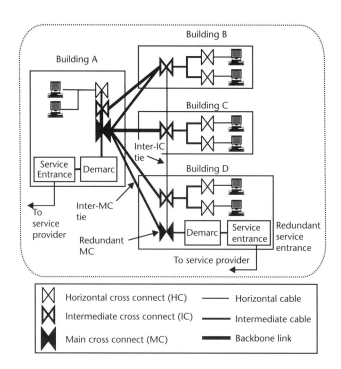

Figure 11.4 Mission-critical campus network design.

the ability of the embedded copper base to adequately support higher LAN speeds remains somewhat questionable. Category 5 systems installed prior to 1995 may encounter problems in supporting Gigabit Ethernet technology. The new Category 5E and Category 6 specifications are designed to supersede Category 5. Category 5E is recommended for any new installations and can support Gigabit Ethernet. Category 6 systems allow greater headroom for cabling deficiencies than Category 5E.

Such deficiencies appear in various forms. Many cabling problems stem from installation in proximity to high-voltage electrical systems and lines. Bad or improperly crimped cables, loose ends, and improperly matched or bad patch cords can create cyclical redundancy check (CRC) errors, making autonegotiation and duplex issues more pronounced. Using voice frequency cables for data transmission can produce signal distortion and degradation.

Noise is unwanted disturbance in network cable and is a common cause of signal degradation. Noise can originate from a variety of sources [11]. There are two fundamental types of noise. *Background noise* is found when there are no devices actively transmitting over a cable. It is attributed to many sources such as thermal noise, semiconductor noise, and even the effects of local electrical equipment. *Signal noise* is noise power resulting from a transmission signal over the cable. The *signal-to-noise ratio* is a measure often used to characterize the level of signal noise with respect to background noise [12]. A high value of this ratio indicates the degree to which audible voice or error-free data will transmit over a cable. The ratio is specified in decibels below 1 mW (dBm).

The industry has arrived at several key test parameters that are often used to certify a copper cable installation and to troubleshoot for some of these problems [13, 14]. The parameters include the following:

- *Attenuation*, sometimes referred to as *loss*, is a problem characteristic in both copper and fiber cable. As illustrated in Figure 11.5, the amplitude of a signal on a medium decreases with distance based on the impedance of the medium. Attenuation is greater at high frequencies than at lower frequencies. Excessive attenuation can make a signal incomprehensible to a receiving device. Attenuation is created by long cable runs, harsh bends, or poor terminations. For these reasons, an Ethernet cable run, for example, must not exceed 90m. In fiber-optic cabling, regenerators must often be used to regenerate the light signal over long distances.

- *Return loss*, which appears as noise or echo, is typically encountered in older cabling plant and is attributed to defective connections and patch panels. It is a measurement of impedance consistency along a wire. Variations in impedance can cause signals to reflect back towards their origin, creating interference, as illustrated in Figure 11.5. Return loss is the summation of all signal energy that is reflected back to the point of signal origin.

- *Cross talk* is a phenomenon that comes in several forms (Figure 11.6) [15, 16]:
  1. *Near-end cross talk* (NEXT) is a condition where electrical signal from one wire leaks onto another wire and induces a signal. NEXT is usually measured at the signal's originating end, where the signal is usually strongest. Contributors to NEXT include defective jacks, plugs, and crossed or crushed wires. NEXT is insufficient for fully characterizing

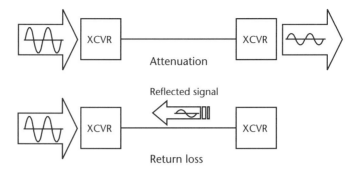

Figure 11.5 Attenuation and return loss.

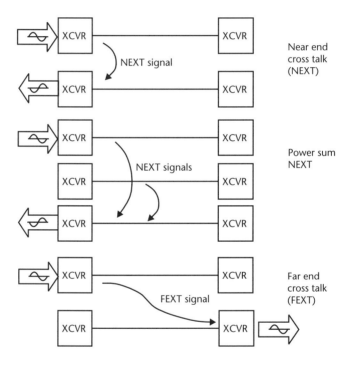

Figure 11.6 Types of cross talk.

performance of cables subject to protocols that simultaneously use multiple twisted pairs of copper wire in a cable, such as a Category 5 cable.

2. *Power-Sum NEXT* (PSNEXT) is a calculated value used to convey the effect of cross talk for protocols that simultaneously use several wire pairs. PSNEXT is used to quantify the relationships between one pair and the other pairs by using pair-by-pair NEXT values. This produces a PSNEXT for each wire pair.

3. *Far-end cross talk* (FEXT) is measure of conditions where an electrical signal on one wire pair induces signal on another pair at the far end of a

link. Because FEXT is a direct function of the length of a link due to the properties of attenuation, it is insufficient to fully characterize a link.

4. *Equal-level far-end cross talk* (ELFEXT) is a calculated value to compensate for the aforementioned length factor by subtracting the attenuation of the disturbing pair from the FEXT value.

5. *Power-sum ELFEXT* (PSELFEXT) is a calculated value designed to more fully describe FEXT characteristics for cables subject to newer protocols that simultaneously use multiple twisted pairs of copper wire in a cable, in the same fashion as PSNEXT.

- *Delay skew* occurs when simultaneous signals sent across several wires are received at different times. Propagation delay is the elapsed time a signal takes to traverse a link. When multiple wires are used to transmit a signal, as in the case of Ethernet, variation in the *nominal propagation velocity* (NPV) of the different wires can make it difficult for receivers to synchronize all of the signals. Delay skew is defined as the range between the highest and lowest propagation delay among a set of wire pairs. Differences in wire insulation among the pairs usually contribute to delay skew.

- *Attenuation–to–cross talk ratio* (ACR) is calculated value that combines the merits of the preceding parameters. It represents the frequency where a signal is drowned out by noise created by NEXT. Because attenuation increases with frequency while NEXT increases irregularly with frequency, ACR is the frequency where the attenuated signal cannot exceed NEXT.

- *Power-sum ACR* (PSACR) is a calculated value used to convey ACR for cables using multiple wire pairs. Both PSACR and ACR are considered the parameters that best characterize a cable's performance.

Outside of poor installation practices, many cable problems are also due to prevailing usage and environmental factors. For example, moisture or chemical exposure can cause corrosion, vibration or temperature extremes can lead to wear, and simple dust can obstruct connector contacts. Higher speed networking technologies such as Gigabit Ethernet are more susceptible to these types of factors. They use weaker high-speed signals that are accompanied by significant noise [17]. For this reason, they employ electronic components that are more sensitive and have tighter tolerances. Thus, usage and environmental factors that may not have been problematic for a 10BaseT network may prove otherwise for a Gigabit Ethernet network.

Upon installation, cabling is passively tested for these parameters using hardware-based tools; the most common is the time-delay reflectometer (TDR). Used for cable and fiber plant (the optical TDR is used for fiber), this device tests for continuity, cable length, and the location of attenuation. Although passive testing is effective, actively testing cable plant with live network traffic can be more definitive in verifying a cable's performance under normal operating conditions.

11.2.3   Fiber-Optic Cable Plant

Optical-fiber cabling, in one form or another, has become a critical component in mission-critical networking. Fiber has found home in long-haul WAN implementations and metropolitan area networks (MANs). As Gigabit Ethernet has been standardized for use with fiber-optic cabling, the use of fiber in conjunction with Ethernet LAN systems has grown more attractive [18]. Yet, many organizations cannot easily justify fiber-optic LAN infrastructure. In addition to cost, the use of fiber-based systems still requires interoperability with legacy copper infrastructure for redundancy if failures occur.

Fiber has advantages over copper in terms of performance. When compared to copper, multimode fiber can support 50 times the bandwidth of Category 5 cable. More importantly, fiber is immune to noise and electromagnetic radiation, making it impervious to many of the aforementioned factors that affect copper. New industry standard fiber V-45 connectors have the look and feel of RJ-45 connectors yielding the same port density as copper-based switches.

Optical TDR (OTDR) devices are used to troubleshoot fiber link problems and certify new installations [19, 20]. Fiber failures do occur due to mishaps in cabling installation and maintenance, much in the same way as copper. Poor polishing, dirty or broken connectors, poor labeling, shattered end faces, poor splicing, and excessive bends are frequently encountered problems. Fiber end face inspection is often difficult to perform, as access to fiber terminating in a patch panel can be tricky and proper inspection requires use of a video microscope. Systems have become available to automatically monitor physical fiber connections using sensors attached to fiber ports.

There are many commonalities between characterizing performance of fiber and copper cables. *Optical noise* can result from amplified spontaneous emissions (ASEs) from Erbium-doped fiber amplifiers (EDFAs). Similar to copper cables, an optical signal-to-noise ratio (OSNR) is used to convey the readability of a received signal. The industry has arrived at several key test parameters that are often used to certify fiber cable installation and to troubleshoot for some of these problems. The parameters include the following [21, 22]:

- *Fiber loss* conveys the attenuation of a light signal over the length of a fiber cable [23]. Simply put, it is how much light is lost as it travels through a fiber. It can be a sign of poorly designed or defective cable. Cable defects can stem from undue cable stress during manufacturing or installation [24].

- *Point loss* is a stepwise increase in loss at some location along a fiber cable. It is usually caused by a twisted, crushed, or pinched fiber.

- *Reflectance* is the amount of light energy that is reflected from a signal source. It usually occurs at mechanical splices and connectors. It is also attributed to air gaps and misalignment between fibers. Reflectance is manifested as a high BER in digital signals. It is analogous to return loss in copper cables.

- *Bandwidth* of a fiber is a function of the properties of the fiber's glass core. It conveys the frequency range of light signals, in megahertz, that can be supported by the fiber.

- *Drift* is variation in light signal power and wavelength resulting from fluctuations in temperature, reflectance, and laser chirp. It conveys the ability of a signal to remain within its acceptable boundary limits.

- *Cross talk* can be manifested in a light signal as the unwanted energy from another signal. In fiber cables, it is often difficult to observe and compute.

- *Four-wave mixing* is the extension of cross talk beyond a given optical signal pair to other signals. The energy can appear in wavelengths in use by other channels.
- *Chromatic dispersion (CD)* results from different wavelengths of light traveling at different speeds within a fiber, resulting in a broadening of a transmission pulse (Figure 11.7) [25]. This causes difficulty for a receiving device to distinguish pulses.
- *Polarization-mode dispersion (PMD)* is caused by modes of light traveling at different speeds. It is typically found in long fiber spans and results from a variety of causes, including fiber defects, improper bending, stress-producing supports, and temperature effects (Figure 11.7).

For fiber, it is important that parameters are measured at those wavelengths at which the fiber will operate, specified in nanometers (nm). They are typically in the range of 850 and 1,300 for multimode fiber and 1,310 and 1,550 nm for single-mode fiber. Fiber should be tested after installation because many of these parameters can change. It is also important to understand the degree of error in any test measures. Error levels can range from 2% to 25%, depending on how the tests are conducted.

11.3   Power Plant

On average, a server system typically encounters over 100 power disturbances a month. Many disturbances go unnoticed until a problem appears in the platform. Studies have shown that about one-third of all data loss is caused by power problems. Power surges and outages, next to storms and floods, are the most frequent cause of service interruptions, particularly telecom service.

Although most outages are less than 5 minutes in duration, they can occur quite frequently and go undetected. Power disturbances usually do not instantaneously damage a system. Many traditional platform problems can be ultimately traced to power line disturbances that occur over a long term, shortening their operational life.

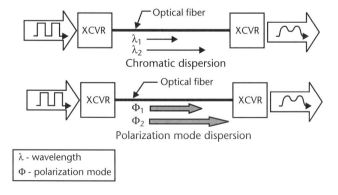

Figure 11.7   Fiber-optic dispersion.

11.3.1 Power Irregularities

Amps (A) are a measurement of current flow and volts (V) are a measure of the amount of work required to move electrons between two points. Power is the amount of current used at a voltage level and hence is the product of volts and current. Power is measured in watts (W) or volt-amps (VA), where 1W = 1.4 VA. Electricity travels in waves that can vary in size, frequency, and shape. Frequency (or cycles), measured in hertz (Hz), is a critical parameter of a power specification. If incorrectly specified, power at the wrong cycle can burn out equipment. Ideally, alternating current (ac) power should be delivered as a smooth sine wave of steady stream current. But this rarely happens due to a host of real-world circumstances.

As electronic equipment grows more sensitive to power irregularities, greater damage and disruption can ultimately result from *dirty power*. A brief discontinuity in power can take systems off-line and cause reboots and reinitialization of convergence processes in networking gear. Some devices will have some *holdup time*, which is the time for an interruption of the input waveform to affect device operation, typically measured in milliseconds [26]. Highly sensitive hardware components such as chips can still be damaged depending on the voltage. Software-controlled machinery and adjustable speed-driven devices such as disk and tape drives can also be affected. Because power travels down cables and through conductors, low-voltage cabling can be affected as well. The following are some of the more frequently encountered power disturbances, illustrated in Figure 11.8:

- *Blackouts* are probably the most noticeable and debilitating of power interruptions. Although they comprise about 10% of power problems, the expected trend is for more frequent blackouts. Inadequate supplies of power to support the growth of data centers and the increased occurrence of sweltering temperatures have contributed to a rising power shortage.

- *Sags or dips* are momentary drops in voltage. Sags of longer duration are often referred to as *brownouts*. They often result from storms and utility transformer overloads. Sags are a major contributor to IT equipment problems.

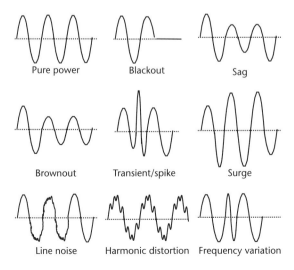

Figure 11.8 Power irregularities.

- *Transients* are high-frequency impulses that appear in the sinusoidal waveform. They are typically caused by a variety of factors. *Spikes,* sometimes referred to as *voltage transient*, are momentary sharp increases in voltage, typically resulting from a variety of sources, including utility companies. Typically, spikes that exceed nominal voltage by about 10% can damage unprotected equipment. Quite often, electrical storms create spikes upon lightning striking power or telecom lines.

- *Surges* or *swells* are voltage increases that last longer than one cycle. They are typically caused by a sudden large load arising from abrupt equipment startup and switchovers.

- *Line noise* can be created by *radio frequency interference (RFI)* and EMI. It can come from machinery and office equipment in the vicinity. It is usually manifested in the form of random fluctuations in voltage.

- *Total harmonic distortion (THD)* is a phenomenon that is usually associated with power plant devices and IT systems. It is the ability of a device to distort the sinusoidal power waveform, the worst being a *square wave*. In a typical electrical system, 60 Hz is the dominant or first harmonic order. Devices can attenuate the waveform and reflect waste current back into an electrical distribution system [27]. To correct for this, power systems will often use filters to trap third-order harmonics [28].

- *Frequency variations* are a change in frequency of more than 3 Hz arising from unstable frequency power sources. They are usually caused by cutting over power from a utility to a generator.

Many of these are also a result of mishaps and human errors by local utilities or within the general locale. Electrical repair, maintenance, and construction accidents and errors by external parties, particularly those residing in the same building, can often impact utility service or power infrastructure.

11.3.2 Power Supply

Commercial ac power availability is typically in the range of 99.9% to 99.98%. If each power disruption requires systems to reboot and services to reinitialize, additional availability can be lost. Because this implies downtimes on the order of 2 to 8 hours, clearly a high-uptime mission-critical facility will require power protection capabilities.

A mission-critical facility should be supplied with sufficient power to support current and future operating requirements. An inventory of all equipment is required along with how much power each is expected to consume. All mission-critical equipment should be identified. The manufacturer specified amps and volts for each piece of equipment should be noted. Manufacturers will often cite maximum load ratings rather than normal operating power. These ratings often reflect the maximum power required to turn up a device versus steady state operation. These ratings will often exceed normal operating power by an average of 30%.

Because ratings convey maximum load, a power plant based strictly on these values could result in costly overdesign. Using current or anticipated operating loads for each system might be a more reasonable approach, although a staggered startup of all systems may be required to avoid power overdraw. The total wattage should be calculated and increased by at least 20% to 30% as a factor of safety, or in accordance with growth plans.

Current power density for data centers can range from 40 to 200W per square foot [29]. Data centers are typically designed for at least 100W per square foot, with a maximum usage of 80% [30]. Quite often, maximums are not realized for a variety of reasons. Heat load, in particular, generated by such high-density configuration can be extremely difficult to manage. Power supply and heat-dissipation technology are gradually being challenged by the trend in compaction of high-tech equipment.

Commercial servers in North America use 120V of ac, with amperage ratings varying from server to server. Whereas today's high-density installations can see on the order of 40 to 80 systems per rack, this is expected to climb to 200 to 300 systems per rack in 5 years. Although advances in chip making will reduce power consumption per system, unprecedented power densities on the order of 3,000 to 5,000W per rack are still envisioned.

Telecom networking systems use –48V of direct current (dc). *Rectifiers* are used, which are devices that convert ac voltage to dc. The negative polarity reduces corrosion of underground cables, and the low voltage is more easily stored and not governed by national electric code (NEC) requirements [31]. Because of the trend in convergence, networking facilities are seeing more widespread use of ac-powered systems. In telecom facilities, ac power is delivered via *inverters*, which are devices that convert ac to dc, drawing power from the –48V dc.

Mission-critical facilities obtain their power from a combination of two primary sources. *Public power* is electrical power obtained from a public utility. *Local power* is electrical power produced internally, often by diesel generators and other types power-generation systems. Also referred to as *distributed generation* or *cogeneration*, local power is often used as a backup or supplement to public power. At the time of this writing, the public power industry is undergoing several trends to improve the cost, production, reliability, and quality of power: deregulation, divergence of the generation and distribution businesses, open-market purchasing, and greater use of natural gas.

Redundancy is key for improved power availability. Power should be obtained from two separate and independent utilities or power grids. The power should arrive from multiple transmission systems along geographically diverse paths through independent utility substations. Two power utility feeds into a facility via diverse paths protects against local mishaps, such as collapsed power poles.

Electrical service should be delivered to a facility at the highest possible voltage. Excess current should be available to enable voltage to reach steady state after an increase in draw following a startup or short-circuit recovery operation.

11.3.3 Power Quality

A mission-critical facility should have a constant supply of clean power, free of any irregularity or impurity. The quality of power entering a facility should fall within American National Standards Institute (ANSI) specifications. *Power-conditioning* devices may be required as part of the facility's electrical plant to ensure a consistent supply of pure power. In general, conditioning keeps the ac waveform smooth and constant so that technical equipment can be impervious to power irregularities. *Line conditioning* devices insert power to minimize brownouts or absorb power to reduce spikes.

Even if incoming power may seem relatively clean, conditioning devices may still be required to protect against irregularities created by internal systems. In fact, the majority of transients come from internal sources such as motor-driven devices like printers, photocopiers, fans, and heating/ventilation/air-conditioning systems. Common fluorescent lighting is also notorious for creating transients. Internally and externally created voltage surges can have different effects on equipment. Internally created transients will usually not instantaneously damage equipment because they occur at lower voltages, but can cause cumulative damage over time. Externally created transients are characterized by short and very high voltages that can cause immediate damage.

11.3.3.1   Surge Protection

Surge protection devices (SPDs) are equipment designed to *clamp* excess voltage. SPDs provide surge protection, suppression, and filtration to guard against transients [32]. Suppression is achieved using *transient voltage surge suppressors* (TVSS), which limit the amount of energy resulting from a transient. They can be used to protect an individual system or cable, or protect other systems from those known to produce transients. There are three types of technologies used in TVSSs:

- *Gas discharge tube (GDT)*. These devices use electrodes inside a pressurized tube containing gases that introduce clamping voltages to smolder an electric arc. GDTs are effective in suppressing very high voltages on the order of 600 to 800V, but they have more difficulty in suppressing surges at lower voltages. Sometimes arcs will persist following a transient and can damage the tube. For these reasons, GDTs are used in conjunction with other suppression technologies.
- *Metal oxide varistor (MOV)*. MOVs are the most commonly found suppression devices. They use zinc oxide fragments under high pressure. They are effective in suppressing short transients at lower voltage levels. However, extremely high voltage transients can increase the clamping voltage to levels that can cause an MOV to fuse shut, possibly damaging the protected systems. For this reason, they are replaced after a known surge or periodically [33]. Because of their inherent caveats, they are less expensive and used in conjunction with other suppression technologies.
- *Silicon avalanche diode (SAD)*. These are semiconductors that emit clamping voltages at almost the same level as the line voltage. For this reason, multiple SADs are often found in a device so that transient energy can be safely dissipated. They are effective in suppressing a wide range of transient voltages and durations, and usually comprise the more expensive devices. Unlike MOVs, they do not degenerate with use.

TVSS manufacturers will often make devices that leverage combinations of these technologies for durability and redundancy so that if an SAD or MOV fails or is damaged, the device can still perform within its ratings. In addition to clamping voltage, SPD efficacy is assessed by other parameters, including response time, or the amount of time required to clamp, and *let-through voltage*, which is the amount of voltage that gets through to equipment [34]. Underwriter Laboratories (UL) specification UL 1449 provides standards on let-through voltage.

SPDs must be strategically placed in accordance with codes and standards (Figure 11.9) [35]. *Primary* surge protection is required at building entrances to withstand extreme over voltages, such as those from lightning (discussed further in this chapter). UL-497 and NEC 800 are the applicable specifications. *Secondary* protection must be provided to protect humans and equipment in accordance with UL-497A and Telecommunication Industry Association/Electronic Industry Association (TIA/EIA)–568 standards. *Isolated* protection is required to handle low-voltage surges introduced from within a building, in accordance with the UL-497B specification.

SPDs are often placed at ac power outlets around desktop areas. Unless data and telephone lines to the same location are protected, the possibility of a voltage potential between protected and unprotected equipment can arise during a surge. This can damage data and telephone equipment. Thus, data and telephone equipment should be protected at their service entry points.

11.3.3.2 Lightning Protection

Unless a resistance path to the ground exists, lightning striking an antenna atop a building can travel into a building and create electrical disturbance, let alone fire and possible injury. A lightning protection system is designed to provide a path to the ground for lightning current to flow. Although many buildings have grounding protection, a sufficiently low-resistance path to the ground is necessary to safely dissipate the lightning energy. Otherwise, a *flashover* can occur, whereby the energy can travel into a building's electrical system and into equipment, creating damage. An unwanted voltage potential can make an unsuspecting person or piece of equipment a conductor as part of a circuit to the ground.

Lightning protection is usually done in accordance with the National Fire Protection Association (NFPA) and ANSI codes. Lightning protection devices should be

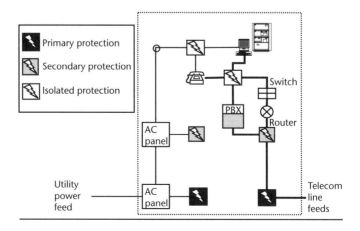

Figure 11.9 Surge protection placement.

placed between the public power supply and the main power distribution panel in a facility to force away lightning energy. Lightning protectors are usually destroyed after a strike and have to be replaced. Although nearly all telecom lines, by law, must have a lightning protector at the service entry point, these devices can still let through voltage into a facility. For this reason, secondary surge protection and proper grounding are necessary.

11.3.3.3   Grounding

Properly grounding protects both equipment and personnel. Grounding protection should be in accordance with local codes and national codes such as ANSI/TIA/EIA-607 Commercial Building Grounding and Bonding Requirements for Telecommunications. Additionally, the Institute of Electrical and Electronics Engineers (IEEE) 142-1991 standard specifies grounding systems and corresponding resistance levels [36]. Although many vendors require separate grounding of their equipment to avoid electrical interference, the NEC requires use of a common grounding system having a properly sized neutral conductor [37].

Grounding devices and potentials in a facility should be tested and identified as such and tied to electrical ground, building steel, racks, cable trays, or other approved equipment. Equipment racks should have access to ground conductors or bus bars and tie back to a main connection at a ground source (Figure 11.10). Using a ground conductor wire gauge that is too small or with frequent bends and grounding to a facility's water supply are common mistakes. Additionally, *ground loops* between two devices connected by cable can create communication interference. Ground loops are stray electrical currents created when a circuit is grounded at more than one point.

Some main ground sources use ground rods, typically made of copper, embedded into the ground. The rods are installed in a grid-like pattern and are connected together with heavy conductors. Depending on the soil resistance, an electrolyte enhancer may have to be used to improve conductivity between the rod and surrounding soil.

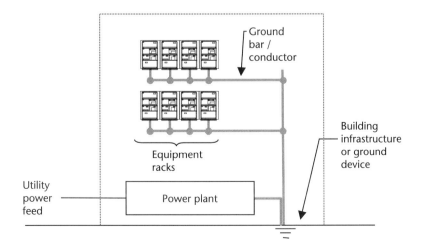

Figure 11.10   Centralized ground source.

11.3.3.4   Corrosion

Corrosion can cause loss of continuity in network connection through various ways. Quite often, metal protection is not a common practice in manufactured components due to fear of altering its engineered dielectric, resistance, and electrical properties. Quite often, manufacturers overlook additional metallic protection of components that can guard against the ongoing effects of humidity, acid rain, and salt within the atmosphere. Although such protection could improve mean time between failure (MTBF), many vendors prefer to simply replace components rather than invest in corrosion protection. New technologies, such as vapor corrosion inhibitors (VCIs) are being introduced on high and low voltage equipment, contacts, switches, circuits, and many other types of components. They protect metals electrochemically and work well in high humidity and salt environments.

If IT infrastructure must reside outside within metallic enclosures or on exterior infrastructure, additional measures need to be taken to protect against corrosion. A phenomenon called *electrolytic corrosion* can occur if dissimilar metals reside in proximity. A voltage potential occurs between the dissimilar metals, resulting in a current between the two materials and producing the possibility of material migrating from one source to the other. Cathodic protection, using donor materials, and applying an equalizing voltage are popular correction mechanisms. When IT systems must reside outside, attention should be given to other metallic infrastructure in the vicinity that is configured as an *acceptor*, absorbing materials from *donor* infrastructure. Without proper protection, metallic donors virtually decay quickly over time, damaging the equipment.

11.3.4   Backup Power

Many firms are finding the need to become more energy self-sufficient. A backup power plant should be part of any mission-critical installation to protect against many of the aforementioned power irregularities. It should be engineered and designed to keep the entire facility operational under specified disturbances. Power protection should been designed from the facility level down to the user system level. Because cost is a limiting factor in how much protection to build, there should be a deep understanding of what conditions will cause the facility to cease operations and for how long.

11.3.4.1   UPS

An *uninterruptible power supply* (UPS) is a device used to provide temporary clean power when a power disruption occurs. They are inserted between a power source and the systems to be protected. UPS devices work with batteries to keep a system running for a brief period of time. This permits cutover to a secondary power source or time for systems to properly shutdown if the disruption is of long duration. Over the years, UPS systems have been incorporated with many additional features, including surge protection and power conditioning.

It is important to selectively decide which systems are to be protected by a UPS, preferably the most critical ones. UPSs can be used to protect virtually any type of equipment, including desktop systems, servers, phone systems, printers, routers, switches, storage systems, security systems, WAN gear, and entire data centers.

They come in all shapes and sizes and can be used in different ways. Centralized UPS systems should be part of a mission-critical facility's power architecture to simultaneously protect a number of systems. As a UPS is a single point of failure, they should be deployed redundantly, preferably in an $N + K$ arrangement.

In general, UPSs operate longer when the attached devices draw less power. The power provided by a UPS should be sufficient to maintain system operation to allow for a graceful, orderly shutdown at minimum. Some UPSs are designed with the intelligence to automatically do this for different platforms and operating systems (OSs). If local power is available (e.g., in the form of a generator), the UPS should bridge the time between the loss of power and when local power is fully operational, which for some gear can take up to 15 or 20 minutes. By doing so, overall availability is improved.

The units come in various form factors, from small stand-alone systems to large-scale rack, panel, or floor mounted systems. Large-scale systems are designed to support entire equipment rooms or data centers. In general, UPSs usually consist of two components: an electronic module that provides the power conversion and a power supply in the form of batteries. Systems with redundant modules can share and balance load between them. The battery supply dictates a UPS's coverage time. Many systems come with their own battery cells or can be connected to a centralized battery supply. There are three general types of UPSs (Figure 11.11) [38, 39]:

- *Standby UPS.* These devices are also referred to as *off-line* UPSs. If ac power drops below an acceptable level, the UPS draws power from the batteries, producing ac power. A rectifier converts incoming ac to dc to keep the battery charged. When power fails, an inverter starts converting the battery's dc current into ac and switches the power source, which can take several seconds to complete, possibly causing systems to shut down. For this reason, they are often used with less critical systems and are less expensive.

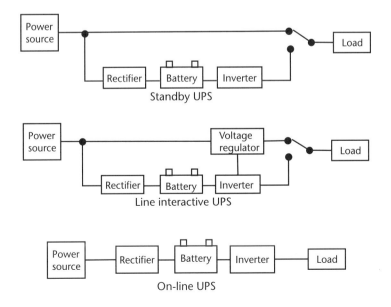

Figure 11.11   Types of UPS.

- *Line-interactive UPS.* These devices work similarly to standby UPSs, except that they regulate the voltage of incoming ac during normal operation. Although some devices do not draw on the battery for voltage regulation to preserve battery life, they can switch to battery power more quickly than standby UPSs when an outage occurs. They are often used for higher end equipment.

- *On-line UPS.* These devices, also referred to as *double* or *dual conversion* UPSs, protect systems during normal operation. They buffer incoming ac power by converting it to dc, running it through a battery, and then converting it back to ac. An inverter continuously converts the dc back to ac, or it is left as dc for telecom systems. When utility power fails, the inverter draws power from the batteries. Buffering the power continuously isolates the load from the power source, avoiding discrete switchovers during disruption. For this reason, on-line UPSs are often used to protect critical platforms and are more expensive.

UPSs use ferroresonant constant voltage transformers (CVTs), which are used to protect UPS battery supplies from transients during charging [40]. Unless harmonically adjusted, CVTs have the potential to produce square waveforms that can damage equipment. To overcome the harmonic distortion that can arise from the dual-conversion process, newer *single conversion* UPSs have evolved. They use a single inverter that only supplies load when the battery is required.

A UPS system should have a power rating that cumulatively meets or exceeds those of the protected systems. They should be designed to withstand 20% to 30% beyond the power load. If the amount of load exceeds the power rating of a UPS, it will either shut down or issue some form of notification that the load has been exceeded [41]. Many designs will go into a *bypass* mode in which it redirects incoming backup power away from the batteries, defeating the purpose of having a UPS altogether [42].

UPSs typically convert about 85% to 95% of all incoming power, rejecting the remaining power as heat. For this reason, choosing UPSs that are operationally efficient at expected operating loads is critical. UPS efficiency is expressed as a *power factor*, whose value is between zero and one, which is used to adjust the UPS power output value. Newer devices employ power factor correction capabilities to reduce the amount of rejected power.

UPSs are designed to momentarily supply power while another backup power source, such as a generator, can resume power delivery. This also provides adequate time for systems they notify to shutdown in an orderly manner.

A UPS should gradually shed load to the backup power source to avoid suddenly loading it. The UPS must be able to gracefully align or *slew* to a generator's frequency; otherwise, undo load instability can occur. Sensing such instability, a UPS could even revert to power delivery using the battery.

Having redundant UPSs for critical systems not only avoids a single point of failure, it also allows a UPS to be serviced without impacting the protected systems. A backup UPS will likely have to support a 100% increase in load with minimal transient voltage. Although some UPSs may have inherent surge-suppressing capabilities, they most likely consist of MOV devices, which are self-sacrificing. For reasons discussed earlier, a separate TVSS should be considered for proper protection.

11.3.4.2 Batteries

Batteries are often the workhorses of a power plant, as most power irregularities are of duration short enough to impact IT operations but not long enough to require generator startup [43]. Batteries that are used in conjunction with UPSs and other backup power plant can be quite expensive. Batteries are typically made from lead oxide acid. Normally, 12V dc batteries are used, requiring acid level monitoring and environmental and hazard protection. They can last from 5, 10, or even 20 years, with a 10-year battery supplying about 30 min of backup time. High operating temperatures, high numbers of discharge cycles, and long discharge periods can shorten battery life [44, 45].

The amount of battery backup required is a function of the amount of load and backup time. Batteries used in conjunction with off-line or line-interactive UPSs have a direct relationship with the local power generation system, such as the generator. They must be chosen to allow time for the generator to start up or enable graceful shutdown of systems in the event of a generator mishap [46]. Batteries are usually deployed in *strings*. The size and number of strings are designed according to the load and backup time. Battery strings should be designed so that they can be recharged in a reasonable amount of time. Battery depletion should be avoided at all costs. Systems should be shut down in the event a backup power source fails in addition to the primary one, so that batteries are not depleted. Depending on the battery size and UPS, a completely drained battery can take anywhere from 2 to 12 hours to recharge [47].

Batteries can also be a weakest link of a power plant. For this reason, alternatives to batteries have recently emerged. Super capacitors, ultracapacitors, and electric double-layer capacitors are battery variants that use oppositely charged carbon plates and an acidic electrolyte to store energy. Flywheel systems, on the other hand, are based on a completely different technology. They use incoming power to wind up a flywheel, storing kinetic energy. The wheel is then spun when required to provide power.

11.3.4.3 Transfer Switches

An automatic transfer switch (ATS) is a device that senses a power irregularity and initializes a generator to supply backup power. During this time, a UPS system supplies auxiliary power. For redundancy, two ATS switchboards should be employed, each with its own utility power feed and surge suppression. Systems having a wide power feed voltage and frequency tolerance are preferred, usually in the range of 95 to 135V and 55 to 65 Hz. This avoids inadvertent activation of a generator or possible equipment damage.

11.3.4.4 Generators

An electric generator is a fuel-driven engine that provides backup power. A generator is activated once as ATS senses a power irregularity. Generators should be matched with UPS and other electrical plant devices based on waveform and frequency. For extreme availability, at least two backup generators should be installed. Generators represent significant capital and operational investment. Like any engine, they require periodic maintenance and test firings to confirm operability.

A typical 750-kW generator with a 1,000-gallon diesel fuel tank can provide 18 hours of power. Tank capacity can vary, with many installations having tanks that can supply 1 to 7 days of power. Generators can provide power indefinitely, as long as they are fueled. Typical fuels include diesel, liquid propane gas, natural gas, or even steam. Refueling contracts should be established with multiple suppliers to guarantee continuous fuel delivery. Multiple fuel tanks are recommended in case one is damaged or the fuel becomes contaminated. Periodic contamination checks of fuel are recommended.

Because emissions, vibration, and noise produced by generators can be environmentally harmful, there are often regulatory procedures and permits that must be met [48]. Cleaner alternatives to the conventional generator are becoming available for local power generation, but can be relatively expensive. Fuel-cell technology combines hydrogen and oxygen to produce energy. Photovoltaic cells convert solar energy to electricity. Newer microturbine technology uses natural gas.

11.3.4.5   Power Fail Transfer Systems

Power fail transfer (PFT) systems are used primarily in conjunction with phone switches [49]. PFT devices sit between a phone switch and incoming phone lines. These systems enable calls to be sent and received through a portion of the phone lines in the event a phone system must reboot.

11.3.5   Power Distribution Architecture

A power distribution architecture integrates power supply, power quality, and backup power plant to reliably circulate power to all systems. A mission-critical power architecture should be organized in a modular fashion, with each module being a self-contained unit. Two levels of redundancy are required: device redundancy and path redundancy. A zoned approach should be used in grouping systems based on their operating requirements and criticality to operations. Critical zones will likely require higher levels of power availability and protection against irregularity. Once zones are identified, logical and physical circuits serving each zone should be defined.

Path redundancy is required so that the failure in power supply, power equipment, or a circuit does not impact IT system operations. Redundant power paths should be designed so that each can be shut down for maintenance while another continues to provide service. Path redundancy is realized through diverse redundant circuits and cable runs between strategic points in the power plant architecture. Redundancy should also be present in the grounding system as well.

Power distribution units (PDUs) are isolation transformers that distribute ac or dc power to individual areas. They can be used to segment delivery of power to individual equipment, racks, or areas. By doing so, power to each area can be shut off without affecting other areas. They are equipped with circuit breakers to guard against inadvertent short circuits that might occur within an area. They also allow electrical maintenance to be conducted in one area while other areas remain operational. The NEC provides standards for safe electrical wiring of facilities. The IEEE standards, which address computing and communication operational issues, assume compliance with NEC standards. The Telecordia G.1502 is a specification for power requirements in technical facilities.

A general rule for a mission-critical power plant is that it must be about 10 times more reliable than the critical loads it protects [50]. A power distribution architecture should be designed so that single failure points are eliminated. Power plant architectures can vary widely depending on a multitude of factors. $N + K$ logic can be applied to either individual components or the overall architecture. Figure 11.12 shows an example of a *2N* distribution architecture with full redundancy starting from the utility feeds to the PDUs and to the end equipment [51]. AC power can be delivered to ac equipment directly or through inverters drawing on dc power (not shown). The following are some key features:

- Separate connections to the power company, each to a different substation on a separate grid, perhaps involving two separate utilities. Each is capable of handling the entire power supply of the facility. The feeds are routed on diverse, independent paths to redundant ATSs.

- In telecom environments, rectifiers are used to change incoming ac to dc before distribution. Multiple redundant rectifiers on each side are supplied from dual redundant ac distribution boards.

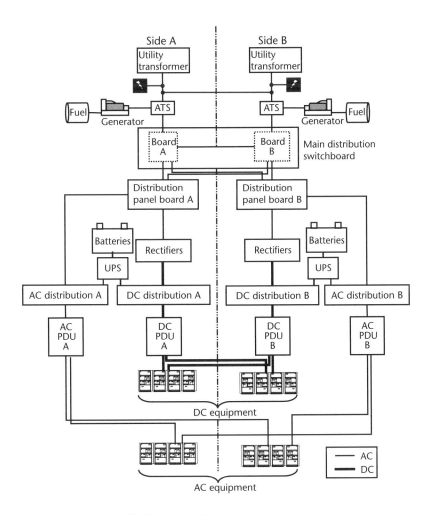

Figure 11.12   The *2N* power distribution architecture.

- On each side, redundant UPS systems can be used and connected on redundant circuits. Multiple banks of batteries can be used such that a UPS can use any battery bank.

- PDUs can be located throughout the facility. Redundant power circuits to each equipment rack or cabinet via redundant PDUs is used. Many computing and networking systems come with dual power feeds to protect against failure. Each feed should connect to a circuit on a separate PDU so that if one circuit fails, the system can still operate.

- Isolation of critical loads on circuits from those operating mechanical and air conditioning systems can avoid power irregularities coming from such equipment. In addition, computing and communication systems should be assigned evenly to a circuit to avoid overload. Branch circuits should be designed so that they are used effectively and extend load isolation to the user level. This means separate circuits for critical servers versus peripheral devices and common office equipment.

- Locating power plant equipment close to the load points can help reduce THD effects, as shorter conductor runs minimize feeder impedance and voltage distortions [52]. On the other hand, locating a technical equipment room too close to a main electrical room creates an opportunity for EMI from the electrical equipment to potentially affect IT system operation.

- As in a networking architecture, scalability is a desired feature in any power plant architecture. A high-availability mission-critical power plant can require exorbitant up-front capital expense. A modular architecture is preferred, allowing power upgrades to be implemented on a pay-as-you-go basis.

Depending on cost, full $2N$ redundancy may not always be possible. In this case, redundancy priority should be given to systems that are mission critical to the operation. For example, workgroup servers that do not run revenue-producing applications may only require backup by a single UPS. Key infrastructure servers such as dynamic hierarchical configuration protocol (DHCP) servers and domain name servers (DNS) or edge networking platforms may require high levels of protection.

11.3.6   Power Management

Centralized monitoring and management of power should be integrated with network management systems. Power management systems provide an array of capabilities that enable graceful transition of systems in the event of power irregularity [53]. They include capabilities such as OS shutdown, system reboot, closing of open files, control of power plant components, usage reporting, quality monitoring, notification, battery monitoring, and interoperability with network management systems.

A power management solution should be scalable and accommodate a variety of electrical and networking systems. Many components such as UPS devices will support SNMP using a standard management information base (MIB), as well as LAN interoperability [54]. Remote monitoring and management of power is highly desirable in situations where staff is locally unavailable to reset or reboot systems.

Many UPS products come equipped with software to alert of power problems, implement administratively defined procedures, and make the management of many small units easier. As mentioned earlier, some UPS management solutions will automatically shut down systems and will route power only to the more critical systems. Some will provide alerts via e-mail, page, and even telephone. Many systems offer remote monitoring and management capabilities via WAN, direct dial-in, and even the Internet. The last item can be worrisome from a security standpoint because attacking an insecure intelligent UPS system through the Internet could result in the UPS notifying servers that power is lost, at which point they are programmed to shut down.

Metering services from utilities provide data on the quantity and quality of power used by their customers. They report fluctuations and interruptions, outages, real-time pricing rates, and load profiles that can be used to manage power consumption and plan and design power plant upgrades.

11.4   Environmental Strategies

The physical environment that houses mission-critical systems must be reliable and safe. Temperature and humidity usually get attention during the physical design stages of building or renovating a facility. As systems are changed or upgraded over time, environmental factors will also change. Facilities should be designed with this in mind, and should take into account growth in systems and system density over the service life of the facility.

Reliance on local building codes usually is insufficient in designing environmental features of a mission-critical facility. Specifications such as the NFPA 75-1999, *Standard for the Protection of Electronic Computer/Data Processing Equipment*, and NEC Article 645-2 contain requirements for information technology equipment rooms. Other comparable specifications can be found in Telcordia Network Equipment and Building Specifications (NEBS) and even military standards.

11.4.1   Air/Cooling

Airflow, temperature, and humidity control are paramount to protecting mission-critical hardware assets. Excessive temperature and humidity can cause central processor units (CPUs) and other system components to fail. Placing systems in unprotected or uncontrolled space, such as general-purpose closets or basements, can be dangerous. Environmentally controlled spaces should be designed to manage the heat generated by numerous systems situated on racks or inside cabinets. They should also control dust and static.

As server, networking, and storage equipment grow more compact, the design of spaces that can provide adequate thermal management and heat dissipation grows more challenging. Servers are becoming flatter and deeper, occupying less space. Many are equipped with their own power supplies. Because redundancy is fundamental to mission-critical operation, they are likely to be deployed in twice the volume.

Traditionally, 0.6W of air conditioning is required to remove heat generated by 1W of computing power, the amount required to run a 20-MHz 80386 chip of

1980s vintage. This implies about 50% more power consumption for cooling [55]. Today's 800-MHz Pentium III chip requires about 75W of power [56]. If this trend continues, a fully equipped rack of servers could approach 18 KW of power in 5 years (as of this writing). Countering this trend are advances in chip technology, which will most likely drive this down by one or two thirds. Nevertheless, the growth in heat load per computing unit will be significant, further exacerbating the challenge of cooling technical spaces.

Technical spaces should have a cooling and ventilation system that is separate from the rest of the facility. Large air conditioning units, often referred to as *chillers*, are designed to discharge cool air into technical spaces. Like any other mission-critical system, redundancy in these units is key. Redundant units deployed in an *N* + *K* arrangement are required. Water-based systems must have redundant water supplies, often in the form of cooling towers. Systems drawing city-based water should employ a secondary source, such as a cooling tower or water delivery service.

Reducing the temperature in a technical space is only one aspect of a cooling process. A stable temperature needs to be maintained throughout the space as well, with a two-degree minimal variation. Although heat alone can damage CPUs, heat cycling (on/off cooling of heat) can also cause component wear as well. Ambient operating temperatures required by every equipment vendor's system should be inventoried in planning the cooling of a technical space. Additionally, ambient moisture is a well-known adversary of electronic systems. Proper humidity control not only protects against excessive moisture, it helps guard against electrostatic discharge (ESD).

Cooling a technical space requires proper airflow throughout the space and through equipment racks. Airflow involves venting hot air and replacing it with colder air. Internal server fans can be ineffective without proper airflow through a rack or cabinet. Flows for dissipating hot air and inserting cold air must be specifically designed. The type of racking and cabinetry play a role in designing airflow.

The presence of a raised floor is another factor affecting cooling [57]. Raised floors are often used to conceal cabling, conduits, and other miscellaneous infrastructure. The under-floor area is also used as a mechanism to channel cool air into equipment racks and cabinets. The under-floor area must be properly sized so that cabling and conduit congestion does not obstruct airflow. Network cabling should be separated from mechanical and electrical infrastructure. Some configurations use multilevel or *stacked* raised floors. In these configurations, the lower level is used for electrical, mechanical, and airflow distribution while the upper level is for computing and network cabling.

Many local building and electrical codes may require *plenum-rated* cable to reside in confined floor or ceiling spaces. If air is channeled through these spaces, these are referred to as *plenum spaces* or *air-plenum returns*. Plenum is a cable sheathing material that produces less toxic vapors when burned than normal sheathing. It is usually more expensive then regular cable sheathing. In place of plenum, some codes may require cabling to be enclosed in a pipe or conduit. Some locations even require both. Figure 11.13 shows several configurations illustrating airflow through equipment racks or cabinetry:

- *Front drawn.* Pulling air from the front area of the equipment, forcing it through the equipment, and then out the rear is a commonly used approach. A problem with this approach is that air in the front is usually colder than staff

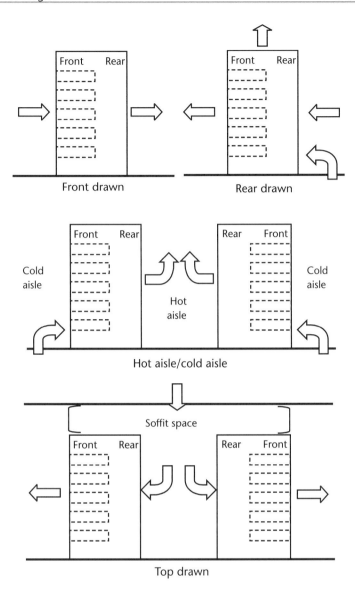

Figure 11.13   Airflow through equipment racks.

can withstand. Another problem is that in configurations where rows are successively aligned in a front-to-back equipment orientation, dirt and impurities emitted from a neighboring aisle can be drawn into the equipment.

- *Rear drawn.* Pulling air in from behind the rack can keep equipment both cleaner and cooler. A cold air source through the access floor or ceiling can avoid some of the problems of the front-drawn scenario. If air is directed from the bottom to the top of the racks, or vice versa, the racks must be nearly enclosed in order to create a chimney effect. In either case, the floor or ceiling may have to be designated as a plenum space.
- *Hot aisle/cold aisle.* This configuration involves using a perforated floor in front of the equipment to allow cold air to come up from the under-floor,

creating a *cold aisle* [58]. To cool equipment, the air is then forced out the back to create a *hot aisle,* which does not have perforated tiles. This works well in configurations where equipment racks face each other. However, it can clash with rack/cabinet implementations that exhaust hot air through the top, sides, or bottom.

- *Top drawn.* This solution involves constructing a soffit above the racks that does not touch the ceiling. The gap between the soffit and the ceiling is just enough so that it avoids creation of a plenum space [59]. Cool air is channeled down through the space and into the cabinets, creating a reverse chimney effect, out to the aisle areas. This solution can clash with cabinet fans designed to pull in air from the front.

To augment these configurations, many rack and cabinets have built-in fans to manage airflow and evenly remove heat. However, solutions should be chosen such that the internal rack/cabinet airflow is consistent with the hot/cold airflows within the technical space. Some solutions even use water-chilled systems, as water can remove about 35 times more heat than an equivalent air volume. For this reason, these systems can be more expensive. From a security standpoint, as in the case of the UPS, unprotected computer-driven intelligent HVAC equipment can tell other equipment that it has failed, forcing unwanted shutdowns.

11.4.2 Fire Protection Planning

Another challenge in designing a mission-critical network facility is creating capabilities to prevent, detect, contain, and suppress smoke and fire to limit injuries, equipment, and facility damage. Fire detection and suppression systems work by using detectors that sense smoke, cut power, and initiate an extinguishing action. The following are some fundamental precautions to use with respect to fire protection:

- *Compartmental design.* The facility should be divided into fire protection zones based on the type of occupancy. Data center, network room, and service entry should be monitored and zoned separately. Equipment spaces should be separated from others by fire-resistant-rated walls, floors, and ceilings with protected (fire-stopped) openings to prevent the spread of smoke or fire to other areas. Compliance with applicable building codes and use of UL-listed systems should be coupled with smart design. Using dedicated air ventilation systems to serve each zone can restrict the spread of smoke once they are shut down. Use of smoke activated dampers between zones can also limit smoke when they are closed. A segmented power distribution plant should be used so that power to all electronic equipment in an affected zone can be readily disconnected.

- *Early detection.* A very early smoke detection apparatus (VESDA) system can immediately identify and report fires to a security desk and local fire departments. New computer-based *addressable* detection and suppression systems can communicate with a single detector and provide flexibility and programmability to accommodate different floor plans and suppression schemes.

- *Suppression.* A fire suppression system can be a single point of failure, and, as such, it should have redundant features. A common approach is to concurrently use two systems, each based on different suppression technology. The following are the two most common types:

  1. A *dry-pipe preaction* suppression system can prevent incidental damage to technical equipment that would otherwise be created from a water-based system. It is a gaseous fire-suppression system, which, upon detection of smoke, is readied and releases nontoxic and noncorrosive heptafluorapropane gas to the affected areas. Halon, once a staple in dry systems, is no longer legal because of its ozone-depletion characteristics.

  2. A *double-action pressurized dry sprinkler* system is often used as a backup to a gas-based system. This is a water sprinkler system that is activated by heat. The pipes going to the sprinkler heads do not contain water until the system is readied, avoiding water damage in situations where pipes inadvertently burst. Sprinkler systems can be activated in only those areas where there is a fire.

As discussed earlier, air-plenum returns are spaces where air flows back to a heating or air conditioning system for ultimate venting throughout the facility. As plenum-rated cables emit less toxic fumes than regular cables do when burned, they enable occupants greater opportunity to evacuate. Plenum is predominantly used in horizontal cabling spans, versus riser cabling, unless the riser space qualifies as an air-plenum return.

11.5 Summary and Conclusions

A mission-critical network facility must provide an environment for continued network operation. The first step to achieving this is through proper location of the facility. Spreading operations across multiple facilities can reduce a firm's dependence on a single location. Network operations should be viewed in a layered hierarchical fashion for facility planning purposes. These layers include the access, distribution, and core layers. Facility location and layout is highly affected by the composition of these layers.

Cable plant is often taken for granted and viewed as a less important element of an overall network operation. However, many network problems result from cabling oversights. Network cabling throughout a facility should entail redundant routing of cables, multiple carrier service entrances, and multiple cross connects. Whether in an MTU or campus setting, diversity in these areas can enhance survivability.

Copper is still the predominant media for LAN implementations. The electrical characteristics of copper make it prone to interferences caused by environmental conditions, electromagnetic effects, surrounding cables, and poor installation practices. There are several well-defined tests that should be satisfied for any installed cable plant. These were reviewed earlier in this chapter.

Fiber-optic cable is widely used for MTU risers and campus backbones. Fiber is immune to much of the electromagnetic phenomena that plague copper—but it is still prone to optical phenomena attributed to poor installation practices, manufacturing defects, and wear. Like copper, a fiber plant requires undergoing several key tests to certify installation and operation.

Power disturbances often go unnoticed and can occur frequently enough to damage systems over time. This chapter covered the many types of power irregularities that can occur. Such disturbances happen for a variety of reasons—many of which can be traced to errors and mishaps. As the demand for power increases, the growth in power disturbances is expected to rise as well. Thus, a mission-critical network facility should have a well-designed power plant that guards against power disturbances and shortages.

To this end, the use of redundancy and power conditioning systems should be intrinsic to the design of an electrical plant. Such systems include SPDs, lightening protection, and grounding systems. Redundancy and backup power supply should be implemented using multiple power utility feeds, UPS, batteries, and generators. A well-designed power distribution architecture integrates these elements into a zoned infrastructure with redundant circuits and devices. Centralized power monitoring of the infrastructure should bridge power management with network management.

Cooling and circulation are often undervalued components in the design of many facilities, yet these two factors are notorious for ultimately causing many system outages. Technical spaces should adequately replace hot air with cold air. Like any other mission-critical components, cooling systems should be deployed with some redundancy inside a facility to protect against system failures and to facilitate repairs. Furthermore, the cooling and circulation patterns within equipment cabinets and frames should be consistent with those of the technical space.

References

[1] Kast, P., "The Broadcast Station as a Mission-Critical Facility," *Broadcast Engineering*, November 2001, pp. 72–75.

[2] Fortin, W., "A Better Way to Cable Buildings," *Communications News*, September 2000, p. 38.

[3] Matthews, J. K., "Maintaining Fiber-Optic Cable at the Demarc," *Cabling Installation & Maintenance*, April 1999, pp. 25–30.

[4] Cerny, R. A., "Working with Fiber," *Broadcast Engineering*, December 1999, pp. 93–96.

[5] Clark, E., "Cable Management Shake-Up," *Network Magazine*, July 1999, pp. 50–55.

[6] Jensen, R., "Centralized Cabling Helps Fiber Come Out of the Closet," *Lightwave*, October 1999, pp. 103–106.

[7] Lupinacci, J. A., "Designing Disaster Avoidance into Your Plan," *Cabling Installation & Maintenance*, April 1999, pp. 17–22.

[8] King, D., "Network Design and Installation Considerations," *Lightwave*, May 1998, pp. 74–79.

[9] Minichiello, A. T., "Essentials of Campus Telecommunications Design," *Cabling Installation & Maintenance*, June 1998, pp. 45–52.

[10] Badder, A., "Building Backbone and Horizontal Cabling—A Quick Overview," *Cabling Business Magazine*, October 2002, pp. 60–64.

[11] Jimenez, A. C., and L. Swanson, "Live Testing Prevents Critical Failures," *Cabling Installation & Maintenance*, June 2000, pp. 51–52.

[12] D'Antonio, R., "Testing the POTS portion of ADSL," *Communications News*, September 2000, pp. 108–110.

[13] Steinke, S., "Troubleshooting Category 6 Cabling," *Network Magazine*, August 2002, pp. 46–49.

[14] Lecklider, T., "Networks Thrive on Attention," *Communications News*, March 2000, pp. 44–49.

[15] Cook, J. W., et al., "The Noise and Cross-talk Environment for ADSL and VDSL Systems," *IEEE Communications Magazine*, May 1999, pp. 73–78.

[16] Goralski, W., "xDSL Loop Qualification and Testing," *IEEE Communications Magazine*, May 1999, pp. 79–83.

[17] Russell, T. R., "The Hidden Cost of Higher-Speed Networks," *Communications News*, September 1999, pp. 44–45.

[18] Jensen, R., "Extending Fiber Optic Cabling to the Work Area and Desktop," *Lightwave*, April 1999, pp. 55–60.

[19] Jensen, R., and S. Goldstein, "New Parameters for Testing Optical Fiber in Premises Networks," *Cabling Installation & Maintenance*, July 2000, pp. 65–70.

[20] Teague, C., "Speed of Light," *Communications News*, June 1998, pp. 44–46.

[21] Breeden, J. P., and D. E. Beougher, "Commissioning a Fiber-Optic LAN," *Cabling Installation & Maintenance*, January 1998, pp. 27–36.

[22] Girard, A., et al., "Critical System Parameters for Network Installation," *Lightwave*, October 2000, pp. 62–70.

[23] Vernon, C., "Testing Small-Form-Factor Fiber-Optic Links," *Lightwave*, October 2001, pp. 98–102.

[24] Straw, A. K., "Standards-Based Factory Testing of Fiber-Optic Cable," *Cabling Installation & Maintenance*, March 1999, pp. 35–40.

[25] Deragon, D. T., "Evolving Requirements for Tomorrow's Optical Network," *Cabling Installation & Maintenance*, April 2001, pp. 45–50.

[26] Sloane, T., "Powering Voice over CATV," *Network Reliability—Supplement to America's Network*, June 2000, pp. 8S–11S.

[27] Katz, A., "A Generator-Friendly Uninterruptible Power Supply," *Cabling Installation & Maintenance*, August 2001, pp. 37–42.

[28] Wong, T. K., "Power Quality in Entertainment Facilities," *Broadcast Engineering*, October, 2002, pp. 40–44.

[29] Neighly, P., "Telecom's Blackout," *America's Network*, April 15, 2001, pp. 28–37.

[30] Campbell, S., "Energy Crisis Hits West Coast," *Computer Reseller News*, January 1, 2001, p. 10.

[31] Stack, F., "Continuous Systems: 24/7 Power Strategies for Converged Networks," *Communications Solutions*, October 2000, pp. 68–72.

[32] Bird, A. O., and J. C. Roth, "Surge Protection Prevents Network Disasters," *Cabling Installation & Maintenance*, April 2000, pp. 23–28.

[33] Schuchart, S. J., "Environmental Protection," *Network Computing*, April, 30, 2001, pp. 65–70.

[34] Anderson, D., "Maximizing the Reliability of Communications Systems: A Total Protection Approach," *Disaster Recovery Journal*, Summer 2001, pp. 72–74.

[35] Cliché, R., "Think Secondary When Defending Against Power Surges," *Cabling Installation & Maintenance*, August 2002 pp. 63–65.

[36] Markley, D., "Even More on Lightening," *Broadcast Engineering*, May 2000, pp. 58–60.

[37] Read, B. B., "How to Protect Your High-Tech Equipment," *Call Center Magazine*, April, 1998, pp. 82–98.

[38] Essex, D., "Guaranteeing Network Power," *Network Magazine*, October 1998, pp. 76–80.

[39] Boston, J., "UPS Systems," *Broadcast Engineering*, September 2000, pp. 78–86.

[40] Grigonis, R., "Power Protection: Not Just a Battery in a Box," *Convergence*, December 2001, pp. 16–28.

[41] Cliche, R., "Uninterruptible Power Supplies Provide a Backup Plan," *Cabling Installation & Maintenance*, April 2002, pp. 44–49.

[42] Katz, A., "Back up Your Backup Power," *Communications News*, July 2000, pp. 50–51.

[43] "Mission Critical IT Systems Spared by Green Solution," *Communications News*, February 2000, pp. 18–20.

[44] Parham, J., and W. Kautter, "Disaster Preparedness for an ISDN Voice Network," *Cable Installation & Maintenance*, November 2002, p. 10.

[45] Katz, A., "Inspect and Replace Aging UPS Batteries," *Communications News*, December 2001, p. 26.

[46] Junkus, J. L., "Carrier-Grade Criteria for Powering Systems," *Communications Technology*, February 2000, pp. 28–30.

[47] Ondrejack, R., "Fade to Black," *Communications News*, July 1999, pp. 36–38.

[48] Boston, J., "UPS & Backup Power Systems," *Broadcast Engineering*, July 2001, pp. 68–71.

[49] Fleischer, J., "Extending Your Call Center's Lifeline," *Call Center Magazine*, April 1999, pp. 102–108.

[50] Bauer, R., "Power and Protection to Fight Off Downtime," *Cable Installation & Maintenance*, September 2000, pp. 25–33.

[51] AT&T Bell Laboratories, *Engineering and Operations in the Bell System*, Murray Hill, NJ: Bell Telephone Laboratories, 1983, pp. 541–549.

[52] Wong, T. K., "Power Quality," Ensuring Equipment Lives Up to Its Promise," *Broadcast Engineering*, pp. 154–158.

[53] Meagher, K., and B. Wyckoff, "Confronting Uncertain Power Conditions," *Cabling Installation & Management*, June 1998, pp. 79–88.

[54] Buratti, J., and I. Simonson, "Keeping up With the Demand for Reliable Power," *Internet Telephony*, March 2000, pp. 84–87.

[55] Angel, J., "Energy Consumption and the New Economy," *Network Magazine*, January 2001, pp. 50–55.

[56] Fixmer, R., "In Hot Pursuit of Energy Efficiency," *eWeek*, January 21, 2002, pp. 41–43.

[57] Dunn, P., "Network Cabinet Design Considerations," *Cabling Installation & Maintenance*, December 2000. pp. 35–42.

[58] McLaughlin, P., "Keeping Your Cool When Your Racks Heat Up," *Cabling Installation & Maintenance*, July 2002, pp. 33–38.

[59] Crumb, D., "Planning for Cooling," *Broadcast Engineering*, September 2001, pp. 62–64.

CHAPTER 12

Network Management for Continuity

This chapter reviews methods and techniques involved in managing mission-critical networks. Network operations and management typically consumes about half of the costs and resources of owning a network. Operation costs typically include installation, training, troubleshooting, and maintenance. The greatest costs are related to troubleshooting—IT managers spend about half of their time trouble-shooting network problems. Then there are management costs associated with administering users, applications, and equipment. The gradual shift of local area network (LAN) traffic to the wide area network (WAN) is evident in the increasing operations costs related to WAN management.

In these volatile economic times, company management is under pressure to improve revenues while maintaining cost effectiveness. As the diversity and complexity of networking technologies grow, so does the complexity in resolving network problems. Consequently, the network manager is faced with the goal of minimizing downtime by reducing the possibility of disruptions and by creating the appropriate response mechanisms.

The network manager must also create the mechanisms to maintain satisfactory performance, manage existing capacity, and anticipate future capacity needs. A related goal is to monitor capacity utilization for effective use of the network resources, thereby avoiding costs of overengineering. Network administrators will work with network managers to determine how the network is performing and how it can be improved. As the performance of a network increases, so does the cost. This is one of the fundamental laws of mission-critical network design. One of the goals of any enterprise is to provide the optimal service for the amount of money invested, to the extent that the enterprise can remain productive.

12.1 Migrating Network Management to the Enterprise

Until the 1980s, many enterprise network managers heavily engaged local exchange carriers (LECs) and interexchange carriers (IXCs) for wide area voice and data management. The monopolistic telecommunications environment was accompanied by a computing industry that was mainframe oriented and dominated by a few large suppliers. Enterprise network management problem solving and planning decisions used to rely heavily on outside entities. This is no longer the case. Today's environment requires firms to create and manage their own networks and engage the services of many carriers and vendors. This has led to many enterprises having heterogeneous multiservices networks comprised of legacy and contemporary systems, applications, and protocols.

This complex environment presents enormous opportunity for human error, which is typically the root cause of many network mishaps. Network management staff are usually their own worst enemies. Even mishaps associated with users or service providers ultimately point to oversights in network management. In the end, network managers must develop company practices, procedures, and policies to protect against human error while controlling and reducing costs in each area. These areas include such items as:

Communications management:

1. External network access requirements;
2. Network architecture and access controls;
3. Network services and operation coordination;
4. Network auditing, monitoring, reporting, and incident response practices.

Configuration management:

1. Configurations for critical systems such as firewalls, switches, routers, domain name server (DNS), authentication, authorization and accounting (AAA) servers, Web servers, private branch exchanges (PBXs), handsets, and other voice-related elements;
2. Configuration-management practices.

Security management:

1. Physical security controls;
2. Security technical policies, practices, and documentation;
3. Security awareness and training;
4. Authentication controls;
5. Risk-management practices;
6. Intrusion prevention;
7. Incident management.

System management:

1. System operation procedures and documentation;
2. System maintenance.

Software management:

1. Application procurement and controls;
2. Application development and controls.

Firms must choose what level of network management they will implement internally. There are typically several approaches. Extensive manual and poorly documented processes typically exemplify an informal approach, where no enterprise network management system is in place, and managers typically manually operate many systems using disparate management tools. This approach is usually characteristic of small to medium-sized firms. Larger firms will use a more controlled approach. Often, processes and infrastructure are well documented, but there is a mix of manual and automated tools. Large firms, typically major corporations or government bodies, will have highly automated tools and processes at the enterprise level [1].

In recent years, many network management tools have come to market with the promise of providing end-to-end solutions in many of these areas. In reality, many of these tools represent only point solutions for a subset of the areas and are often difficult to integrate together. Compounding this problem is the fact that network management is shifting gradually to enterprise service management. This implies bringing network, systems, applications, and traffic management towards a common enterprise model. Such models address managing various services that require different protection, restoration, and performance management schemes, depending on their importance to the firm.

Other than using network management tools, many firms decide to outsource a portion or all of their network management to another party. Outsourcing can involve either a specific task, such as network, access, or security monitoring, or the responsibility of managing a firm's entire network—a model characteristic of total enterprise outsourcing. A comprehensive outsourcing service should monitor every node in the firm's network and identify, analyze, and resolve problems.

It is usually easier to establish a comprehensive network management program upon first implementation of a network, versus after the network has been running for some time. The process usually involves an exhaustive inventory of systems and applications. Those network elements that will be *peered into* for monitoring are identified. These are usually such elements as frame relay devices, permanent virtual circuits (PVCs), digital service units (DSUs)/channel service units (CSUs), servers, switches, and routers.

Such elements will require simple network management protocol (SNMP) agents and remote monitoring (RMON) capabilities within them. Many network management tools rely on remote agents to supply information to a wider scale monitoring system. In general, agents are software processes that run on network components and report status information to a network management server. The network management server generally accumulates data from the agents and computes statistics on the network. Status, alarm, and performance information is then displayed on a workstation for viewing by the network manager. The firm's network may have to transition to these standards [2]. This often implies linking legacy system network architecture (SNA) systems to components with such capabilities. It may also involve converting dedicated point-to-point circuits to WAN circuits.

12.2   Topology Discovery

Mission-critical network management requires having a complete view of a network topology, for the simple reason that one cannot manage what cannot be seen. This includes both voice- and data-related elements and traffic. The best network management tools feature graphical displays of network topology used in conjunction with measurement collection, monitoring, and configuration capabilities. Many tools can discover many kinds of elements, including applications.

The better network management tools can accurately determine layer 3 and layer 2 elements at the outset, on a continuous basis so that network changes are readily identified. This includes not only an inventory, but mapping applications and operating systems (OSs) to systems, servers, and desktops. Initially, an autodiscovery process on an active network can be slow and may create a heavy load, but many products are more benign than this. Afterwards, the process should be done on a daily basis during off hours. Many systems allow managers to define the frequency of discovery.

Once a topology is obtained, it should be validated. A common problem with discovery is reliance on all elements having registered SNMP system identifiers. Often, this is not the case, and the discovery must be supplemented with manually gathered information or through the use of other tools. Topologies can often overlook virtual items, such as paths and virtual addresses. This involves obtaining configuration information from routers and switches and reviewing network documentation.

Mission-critical elements and services, such as edge routers, WAN connectivity, and switches should be noted. A network management tool will query each element at a regular interval, which can be specified by the network manager. Tools typically use SNMP management information base (MIB) definitions of metrics for performance, availability, and status of each device. SNMP traps from devices that support critical services should be defined and monitored. For WAN connections, traps and performance information should be obtained from the carrier's network. Network performance should be monitored closely for about a week in order to understand the traffic and threshold patterns.

12.3   Network Monitoring

Much time is often wasted tracking and troubleshooting problems with undetermined root causes. Network-management tools must coordinate monitoring network infrastructure and applications so that root causes and effects of software or hardware failures can be readily identified across multiple components and domains [3]. Polling and instrumentation are the two most important characteristics if a network-monitoring tool. Many tools employ real time intelligent polling in conjunction with agents or probes.

In network monitoring, more is not necessarily better. Enough information should be obtained so that decision making can be done quickly and easily, with minimal noninterfering data collection overhead. A selective approach to gathering measurements is preferred over collecting inordinate amounts of raw data. A good network-monitoring tool should have intelligent agents with local processing and storage to reduce information prior to reporting it to the monitoring console [4, 5]. Such agents should also be able to acquire knowledge about the devices they are monitoring and apply that knowledge to problem solving. Their actions should be governed by rules specified by the network manager. Agents that can initiate such actions can reduce the mean time to recovery (MTTR).

Agents should be as least invasive as possible when collecting data. They should consume less than one percent of a system's resources so that performance is not degraded. They should also allow incorporation and definition of user-defined agents if needed with those widely used. Typical metrics obtained include server I/O information, system memory, and central processing unit (CPU) utilization and network latency in communicating with a device. In general, SNMP GET and TRAP commands are used to collect data from devices. SET commands are used to remotely configure devices. As this command can be quite destructive if improperly used, it poses a high security risk and should be utilized judiciously [6, 7]. Many devices use SNMP MIP, but manage via Java applets and secure socket layer (SSL)–management Web servers.

A network itself is a single point of failure for network management. Network monitoring implies communication with the network elements. Often times, SNMP commands are issued *inband*, meaning that they are sent over the production network. If the network is down, then SNMP is of little use. Often, an agent can indicate if communication with any other nodes has failed or if it is unacceptable. Some agents can try to fix the problem on their own. If communication is lost, some agents can communicate via other nodes. Some can even assume control server tasks if communication with the server is lost. However, without the ability to communicate to a failed node, nothing beyond the problem notification can be done. An alternative approach is to communicate to a device via an out-of-band mechanism, typically through a secure dial-up connection or SSL over dial-up through a Web server.

Many network monitoring solutions focus on devices. Often, information alerts regarding a device may not indicate what applications, and ultimately services and transactions, are going to be impacted. Network-monitoring tools should be able to help identify the impact of device problem by application. This requires the identification of resources and processes required for the operation of an application, including such items as databases and storage, connectivity, servers, and even other applications. Application-performance problems are often difficult to identify and usually affect other applications.

Sampling application transaction rates is an approach that can help identify application performance problems. Sampling rates can vary by application. Sampling at low rates (or long intervals) can delay problem identification and mask transient conditions. Sampling too frequently can cause false alarms, particularly in response to transient bursts of activity.

Many monitoring tools focus on layer 3, primarily Internet protocol (IP)–related diagnostics and error checking. The Internet control message protocol (ICMP) is widely used for this purpose. Router administrators can issue a PING or TRACEROUTE command to a network location to determine if a location is available and accessible. Other than this, it is often difficult to obtain status of a logical IP connection, as it is a connectionless service.

Newer tools also address layer 2 problems associated with LANs, even down to the network interface card (NIC). Some enable virtual LAN (VLAN) management. They allow reassigning user connections away from problematic switch ports for troubleshooting.

Proactive monitoring systems can identify and correct network faults in critical components before they occur. Intelligent agents should be able to detect and correct an imminent problem. This requires the agent to diagnose a problem and identify the cause as it occurs. This not only makes network management easier, but also significantly reduces MTTR. Furthermore, there should be strong business awareness on the part of network management to anticipate and prepare for special events such as promotions or new accounts. Overall, a proactive approach can avert up to 80% of network problems.

Tracking and trending of performance information is key to proactive fault management. A correlation tool can use different metrics help to identify patterns that signify a problem and cause. Cross correlation between CPU utilization, memory utilization, network throughput, and application performance can identify the leading indicators of a potential faults. Such information can be conveyed to the intelligent agents using rules that tell them how and when to recognize a potential event. For example, an agent performing exception monitoring will know, based on the pattern of exceptions, when an extraordinary event has or is about to take place.

Probe devices can provide additional capabilities beyond SNMP and intelligent agents. Probes are hardware devices that passively collect simple measurement data across a link or segment of a network. Some devices are equipped with memory to store data for analysis. Probes should have the capacity to collect and store data on a network during peak use. Many probes are designed to monitor specific elements [8]. For example, it is not uncommon to find probes designed specifically to monitor frame relay or asynchronous transfer mode (ATM) links (Figure 12.1). They are often placed at points of demarcation between a customer's LAN and WAN, typically in front of, in parallel with, or embedded within a DSU/CSU to verify WAN provider compliance to service levels.

12.4   Problem Resolution

Automated network-management tools by themselves are not enough to assure satisfactory network availability and performance. A systematic and logical approach in responding to events can also be an effective tool. First and foremost, it is important to identify the most critical portions of a network and potential points of failure. This helps to outline the entire network-management effort and identify those areas within an IT infrastructure where attention should be most directed. A systematic approach should entail the following steps:

1. *Event detection.* Detection is the process of discovering an event. It is typically measured as the time an adverse event occurs to the time of

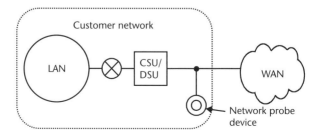

Figure 12.1   Use of probe example.

becoming aware of it. Recognizing certain behavioral patterns can often forewarn that a fault has or is about to occur. Faults should be reported in a way that discriminates, classifies, and conveys priority. Valid faults should be distinguished from alerts stemming from nonfailing but affected devices. Quite often, a device downstream from one that has exhibited an alarm will most likely indicate an alarm as well.

The magnitude and duration of the event should also be conveyed. Minor faults that occur over time can cumulatively signify that an element is about to fail. Because unexpected shifts in an element's behavioral pattern can also signify imminent failure, awareness of normal or previous behavior should be conveyed.

Automating the process of managing alarms can help maintain operational focus. Furthermore, awareness of the connectivity, interdependencies, and relationships between different elements can aid in alarm management. Elimination of redundant or downstream alarms, consolidation, and prioritization of alarms can help clear much of the smoke so that a network manager can focus on symptoms signifying a real problem.

2. *Problem isolation.* Isolation is the process of locating the precise point of failure. The fault should be localized to the lowest level possible (e.g., subnetwork, server, application, content, or hardware component). It should be the level where the least service disruption will occur once the item is removed from service, or the level at which the defective item can cause the least amount of damage while operational. A network-management tool should provide the ability to window in, if not isolate, the problem source. An educated guess or a structured search approach to fault isolation is usually a good substitute if a system is incapable of problem isolation.

3. *Event correlation.* Event correlation is a precursor to root cause analysis. It is a process that associates different events or conditions to identify problems. Correlation can be performed at different network levels (Figure 12.2). At the *nodal level*, an individual device or network element is monitored and information is evaluated to isolate the problem within the device. At the *network level*, several nodes are associated with each other. The group is then evaluated to see how problems within that group affect the network. Nodes or connections within that group are interrogated. The *service level* associates applications with network elements to determine how problems in either group affect each other. Faults occurring at the service level usually signify problems at the network or node levels [9, 10]. Some network-management tools offer capabilities to associate symptoms with problems. These tools rely on accurate relationship information that conveys application dependencies on other elements.

4. *Root cause analysis.* It is often the case that many network managers don't have the time to exhaustively research each problem and identify root cause. Most spend much of their time putting out fires. Root cause analysis, if approached properly, can help minimize the time to pinpoint cause. It goes without saying that merely putting out the fire will not guarantee that it will happen again. Root cause analysis is designed to identify the nature, location, and origin of a problem so that it can be corrected and prevented

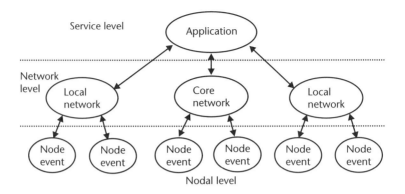

Figure 12.2   Event correlation at different levels.

from reoccurring. It is unlike correlation, which refers to associating events to identify symptoms, root cause analysis attempts to identify the single point of failure. Root cause analysis software tools are available to help automate the process [11]. The process is actually quite simple:

- *Collect and analyze the information.* This means collect all of the valid and correlated symptoms from the previous steps. Information should include:
  - Events and values of metrics at the time of failure;
  - How events or metrics changed from their normal operating or historical behavior;
  - Any additional information about the network; in particular, any recent changes in equipment or software.
- *Enumerate the possible causes.* The process of elimination works well to narrow down the potential causes to a few. Although many causes result mainly from problems in design, specification, quality, human error, or adherence to standards, such problems are not easily or readily correctable. A cause in many cases may be broken down into a series of causes and should be identified by the location, extent, and condition causing the problem.
- *Test the cause.* If possible, once a probable cause is identified, testing the cause to obtain the symptoms can validate the analysis and provide comfort that a corrective action will mitigate the problem. This might involve performing nondisruptive tests in a lab or simulated environment. Testing the cause is usually a luxury and cannot always be easily performed.

5. *Corrective action.* A recommendation for corrective action should then be made. It is always wise to inform customers or client organizations of the root cause and corrective action. (Contrary to what many may think, this reflects positively on network management and their ability to resolve problems.) Corrective action amounts to identifying the corrective procedures and developing an action plan to execute them. When implementing the plan, each step should be followed by a test to see that the expected result was achieved. It is important to document results along the way, as users may become affected by the changes. Having a formal process that notes what was changed will avoid addressing these effects as new problems arise.

The results of the analysis should be kept in a log for future reference. If the same problem reappears at a later time, a recovery procedure is now available. It is often wise to track the most common problems over time, as they can often point to defects in a network design or help in planning upgrades. One will find that some of the most common problems fall into the following categories:

- *Memory.* Inadequate memory is often the culprit for poor system performance and outages. When memory is exhausted, a system will tend to swap content in and out from disk, degrading system performance. This can often mislead one to believe that the problem lies in another area, such as inadequate bandwidth or a database problem.
- *Database.* Bona fide database problems usually materialize from poorly structured queries and applications, rather than system problems.
- *Hardware.* A well-designed software application can still be at the mercy of poorly designed hardware. A hardware component can fail at any time. Preventive maintenance and spares should be maintained for those components with high failure expectancy.
- *Network.* A bottleneck in a network can render any application or service useless. Bottlenecks are attributed to inadequate network planning.

12.5   Restoration Management

Restoration management is the process of how to manage a service interruption. It involves coordination between non-IT and IT activities. Restoration management allocates the resources required to restore a network to an operational state. This means restoring those portions of the network that were impacted by an outage. Restoration is that point where a network continues to provide operation, not necessarily in the same way it did prior to the outage but in a manner that is satisfactory to users. It is important to note that an IT environment can restore operation even if a problem has not been fixed. Once the network provides service it is restored, the remaining activities do not necessarily contribute to the overall restoration time.

This point is paramount to the understanding of mission critical—a network is less likely to restore service the longer it is out of service. Restoration management can involve simultaneously coordinating and prioritizing contingency and several recovery efforts, with service restoration as the goal. The following are several steps that can be taken to manage a restoration effort:

1. *Containment.* When an outage has occurred, the first step is to neutralize it with the objective of minimizing the disruption as much as possible. Some refer to this stage as *incident management.* Regardless of what it is called, the appropriate problem resolution and recovery efforts should be put in motion. Efforts may involve hardware, communications, applications, systems, or data recovery activities. A determination should be made as to the severity level of the problem. There is no real standard for severity levels—a firm should use what works best. For each level, a list

of procedures should be defined. These should identify the appropriate failover, contingency, recovery, and resumption procedures required to restore service. Instantaneous restoration is a job well done.

Today's enterprise networks are deeply intertwined with other networks, typically those of customers, suppliers, or partners. Consequently, their plans should become relevant when a planning a restoration process. The tighter the level of integration, the more relevant their plans become. At this stage, it is important to identify all affected or potentially affected parties of an outage, including those external to the organization.

2. *Contingency.* When a problem occurs, a portion of a network has to be isolated for recovery. A firm must switch to a backup mechanism to continue to provide service. This could well be a hot or mirrored site, another cluster server, service provider, or even a manual procedure. Much discussion in this book is devoted to establishing redundancy and protection mechanisms in various portions of a network with goal of providing continuous service. Redundancy at the component level (e.g., hardware component, switch, router, or power supply), network level (e.g., physical/logical link, automatic reroute, protection switching, congestion control, or hot site), and service level (e.g., application, systems, or processes) should in some way provide a contingency to fall upon while recovery efforts are in motion.

3. *Notification.* This is the process that reports the event to key stakeholders. This includes users, suppliers, and business partners. A determination should be made if stakeholders should be notified in the first place. Sometimes such notifications are made out of policy or embedded in service level agreements. A stakeholder should be notified if the outage can potentially affect their operation, or their actions can potentially affect successful service restoration. For an enterprise, the worst that can happen is for stakeholders to learn of an outage from somewhere else.

4. *Repair.* Repair is the process of applying the corrective actions. These can range from replacing a defective component to applying a software fix or configuration change. The repair step is the most critical portion of the process. It involves implementing the steps outlined in the previous section. It is also a point where errors can create greater problems. Whatever is being repaired, *hot* replacement should be avoided. This is the practice of applying a change while in service or immediately placing it into service. Instead, the changed item should be gradually placed in service and not be committed to production mode immediately. The component should be allowed to share some load and be evaluated to determine its adequacy for production. If incorrect judgment is made that a fix will work, chances are the repaired component will fail again. Availability of spares or personnel to repair the problem is implicit in the repair process.

5. *Resumption.* Resumption is the process of synchronizing a repaired item with other resources and operations and committing it to production. This process might involve restoring data and reprocessing backlogged transactions to roll forward to the current point in time (PIT).

12.6   Carrier/Supplier Management

Suppliers—equipment manufacturers, software vendors, or network service providers—are a fundamental component to the operation of any IT environment. The more suppliers that one depends on, the greater the number of problems that are likely to occur. In this respect, they should be viewed almost as a network component or resource. For this reason, dependency on individual suppliers should be kept to a minimum. This means that they should be used only if they can do something better and more cost effectively. It also means that organizations should educate their suppliers and service providers about their operational procedures in the event they need to be engaged during a critical situation.

Organizations should take several steps when evaluating a service provider's competence, particularly for emergency preparedness. Their business, outage, complaint, response, and restoration history should be reviewed. They should also be evaluated for their ability to handle mass calling in the event a major incident has taken place. A major incident will affect many companies and competing providers. Access to their key technical personnel is of prime importance during these situations. Providers should also have the mechanisms in place to easily track and estimate problem resolution.

When dealing with either service providers or equipment suppliers, it is a good idea to obtain a copy of their outage response plans. It is quite often the case that redundant carriers meet somewhere downstream in a network, resulting in a single point of failure. If a major disaster wipes out a key POP or operating location, one may run the risk of extended service interruption. With respect to carriers, plans and procedures related to the following areas should be obtained: incident management, service-level management, availability management, change management, configuration management, capacity management, and problem management.

A good barometer for evaluating a service provider's capabilities is its financial stability. A provider's balance sheet usually can provide clues regarding its service history, ubiquity, availability, levels of redundancy, market size, and service partnerships—all of which contribute to its ability to respond when needed. In recent years, insolvency is quite prevalent, so this knowledge will also indicate if their demise is imminent. Whenever possible, clauses should be embedded within service level agreements (SLAs) to address these issues.

A determination has to be made as to what level of support to purchase from a supplier. Many suppliers have many different types of plans, with the best usually being 24 x 7 dedicated access. A basic rule to follow is to buy the service that will best protect the most critical portions of an operation—those that are most important to the business or those that are potential single points of failure.

Some protective precautions should be taken when using suppliers. Turnover in suppliers and technology warrants avoiding contracts longer than a year. For critical services, it is wise to retain redundant suppliers and understand what their strengths and weaknesses are. Contract terms with a secondary supplier can be agreed upon, but activated only when needed to save money. In working with carriers, it is important to realize that they are usually hesitant to respond to network problems that they feel are not theirs. A problem-reporting mechanism with the carrier should be established up front. There should be an understanding of what circumstances will draw the carrier's immediate attention. Although such mechanisms are spelled out in a service contract, they are often not executed in the same fashion.

12.7   Traffic Management

Traffic management is fast becoming a discrete task for network managers. Good traffic management results in cost-effective use of bandwidth and resources. This requires striking a balance between a decentralized reliance on expensive, high-performance switching/routing and centralized network traffic management. As diversity in traffic streams grows, so does complexity in the traffic management required to sustain service levels on a shared network.

12.7.1   Classifying Traffic

Traffic management boils down to the problem of how to manage network capacity so that traffic service levels are maintained. The first step in managing traffic is to prioritize traffic streams and decide which users or applications can use designated bandwidth and resources throughout the network. Some level of bandwidth guarantee for higher priority traffic should be assured. This guarantee could vary in different portions of the network.

Traffic classification identifies what's running on a network. Criteria should be established as to how traffic should be differentiated. Some examples of classification criteria are:

- Application type (e.g., voice/video, e-mail, file transfer, virtual private network [VPN]);
- Application (specific names);
- Service type (e.g., banking service or retail service);
- Protocol type (e.g., IP, SNMP, or SMTP);
- Subnet;
- Internet;
- Browser;
- User type (e.g., user login/address, management category, or customer);
- Transaction type (primary/secondary);
- Network paths used (e.g., user, LAN, edge, or WAN backbone);
- Streamed/nonstreamed.

Those classes having the most demanding and important traffic types should be identified. Priority levels that work best for an organization should be used—low, medium, and high can work fairly well. Important network traffic should have priority over noncritical traffic. Many times, noncritical traffic such as file transfer protocol (FTP) and Web browsing can consume more bandwidth.

The distributed management task force (DMTF) directory-enabled networking (DEN) specifications provide standards for using a directory service to apply policies for accessing network resources [12]. The following is a list of network traffic priorities with 7 being the highest:

- Class 7—network management traffic;
- Class 6—voice traffic with less than 10 ms latency;
- Class 5—video traffic with less than 100 ms latency;
- Class 4—mission-critical business applications such as customer relationship management (CRM);
- Class 3—*extra-effort* traffic, including executives' and super users' file, print, and e-mail services;
- Class 2—reserved for future use;
- Class 1—background traffic such as server backups and other bulk data transfers;
- Class 0—*best-effort* traffic (the default) such as a user's file, print, and e-mail services.

12.7.2   Traffic Control

For each traffic class, the average and peak traffic performance levels should be identified. Table 12.1 illustrates an approach to identify key application and traffic service parameters. The clients, servers, and network segments used by the traffic, should also be identified if known. This will aid in tracking response times and identifying those elements that contribute to slow performance. Time-of-day critical traffic should be identified where possible. A table such as this could provide the requirements that will dictate traffic planning and traffic control policies.

This table is by no means complete. A typical large enterprise might have numerous applications. Distinguishing the most important traffic types having special requirements will usually account for more than half of the total traffic.

Table 12.1   Application Classification Example

Application	Host Location(s)	Type	Traffic Class	Priority	Bandwidth (G = Guaranteed; B = Best Effort)	Delay	Availability	Access Locations (Users)
Emt 1.1.2	ch01 at01 ny01	ERP	4	H	1 Mbps (G)	40 ms	99.99%	CH (40); AT (20); NY (100)
Ora 3.2	ny01	DBMS	4	H	1 Mbps (G)	25 ms	99.99%	CH (40); AT (20); NY (100)
User	ch01 at01 ny01	Miscellaneous	0	M	4 Mbps (B)	100 ms	99.5%	CH (40); AT (20); NY (100)
Gra 2.1	Ny01	Web site	2	M	1 Mbps (Peak) 56 Kbps (Min)	150 ms	99.5%	CH (40); AT (20); NY (100); Internet

Additional information can be included in the table as well, such as time of day, protocol, special treatment, and budget requirements. Tables such as this provide the foundation for network design. McCabe [13] provides an excellent methodology for taking such information and using it to design networks.

Traffic surges or spikes will require a dynamic adjustment so that high-priority traffic is preserved at the expense of lower priority traffic. A determination should be made as to whether lower priority traffic can tolerate both latency and packet loss if necessary. Minimum bandwidth or resources per application should be assigned. An example is streamed voice, which although not bandwidth intensive, requires sustained bandwidth utilization and latency during a session.

Overprovisioning a network in key places, although expensive, can help mitigate bottlenecks and single points of failure when spikes occur. These places are typically the edge and backbone segments of the network, where traffic can accumulate. However, because data traffic tries to consume assigned bandwidth, simply throwing bandwidth at these locations may not suffice. Intelligence to the backbone network and at the network edge is required. Links having limited bandwidth will require controls in place to make sure that higher priority traffic will get through when competing with other lower priority traffic for bandwidth and switch resources. Such controls are discussed further in this chapter.

Network traffic can peak periodically, creating network slowdowns. A consistent network slowdown is indicative of a bottleneck. Traffic spikes are not the only cause of bottlenecks. Growth in users, high-performance servers and switch connections, Internet use, multimedia applications, and e-commerce all contribute to bottlenecks. Classic traffic theory says that throttling traffic will sustain the performance of a network or system to a certain point. Thus, when confronted with a bottleneck, traffic control should focus on who gets throttled, when, and for how long. *Traffic shaping* or rate-limiting tools use techniques to alleviate bottlenecks. These are discussed further in Section 12.7.3.1.

12.7.3 Congestion Management

Congestion management requires balancing a variety of things in order to control and mitigate the congestion. Congestion occurs when network resources, such as a switch or server, are not performing as expected, or an unanticipated surge in traffic has taken place. The first and foremost task in congestion management is to understand the length and frequency of the congestion. Congestion that is short in duration can be somewhat controlled by switch or server queuing mechanisms and random-discard techniques. Many devices have mechanisms to detect congestion and act accordingly. Congestion that is longer in duration will likely require more proactive involvement, using techniques such as traffic shaping. If such techniques prove ineffective, then it could signal the need for a switch, server or bandwidth upgrade, network redesign, path diversity, or load control techniques, such as load balancing.

Second, the location of the congestion needs to be identified. Traffic bottlenecks typically occur at the edge, access, or backbone portions of a network—points where traffic is aggregated. They can also occur at devices such as servers and switches. This is why throwing bandwidth at a problem doesn't necessarily resolve it. Latency is caused by delay and congestion at the aggregation points. Throwing bandwidth could aggravate problems further, as data traffic, thanks to transmission control protocol (TCP), will try to consume available bandwidth. Adding bandwidth could result in a device becoming overwhelmed with traffic.

The next task is to redirect traffic around or away from the congestion. At this point, the exact source of the congestion may not yet be known; however, our mission-critical strategy is to keep service operational while recovering or repairing a problem. If congestion is building at a device, then the recourse is to redirect traffic to another device. Load balancing, discussed earlier in this book, can be an effective means to redirect traffic. For Internet environments, a less dynamic approach is changing the IP address resolution, which may require DNS changes. A secondary address pointing to the backup site would have to be prespecified to the DNS provider. If the backup site is on a different IP subnet than the primary site, instituting address changes can become even more difficult.

When making address changes, there must be some assurance that the backup site has the ability to service the traffic. This not only means having the adequate resources to conduct the necessary transactions, it also implies that the network connectivity to that location can accommodate the new traffic. The site serves no purpose if traffic cannot reach it.

12.7.3.1   Traffic Shaping

Traffic shaping is a popular technique that throttles traffic for a given application as it enters a network through a router or switch. There are several approaches to shaping that are discussed further in this chapter. One popular technique is based on the *leaky bucket* buffering principle of traffic management, which is intended to throttle traffic bursts exceeding a predefined rate. This concept is illustrated in Figure 12.3. A packet is assigned an equivalent number of tokens. As packets enter

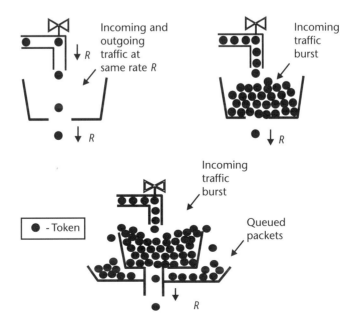

Figure 12.3   Traffic shaping illustration.

the network through a device, the device fills the bucket at a rate $R = B/T$, where R is the rate at which the bucket fills with tokens, B is the burst size equivalent to the size of the bucket, and T is the time interval that traffic is measured. R is in essence the defined rate over a period of time that a certain number of bits can be admitted into the network for a given application. B and T are administratively defined [14].

The device fills the bucket at rate R. As packets enter the network, the equivalent number of tokens is leaked from the bucket. As long as the bucket fills and leaks at the same rate R, there will be enough tokens in the bucket for each packet, and tokens would not accumulate further in the bucket. If traffic slows, then tokens accumulate in the bucket. A burst of traffic would drop a large number of tokens into the bucket. Traffic is still admitted as long as there are tokens in the bucket. If the bucket fills to the maximum size B, then any new tokens are discarded or queued. When packets are dropped, they must be retransmitted, else data will be lost.

Critical to the use of traffic shaping is the proper specification of the R, B, and T parameters. Enough burst capacity B should be assigned so that there is enough to handle more critical applications during periods of peak load.

12.7.4   Capacity Planning and Optimization

Capacity planning, more often than not, is performed on an incidental basis versus using a systematic, analytical approach. Much of this is attributed to several factors. First, declining system and memory unit prices have made it attractive to simply "pad" a network with extra protective capacity and upgrade on the fly. Second, the rapid pace and change with respect to technology and services makes extensive capacity planning a difficult and intensive effort, rendering it a backburner activity. Many IT organizations find too little time for planning and usually make it low priority.

Here lies the crux of the problem. Capacity planning should be treated as a business activity. It is the process of determining the IT resources and capacity required to meet business direction and growth. It should address how investment in IT will affect the firm's service levels to satisfy customer needs. It should also define how to effectively use the resources to maximize revenue.

Capacity planning and optimization are two complementary iterative processes. Capacity planning determines the topology and capacity of a network subjected to demand. The topology specifies how network elements will connect and interact with each other. Capacity will specify how they are to perform. Performance is defined using such variables as network load, peak and average bandwidth, storage, memory, and processor speeds.

Central to capacity planning is the estimation and forecasting of anticipated network load. There are many mathematical modeling approaches available, but in the end none compare to human judgment. Models are only as good as the assumptions behind their specification and the data they process. Yet models are often preferred as providing an objective view of network load and are used to confirm judgments already made. It is quite often the case that a model user knows what answers they expect to obtain from a model.

Incomplete data or too much data are often additional inhibiting factors. Network complexity often makes it difficult to accurately model an environment. Using simplifying assumptions to model behavior can lead to erroneous results. For example, use of linear regression is insufficient to model IT systems that in general exhibit nonlinear and exponential characteristics. A straight-line model will often wrongly specify response times for a given load pattern (Figure 12.4). Queuing theory shows that CPU throughput can saturate at different load levels. In fact, server response times tend to increase exponentially once a system's utilization exceeds 50%. This can affect the decision of whether to add systems to handle the estimated load [15].

Similarly, models must be specified for each type of component in an IT environment. This includes models for storage capacity, memory, and CPU utilization. Further complexity is encountered when it is realized that all of the components interact with each other. To make life easier, many types of capacity-planning tools have emerged on the market. One will find that no one tool adequately addresses all aspects of an IT environment. Many represent point solutions, with emphasis on a particular function or group of functions. In any case, the network manager must define what data, statistics, trends, systems, and models are appropriate and sufficient to characterize the environment they manage. Planning tools fall into three general categories [16]:

1. *Long-term* planning tools collect historical data and compute trends over time. This usually requires much data to arrive at a meaningful result. For this reason, they are not very popular. Today's environment prefers short-term solutions based on minimum data input.

2. *Real-time* tools collect data in real time and work within a short-term planning horizon, on the order of 90 days. Because they address immediate problem solving, they are more popular. Many view them as performance-management tools, without any long-term planning capabilities.

3. *Predictive modeling* tools simulate traffic over a computerized model of an operating network. They are effective in gaining insight using "what if" scenarios. The problem with these tools is accuracy. When using a computerized model, it is prudent to calibrate the model based on the current network operation. The model should be able to simulate the current network using current load patterns to match the existing network behavior [17].

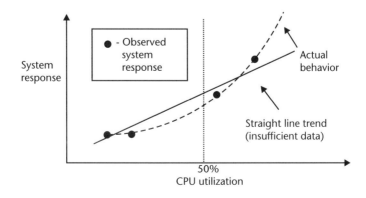

Figure 12.4   System utilization modeling example.

When choosing a capacity planning software tool, look for the following characteristics:

- The tool requires minimal data for analysis.
- It has the ability to isolate applications that consume bandwidth.
- It has the ability to identify underused resources.
- It has the ability to model "what if" scenarios.
- It provides practical guidance on network optimization.

The last item is particularly important when it comes to models. Many will define optimal solutions that are impractical to deploy or whose results can change significantly with variations in traffic patterns. Optimization is the process of tuning a network for capacity, speed, and performance. The objective is to neither under-provision nor overprovision, neither too soon nor too late. When unsure as to how to optimize a network, a rule to follow is to maximize use of the existing capacity and then add more capacity only in those areas where required. This requires prioritizing those applications that are most critical and allocating bandwidth, CPU, software, storage, and switching resources among them.

As said several times already, throwing bandwidth at a problem does not necessarily fix it. Broadband links do not assure speed and service reliability. Bandwidth currently exceeds CPU speeds, so impedance matching is required throughout a network. Overreliance on protection capabilities of switches and routers can prove fatal. For instance, routing protocols based on open shortest path first (OSPF) will assign traffic to least-cost links, without regard to traffic load. Traffic accumulating across all routes using a particular link can easily exceed the link's capacity. To get around this, routing approaches that address traffic load can be used. These are discussed in the next sections.

12.7.4.1 MPLS Planning

Multiprotocol label switching (MPLS) provides an opportunity to avoid this particular problem. MPLS is a combined connectionless and connection-oriented packet forwarding mechanism. As congestion increases, MPLS packets experience delay just as any other mechanism. However, MPLS traffic engineering can ensure that label switched paths (LSPs) can be defined so that congestion is minimized and service levels are met. This is because MPLS assures that every packet in a flow travels the same carefully engineered route for predicted traffic.

To engineer for MPLS traffic, the flow of traffic between an origin and destination is made to follow a specific LSP. This is unlike the best-effort provisioning that is characteristic of TCP/IP networks, where every packet travels a different route and is then reassembled at the destination. Furthermore, IP routing fails to establish routes based on any one suitable metric. On the other hand, MPLS traffic can be engineered based on the maximum bandwidth of each network link. The sum of the engineered bandwidth over all LSPs using a link should not exceed the maximum link bandwidth.

Engineered LSPs make use of constraint-based routing (CBR) to avoid exceeding the link bandwidth. CBR assigns excess traffic to other paths based on a predefined load ratio. This in essence redirects and spreads traffic among multiple LSPs, which is particularly helpful in cases where a traffic surge occurs or a network link or device fails. The quality of service (QoS) and traffic parameters, such bandwidth, delay, and loss can be defined for a particular service class. High-priority traffic is assigned to LSPs, which can provide the resources to satisfy the required QoS of the service [18, 19].

LSPs are established and removed using the constrained routing label distribution protocol (CR-LDP), which is an extension of LDP. LDP was originally designed to set up LSPs service flows on a hop-by-hop basis, versus an engineered basis. The combined LDP and CR-LDP protocol runs over TCP.

12.7.4.2   RSVP Planning

Resource reservation protocol (RSVP) is an Internet Engineering Task Force (IETF) standard that has been used for many years. RSVP with traffic engineering extensions (RSVP-TE) expands the RSVP protocol to support label distribution and explicit routing for different services. Like CR-LDP, RSVP-TE can set up traffic engineered LSPs based on QoS information and automatically allocate resources. Unlike CR-LDP, it is not confined to only that portion of a network that uses MPLS. RSVP runs over IP, with RSVP-TE invoked using user datagram protocol (UDP), making it usable across an entire network end to end [20]. This avoids interoperability issues at network boundaries and allows the engineering of traffic flows between customer premises.

Although it uses UDP, RSVP-TE has included enhancements to enable it to run with the same reliability as with TCP, allowing recovery from packet loss in certain instances. There is much debate as to whether RSVP-TE or CR-LDP is better. RSVP-TE may have an edge because of its inherent compatibility with IP [21].

12.8   Service-Level Management

Until recently, many firms implemented network management using a piecemeal approach, collectively managing the health of various network devices. This approached has changed somewhat in recent years. Instead of focusing solely on processor downtime and throughput, today's approach focuses on maintaining levels of service for end users. Service-level management (SLM) is the process of guaranteeing a specified range of service level. These levels are usually contained in SLAs, which are contracts between two internal organizations or between a service provider and a customer.

Using some of the aforementioned techniques, services can be classified and service levels can be defined, based on the services performance requirements. For each level, an objective is defined which in essence is a metric that will help characterize whether the service is achieving the desired level. Typically, three basic classes of metrics are defined: availability, reliability, and response time. Also specified is an understanding as to where these measures would take place, either on a device, segment, link, or end-to-end basis.

SLM has given rise to many software tools designed to automate the SLM process. Used in conjunction with some of the previously mentioned capacity-planning tools, SLM tools allow firms to proactively provision a service from end to end in an existing network. They also provide some of the reactive features that respond to network events. The better SLM tools offer detailed monitoring in addition to the service level. They can monitor a variety of network elements, including servers, LANs, applications, and WAN links for compliance. They also assist in fault isolation, diagnostics, problem escalation, performance assessment, and trending. Some can provide reports and measurements in a format usable by other applications, such as billing.

12.9   QoS

QoS is a global term for a variety of techniques designed to prioritize traffic and enforce the policies to manage latency and bandwidth requirements. QoS is designed to give the most mission-critical traffic preferential treatment as it traverses a network. When users encounter long response times and time outs while using important applications, it is indicative of poor QoS. QoS has become the mechanism to ensure that important applications perform satisfactorily for users.

QoS requires continuous real-time monitoring as well as careful traffic engineering and planning in order to build in required performance. The goal of QoS is to ensure that *all* traffic in a given network meets their required service levels. In times of congestion, it should assure that the most essential traffic obtains the resources it needs to perform satisfactorily.

A secondary benefit of implementing QoS in a network is the ability to optimize use of a network, forestalling the expense of adding resources. For example, it can defer the costs of adding bandwidth and CPU to a network. In light of declining bandwidth, server, and storage costs, this may not seem compelling. But such costs can be significant for large enterprise networks, particularly in an economically depressed environment.

QoS is best used in large multiservice networks, where bandwidth is up for grabs. The last mile, which is the most vulnerable part of network, is usually the one portion of a network that most benefits from implementing QoS.

12.9.1   Stages of QoS

In this section, we will try to explain QoS according to the logical sequence of steps that are involved. Then, we will discuss how QoS is applied to the more popular networking technologies and environments (as of this writing). The following sections describe the steps in developing QoS criteria [22].

1. *Traffic classification.* In order for a network to enforce QoS for a service, it must be able to recognize traffic flows supporting that service. A *flow* is a conversation between an origination and destination identified by layer 3 addresses and layer 4 ports, which identify application services. Network devices must inspect each packet they receive, looking for *marks* indicating the class of service of the packet. Once a mark is recognized, the appropriate performance mechanisms are applied to the packet. As of this writing, there is no standard way of classifying traffic per se. One version was discussed earlier in this chapter. The IETF is trying to establish a common classification

scheme for use by routers. The following are two known standards developed by the IETF for classifying data traffic:

- *Intserv.* Integrated services (Intserv), sometimes referred to as IS, is an approach that uses RSVP to reserve network resources for particular traffic flows. Resources such as bandwidth and latency are reserved for a traffic flow. Because of the overhead involved in maintaining the status of every flow in the network, Intserv is not very practical [23].

- *Diffserv.* Differentiated services (Diffserv) is an alternative approach that has found greater acceptance [24]. IP packets are classified at the network edge using the IP version 4 (IPv4) type of service (TOS) field or the IP version 6 (IPv6) traffic class field to classify a service (these fields are discussed in the next section). Based on these fields, the QoS treatment or per-hop behavior to be applied on a per-packet basis can be specified. Once the TOS is established, the packet is queued and buffered along the way, using a mechanism called weighted random early detection (WRED), which is described further in this chapter. For MPLS networks, the class of service (CoS) field in the MPLS header is also set accordingly at the ingress LSR. This field is used to classify and queue packets as they travel through the MPLS portion of a network.

2. *Marking.* Marking is the process of coding the service classifications within packets so that network devices can identify theme. Both classification and marking should be performed at the network edge. There is usually a chance that a packet will be marked again as it traverses a network. If at all possible, it should be marked with the class. In layer 2 networks, frames are tagged using the IEEE 802.1p standard (this is discussed further in Section 12.9.3.1). For layer 3 IP networks, the TOS byte, an 8-bit field in the IP header, is coded with one of the following possible values:

- The *differentiated services code point (DSCP)* populates the first 6 bits of the TOS byte. DSCP specifies the per-hop behavior that is to be applied to a packet. Not all equipment vendors yet support DSCP.

- *IP precedence* is a 3-bit field in the TOS byte that is populated with IP precedence. Values 0 (default) to 7 can be assigned to classify and prioritize the packet. IP precedence is being phased out in favor of DSCP.

- The *type of service field (ToS)* is a code with values from 0 to 15 that populates the TOS byte to convey whether the packet requires any special handling. The ToS field is also being phased out in favor of DSCP.

3. *Policing.* Policing is the process of enforcing the treatment of packets based on their classification and prevailing network conditions. Incoming and outgoing packets from a network are policed using various mechanisms. Policing enforces the prioritization of traffic derived from the previous two steps. During congestion, low-priority traffic is throttled in favor of higher priority services. The following are several mechanisms that are used [25]:

- *Traffic shaping.* Traffic shaping was discussed earlier in this chapter. As congestion is detected, the volume and rate of incoming and outgoing packets for particular flows are reduced. Packets and can be either discarded or queued. Typically, an application host, router, switch, or firewall can apply the mechanism at the TCP level [26]. This prevents non-critical traffic from overwhelming a network during congestion. When applying traffic shaping, low-priority traffic should not be starved out completely. Instead, it should be scaled back by allowing it to drip out onto the network. Often, many devices focus only on traffic ingress to a network, versus controlling it throughout the network. For traffic shaping to be effective, it should be used uniformly across a network. It can work well to apportion bandwidth for egress traffic by providing more bandwidth for users requesting important applications.

• *TCP rate shaping.* Another related traffic-shaping technique is *TCP rate shaping*, sometimes referred to as TCP *window sizing* [27]. This mechanism adjusts the TCP window size to control the rate at which TCP-based traffic is transmitted. If the TCP window is full, a host pauses transmission. This has the effect of slowing traffic flow between two devices.

• *Queuing.* Queuing is the means whereby traffic shaping is accomplished. Queuing is a method of dictating the order in which packets are issued to a network. Various strategies can be used in conjunction with some of the policing techniques under discussion. Queuing based on the service class is preferred, as it can be used to assure service levels for critical traffic. Queuing should be performed in core routers and edge devices to assure consistent traffic prioritization. Heavy queuing of lower priority streamed traffic can introduce enough latency and jitter, making it useless. Latency-sensitive traffic should be prioritized appropriately so that it can achieve the specified level of service for its class.

  Packet dropping (also referred to as *tail dropping*) occurs when a queue reaches its maximum length. When a queue is full, packets at the end of the queue prevent other packets from entering the queue, discarding those packets. When a packet drop occurs, it results in the far end device slowing down the packet transmission rate so that the queue can have time to empty. The following describes two popular queuing schemes:

  – *Weighted far queuing* (WFQ) creates several queues within a device and allocates available bandwidth to them based on administratively defined rules. Weights can be assigned to each, based on the volume of the traffic in queue. Lower weights can be assigned to low-volume traffic so that it is released first, while high-volume traffic uses the remaining bandwidth. This avoids queue starvation of the lower weighted traffic. Queue starvation is a term used to denote situations arising from inadequate queue space, resulting in undelivered, or discarded, packets.

  – *Priority queuing* assigns a queue priority, usually from high to low. Queues are served in priority order, starting with the high-priority queue first, then the next lower priority queues in descending order. If a packet enters a high-priority queue, while a lower priority queue is being serviced, the higher priority queue is served immediately. This can ultimately lead to queue starvation.

• *Random early detection.* Random early detection (RED) is a form of congestion control used primarily in routers. It tracks the packet queue within the router and drops packets when a queue fills up. RED was originally intended for core Internet routers. It can result in excessive packet dropping, resulting in unwanted degradation in application performance. Excessive unwanted packet loss can result in unnecessary retransmission of requests that can congest a network.

- *Fair bandwidth.* Fair bandwidth, sometimes referred to as *round robin*, is a strategy that simply assigns equal access to bandwidth across all services. Although it may seem crude, it is in fact the most prevalent QoS mechanism in use today. Most LAN and Internet usage is done in a best-effort environment, whereby all traffic has the same priority. The net effect of this is that applications that require significant bandwidth or latency to function properly (e.g., video teleconferencing or voice over IP) will be short changed in order to provide bandwidth to other users.

- *Guaranteed delivery.* Guaranteed delivery is the opposite of fair bandwidth. It dedicates a portion of bandwidth for specific services within a network, based on their priority. Other, less important applications are denied bandwidth usage, even during congestion.

12.9.2   QoS Deployment

QoS management uses policing devices and software tools to administratively specify the QoS parameters and rules. This represents the greatest hurdle in implementing QoS. Many solutions manage devices on a one-by-one, versus networkwide, basis. The challenge for most network managers is too integrate, either technically or manually, a set of tools into a coherent network-management system.

QoS devices calculate QoS statistics from real-time raw data received from the network. The administrative tools allow the manager to specify the availability and performance parameters. When starting out, it is best to deploy QoS on a gradual basis, starting with the most basic QoS needs. This allows a manager to test and learn how such devices enforce QoS. One of the first and most basic services that will likely require QoS is voice over IP (VoIP), which needs guaranteed latency to function properly.

As QoS requirements grow, managing QoS on a device basis could become unwieldy. A central repository for router or switch definitions across a network will be required. Using a directory service, such as NDS or Active Directory, is an effective way to define and retain traffic priorities. The DEN recommendations that were earlier described can be a good starting point.

The following are some approaches to how QoS is implemented within different kinds of network devices. Each has different trade-offs regarding the level of control and the effect on network performance:

- *Traffic shapers.* These are appliances specifically designed to perform the traffic-shaping function. Many have found use in conjunction with managing access links between a router and a WAN or ISP network. They can be situated on the outside of an edge router to control traffic destined to a WAN. Sometimes they are placed inside an edge network just before the router to apply policy decisions to local traffic. Some devices can be configured to treat packets based upon a variety of parameters in addition to service type, such as the type of protocol, application IP sockets, and specific pairs of IP addresses.

- *Routers*. Routers, also known as layer 3 switches, are viewed as the most appropriate location to police QoS. However, to perform this function in real time requires additional packet processing overhead to classify each packet. This can add further delay to the routing function. The service time per packet and consequently the number of queued packets can increase rapidly with load. Routers also tend to be inflexible in reallocating resources to services when conditions change. Furthermore, router settings, particularly at the edge, must be coordinated with those of the WAN service provider or ISP.

- *Load balancers*. Earlier discussion in this book showed how load balancers could play a major role in controlling traffic. They can, in fact, serve as an appropriate place for policing because they can alleviate bottlenecks in the edge that could otherwise make all core router QoS mechanisms ineffective. As of this writing, some load balancer products are just beginning to incorporate QoS capabilities, in the same way they have taken on firewall and security features. It is currently unclear as to how load balancers can be used in conjunction with other QoS devices to assure end-to-end policing. Some view them assuming a passive role, checking and certifying that a packet's QoS settings are unchanged from their intended settings. Others see them as taking a proactive role, explicitly changing settings during congestion.

- *Caching devices*. From past discussion, we noted that caching devices are used to direct user requests to a device that can locally satisfy a request for static content, with the net effect of preserving Web server performance. A caching device can be used in conjunction with a policing device, such as a traffic shaper, by controlling and containing service traffic flows representing repetitive content requests and servicing them locally.

12.9.3 QoS Strategies

QoS can take on different roles or can be leveraged differently in situations beyond an IP network. The following sections describe some common situations warranting special consideration in deploying QoS.

12.9.3.1 Ethernet

Because Ethernet dominates a majority of enterprise LAN environments, the IEEE has devised some mechanisms to address QoS. The IEEE 802.1p standard has defined a 3-bit value that assigns eight priority class values to LAN frames. The value is used to *tag* Ethernet frames with certain priority levels. The value is inserted in an IEEE 802.1Q frame tag. (The IEEE 802.1Q specifies virtual LAN standards.)

The priority tag is used in a similar manner as IP precedence, but at layer 2. Some routers will use this tag to define an IP precedence or DSCP value to be placed within the IP header. Use of the tags requires having a hub or switch with the ability to recognize and set the tag values. The device should also have queuing capacity to store packets that are queued. Many Ethernet edge switches come equipped with two queues, while backbone switches feature four queues.

12.9.3.2 LAN/WAN

QoS can be used in conjunction with WAN access points. LAN/WAN interface points are typically prone to congestion and can be single points of failure if not properly architectured [28]. Traffic from within the LAN can accumulate at these points and be placed on WAN links with typically far less bandwidth than that available on the LAN. Furthermore, protocols can differ as a packet travels from a LAN host through a WAN network, and then to a host on a destination LAN.

When monitoring WAN performance, it is desirable to monitor from network edge to network edge so that local loop conditions can be detected, versus monitoring between carrier points of present (POPs). Monitoring should be done via devices attached to WAN interfaces, such as routers, intelligent DSU/CSUs, or frame relay access devices (FRADs).

For an enterprise, QoS policing could be used to assure WAN access for mission-critical applications, avoiding the need to upgrade expensive access or WAN bandwidth. If at all possible, the LAN QoS priority levels just discussed should be mapped to the WAN provider's QoS offerings. Traffic shapers placed in front of an edge router should have the ability to discern local traffic from that destined for the WAN. Some traffic shaping devices can recognize forward explicit congestion notifications (FECNs) and backward explicit congestion notifications (BECNs), which typically indicate WAN congestion.

12.9.3.3 RSVP

RSVP is a protocol that can be used to facilitate QoS. Applications can use RSVP to request a level of QoS from IP network devices. In fact, RSVP is the only protocol under consideration by the IETF to implement Diffserv. RSVP-TE, with the ability set up traffic-engineered paths, enables users to define such paths based on QoS attributes. QoS signaling within RSVP is limited only to those traffic flow services offered by the RSVP protocol, namely the guaranteed service (GS) and controlled load service (CLS). Both are designed to ensure end-to-end service integrity and avoid interoperability issues at network boundaries.

12.9.3.4 CR-LDP

CR-LDP is another means of implementing QoS. It provides several QoS parameters to dictate IP traffic flow. Traffic-engineering parameters can be signaled to devices to reserve resources based on QoS parameters. Latency can be specified using parameters that convey frequency of service. LSPs can be assigned according to a specified priority if needed. To assure QoS consistency at the edge network, CR-LDP QoS settings must correspond to the service-class parameters defined by the WAN. Frame relay, for example, defines default, mandatory, and optional service classes that specify different delay requirements that can be mapped to CR-LDP QoS parameters.

12.9.3.5 ATM

The CR-LDP QoS model is actually based on the ATM QoS model, which is quite complex. Conforming to ATM QoS requirements at the network boundaries has contributed to the problems in providing end-to-end QoS for IP applications over ATM backbones. A better approach is to adhere to the IP QoS approach using Diffserv and RSVP.

12.9.3.6   SONET

Synchronous optical networks (SONET) are precisely engineered with implicit mechanisms to assure QoS. The meaning of QoS when applied to SONET differs from that which has been discussed thus far. Because SONET is a layer 1 technology, the QoS of the higher layer traffic is left up to those layers. However, SONET does have some features to help assure QoS for the higher layers.

Several metrics are used to characterize QoS. An errored second (ES) is encountered when a transmission error is identified, such as a violation of bit error rate (BER) thresholds. A severely errored second (SES) is defined when multiple errors occur. Ten sequential SESs lead to unavailable seconds (UAS). UAS is used as a key parameter to characterize the availability of a SONET circuit. Because APS helps prevent UAS, continued UAS could indicate a significant problem [29].

Most SONET equipment includes capabilities to monitor UAS by examining overhead administrative bits carried by a SONET signal. An alarm is usually generated when UAS exceeds a specified threshold along a circuit's path. It is important to check with the equipment vendor to see how UAS is measured, as vendors may use different approaches. If integrating a multivendor network, UAS should be measured in a consistent fashion throughout the network.

12.9.3.7   Streaming Services

Streaming services such as VoIP and video typically involve time-sensitive traffic. Although many of the previously discussed IP-related QoS mechanisms are applicable, they are not sufficient. VoIP traffic requires specific bandwidth, jitter, and latency while traversing a network. It also requires examining packet information at layers above layer 3, such as required by the H.323 and IETF session initiated protocol (SIP) videoconferencing standards. Such services require priority handling across intermediate network devices to slow or speed up packet delivery to the destination so that the end user sees a constant rate.

Vendors are actively addressing this issue because VoIP will likely be one of the premier services requiring QoS. Additional QoS capabilities are being embedded into equipment to address these services. VoIP traffic will most likely require higher priority than instant messaging or Web surfing traffic, for example. In addition to QoS, capacity planning is required to identify and reserve the bandwidth required to support these services.

12.10   Policy-Based Network Management

Policy-based network management (PBNM) is the term describing a class of approaches designed to centrally classify, prioritize, and control traffic in a network. A network manager establishes a set of rules or policies that dictate how network resources are to be used for different traffic classes.

SLM is realized through software tools called policy managers. Originally intended for defining queuing mechanisms on routers, these tools are gradually being transformed into PBNM tools that can be used to actively manage and monitor the application of QoS mechanisms in a network [30]. Figure 12.5 illustrates, at a conceptual level, the relationship between PBNM and some of the other QoS components previously discussed. PBNM is also being positioned to include other aspects of network management, encompassing different protocols and protocol layers. Areas such as security, DNS, and VLAN are also covered.

Many of these tools are still in their infancy and require substantial interoperability with various OSs and devices. This is due in part to lack of a standardized approach to configure and control information across various vendor devices and applications. As discussed earlier, the DEN effort is representative of many approaches to arrive at a standard. A directory-related approach like DEN is preferred, but implementations are still varied. The IETF has been actively working on

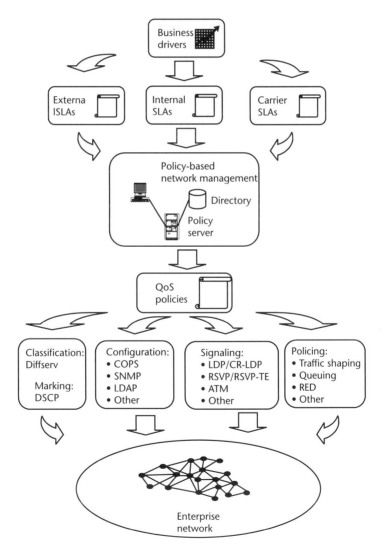

Figure 12.5   Policy-based network management.

a standardized PBNM approach. PBNM can be broken down into three basic functions [31]:

1. *Configuration* is the provisioning of network devices, such as switches or routers, with parameters that dictate how packets are treated for different types of traffic. Several tasks are involved:
   - *Policy-based configuration* is the task of identifying and selecting how devices are to be configured. These rules control the distribution of policies among devices. Policies can be distributed and invoked by device type, geographic location, time of day, or special events.
   - *Configuration of policy* is the task of setting the parameters on selected devices. For example, parameters can be set for QoS, security, or DNS.
   - *Policy-based configuration of policy* is the task of specifying how to determine the types of configuration parameters for a given device.
2. *Enforcement* is the policing of the provisioned policies.
3. *Verification* is the auditing of a network to see if a policy has taken place and to monitor its overall effect. A network traffic patterns change, so do policies that are being implemented.

Policy rules are grouped into several categories. *Conditions* are those events that cause a policy to take effect. They can range to almost anything within the scope of the invoking device; *actions* dictate what is to be done when a condition takes place; *roles* define how an action is executed. A role classifies a device or interface as being a core, gateway, or access device. Policies are then implemented based on the role of the device.

So, for example, a condition may explicitly state the required characteristics of the traffic entering a device for an action to be performed, in the same way an IF-THEN condition is used in software programming. Such characteristics can include the router port, originating media access control (MAC)/IP address, hypertext transfer protocol (HTTP) uniform resource locator (URL), or TOS value. Traffic with the characteristics that match the condition is then treated according to the predefined actions (e.g., queued, TOS changed, or shaped according to certain rules).

All of this may seem somewhat esoteric at this point, and this is precisely why standardized practices are dragging along. Implementing a PBNM system requires using various protocols to allow network devices, policy servers, and applications to communicate and interoperate with each other. Diffserv capabilities will likely be used to specify the TOS field in the IP header. RSVP will be used to enable applications to communicate their QoS requirements and obtain a TOS, based on the PBNM policies.

Although Diffserv and RSVP represent accepted standardized mechanisms for real-time implementation, there are still some differences in approach regarding device configuration. The following are those that are currently being considered [32]:

- *Common open policy serve* (COPS) is a protocol that has two forms. One form, called COPS-PR, is used to provision devices, namely routers, by another device. Another form, COPS for dynamic QoS (COPS-QoS) is a subset of COPS-PR that is used to obtain and implement QoS policies defined by another device. COPS is gradually being positioned as the protocol for PBNM device configuration and management. COPS has a well-defined parameter set that has the potential to be understood by an array of devices.

- *Telnet* is a protocol used to access the command-level interface (CLI) of a device, allowing a manager to input device commands, usually for the purpose of troubleshooting. This requires knowing the command line syntax of a device. Because CLIs can vary widely among different devices and device versions, it is impractical to implement Telnet as a standard protocol for configuring PBNM.

- *SNMP* is a protocol that is limited to utilizing a device's MIB variables. PBNM configuration will likely require specification of variables beyond those within the MIB.

- *Local directory application protocol* (LDAP) is used to access user and resource data from stored directory information. It can be used to store and distribute policy information, but this implementation can vary by vendor. No common directory format is currently available. Use of LDAP is premised on the belief that a system's user directory is a central repository of information required for PBNM implementation. Directories will typically define the user types for controlling file or device access based on organization. It is felt that this information can also be leveraged to define user traffic policies as well.

- *HTTP* and the *Java* programming language are the most common mechanisms in use today to implement PBNM configuration. HTTP is an application protocol that can govern file exchanges over the Web. Java is popular because it can be used to create configurations among servers and clients in a network.

12.11   Service-Level Agreements

Service-level agreements (SLAs) are contracts between a network provider and a user. The SLA specifies what service level the network provider will guarantee a network user. There is no standard for an SLA—its use varies by context. In today's world, there are two forms of an SLA.

- An *external* SLA is made between a customer and a service provider, such as an LEC, WAN provider, or ISP. The SLA consists of two fundamental parts. First, there is a *service level objective*, which is a condition required by the SLA customer that the network manager is to provide. The second part is a *penalty* that the network provider will face if it fails to meet the service-level objective. Not surprisingly, penalties have been the subject of much dispute and have been the problem leading to a slow adaptation of SLAs in practice. Many carrier SLAs, for instance, have failed to place any stringent penalties on carriers when failing to comply with an SLA.

   There have been many creative approaches to penalties. External SLAs often specify a straight fixed dollar amount to be paid or credited by the network provider. Some credit the customer just for the time service was inadequate based on the negotiated service rate. Neither case addresses the issue of potential business loss that a customer can entail in light of an SLA delinquency, particularly in the case of carrier SLAs. However, growing reliance on networking for e-commerce has caused many enterprises to take a stronger stand in specifying stricter penalties in their SLAs.

- An *internal* SLA is defined within an enterprise by an IT organization to guarantee and assign service levels to customer organizations or even end customers. Unlike external SLAs, internal SLAs can be used as a means for management to communicate business needs and subsequently network needs to their IT organizations. Internal SLAs may require more periodic review by both parties due to frequent or unexpected requests from internal business units arising from business shifts. An IT organization must in turn have a more flexible and expedient mechanism in place to specify policy changes into their network or create new network services.

In either case, SLAs are seen as the driver for PBNM. They are documents that network managers use to program policy rules and parameters into the PBNM systems. SLAs are transforming from specifying aggregate measurements averaged over a month to more detailed and granular measures. This is being driven by the need to maintain performance on a transaction basis. Network managers are being pressured into translating business-related metrics such as revenue and downtime into network-related metrics, such as latency and bandwidth. These metrics can then be incorporated into PBNM policies.

This means that software tools must be in place that can conveniently take SLA requirements and convert them into policies. Although some tools are appearing on the market to perform and automate this task, it is expected there is still some manual effort involved and such tools are not all encompassing. Further, it takes some strategy to decide what SLA criteria an IT organization is capable of complying with. Many managers will take a conservative approach and guarantee service levels well below their operational norm for only conditions that they can monitor.

SLAs should be meaningful and should not be implemented just as a formality. Service-level objectives should be selected that best characterize a network service. For each service-level objective, metrics that best exemplify how that that objective is being met should be identified. The volume of service provided, the target levels of service quality, how they are to be measured, and associated restrictions, assumptions, penalties, measurement criteria, reporting, and enforcement need to be specified. The last area is typically one of contention. A customer will need to prove that a service-level target was indeed violated by a network provider. Both parties would need to agree upon what mechanism will be used as the basis. The SLA is valid only if both parties agree and adhere to the enforcement mechanism.

12.11.1   Carrier/Service Provider Agreements

Carrier SLAs are rapidly growing in importance among firms. A carrier SLA is a key component of a mission-critical network. The following are some precautions to follow when formulating carrier SLAs:

- Because a carrier SLA can be a single point of failure, multiple SLAs with multiple carriers should be developed.

- An SLA may only be reflective of a carrier's normal operating conditions. If widespread disasters or extraordinary events occur, the SLA terms may not be applicable. Contingency plans for these circumstances should be spelled out.

- Problem management and escalation procedures should be spelled out in an SLA. Escalation should be based on the level of impact of a network problem on a customer's operation.

- Service-provisioning requirements should be specified. They should be designed to assure confidence that services would be installed when needed, especially when outages occur.

- SLAs should also indicate carrier requirements in the event the customer faces a significant outage. This is an insurance premium a customer will pay to capture the carrier's attention and resources during trying circumstances. Recovery metrics and objectives previously discussed in this book are applicable.

- SLAs are no guarantee that an outage will not take place. Good mission-critical design practices as described in this book provide the best protection against outages.

- A carrier's advertised performance should not be the basis of an SLA. A carrier's advertised guarantee will reflect overall network performance, averaged over many customers. An SLA should assure the performance of an individual company's traffic volume and pattern. After all, a carrier will maintain one customer's performance and the expense of other customers.

- The interval, frequency, and network boundaries of service-level metrics should be spelled out. Metrics averages over long time intervals, like a month, can mask substandard performance. Availability, latency, packet loss, installation time, and other relevant metrics should convey end-to-end conditions as much as possible.

- The management tools that will be used to monitor and enforce the SLA should be identified. Instrumenting, a term used to signify the use of special measuring instruments (i.e., probes), should also be spelled out. Unlike instruments, management tools can apply corrective actions and identify root cause. Both parties must agree upon these.

- SLA terms should call for immediate, penalty-free exit for the customer in the event a carrier undergoes bankruptcy, undergoes a merger, or is acquired by another organization.

The topic of SLAs is an extensive one. It likely requires far more treatment than is provided here. It involves discussion of technical, financial, and legal issues, surpassing the limitations of this book. The author hopes to have covered some essential thinking and hopes that the reader will follow through using other information sources.

12.12   Change Management

Improper modifications of system configurations are often the blame for many network outages. Change management is the process of controlling who can make changes to systems, when they are made, and how they are made. When implementing or operating an enterprise network, haste makes waste. Quite often, business pressure will require an IT organization to perform sweeping change to an environment without understanding the impact of the change.

Network changes and configuration of equipment should be done in phases. New features, applications, systems, and locations should be added gradually to an environment. No feature should be incorporated into a network unless it can be operationally supported. For example, many telecom providers will not introduce any new service features in a network unless their operations and support systems (OSSs) have been modified to support the feature. The same practice should likewise be followed in an enterprise network.

More often than not, many IT organizations conduct change management in a reactive mode—changes are made as items break. Instead, change management should be conducted proactively. Changes should be made to prevent mishaps, add new users, or add new features. The impact of a change should be understood and documented prior to making the change. This should include a checklist itemizing all items that will be changed and the effect each has on systems or applications in the network. Performing the change first in a lab or test environment is preferred, versus making changes on a live network.

As new equipment is incrementally installed in a network, interoperability with legacy systems and protocols should be maintained. The new system should operate in parallel until there is confidence that it can function properly. Live traffic should be migrated gradually to the new system. The older system should be decommissioned once the new system can handle live traffic load for a period of time without problem. A back out plan should be devised so that if a live change has negative effects, the change can be removed and the environment can be rolled back to the state it was prior to the change.

This strategy can be applied at the device, application, service, or geographical levels. Not only does it assure that a change will work, it also ensures minimal disruption of live service. Many find that a problem with this strategy is the time and expense it requires. An alternative strategy is to flash cut a change into a live network during off hours to minimize disruption and then back out if necessary. Although this approach can be effective, the evolution to 7 x 24 operation could eventually make this approach dangerous. Setting up a test lab environment is an effective way to simulate system changes in a 7 x 24 operation.

12.13   Summary and Conclusions

Managing mission-critical networks boils down to ensuring service levels. It involves managing network communications, configurations, security, systems, and software. Network-management tools can be used to reduce complexity. They invariably consist of a dedicated server that accumulates data, status, alarm, and performance information for viewing by the network manager. Network-management tools must effectively consolidate this information so that the cause and effects of failures and overloads can be readily recognized. The right amount of information should be collected so that decision making can be done quickly and easily, with minimal overhead. A prerequisite for the use of these tools is a complete and accurate view of network topology.

SNMP is a widely use standard in network management. Many network management tools employ SNMP agents—software processes that run on network components and collect and report status information. Proactive agents can identify and correct problems before they even occur. Because SNMP commands are issued in band over a production network, their effectiveness is prone to the very same network problems they discover. Probes can provide an additional means of monitoring. These are hardware devices that passively collect data across network links.

Above all the sophisticated tools, most important is having a systematic and logical process to respond to adverse events. At a minimum, the process should involve some fundamental tasks. These include fault detection, isolation, correlation, root cause analysis, and correction. Likewise, a systematic restoration process should be in place that involves activities related to problem containment, contingency, notification, repair, and resumption.

As the diversity in traffic streams grows, so does traffic-management complexity. At the outset, traffic management requires prioritizing traffic streams and deciding which users or applications can use capacity and resources under different situations. For each traffic class, average and peak performance levels should be identified. When traffic spikes occur, dynamic adjustment is required so that higher priority traffic can be allocated more network resources over lower priority traffic. Simply throwing bandwidth to abate congestion does not suffice—it must be allocated in the right network locations.

Traffic bottlenecks typically occur at points where traffic accumulates, such as servers and switches, particularly those that serve as edge, access, or backbone gateways. Automatic switch-based traffic control mechanisms, such as queuing and buffering, are effective in abating short-term congestion. Longer term congestion likely requires more proactive involvement, using techniques such as such as traffic shaping to rate limit certain types of traffic and redirection to move traffic around or away from the congestion. Traffic shaping is a popular technique that throttles traffic as it enters a network through a router or switch.

Capacity planning determines the topology and capacity of a network. New routing approaches have been devised to leverage capacity planning and traffic engineering. They are intended to route services so that individual link capacities are not exceeded. MPLS traffic engineering predetermines LSPs so that congestion is minimized and service levels are met. RSVP-TE goes a step further by engineering LSPs based on QoS information and extending this capability to IP networks.

SLM is the process of guaranteeing a specified range of service levels, characterized by the QoS. QoS management involves the classification, marking, and policing of traffic. It employs policing devices and software tools to administratively specify QoS parameters and policy rules. QoS can be implemented in different types of network devices. SLM is achieved through software tools called policy or PBNM systems. SLAs drive the policy rules and parameters that are programmed into PBNM systems. SLAs are contracts between two organizations. If drawn correctly, an SLA should quantitatively state service level objectives.

Network changes should be done in phases. New features, applications, systems, and locations should be added gradually, versus all at once. Improper modifications of system configurations are often the blame for many network outages. The potential impact of any change should be well understood and documented prior to making the change.

Network managers are faced with the objective of minimizing downtime by reducing the possibility of disruptions and creating the appropriate response mechanisms. In a mission-critical network, the ultimate network management goal is to ensure reliable and optimal service for the amount of money invested.

References

[1]   Angner, R., "Migration Patterns," *America's Network*, May 1, 2002, pp. 51–52.

[2]   Llana, A., "Real Time Network Management," *Enterprise Systems Journal*, April 1998, pp. 20–27.

[3]   Wilson, T., "B2B Apps Prone to Outages," *Internet Week*, No. 850, February 26, 2001, pp. 1, 45.

[4]   Zeichick, A., "Predicting Failure," *Internet Week*, No. 827, September 4, 2000, pp. 33–38.

[5]   Mandelbaum, S., "In the Saddle with Availability Management," *Enterprise Systems Journal*, September 2000, pp. 50–53.

[6]   Liebmann, L., "SNMP's Real Vulnerability," *Communication News*, April 2002, p. 50.

[7]   Liebmann, L., "SNMP's Real Vulnerability," *Network Magazine*, May 2002, p. 76.

[8]   Cooper, A., "Network Probes Provide In-Depth Data," *Network World*, July 3, 2000, p. 41.

[9]   Freitas, R., "Event Correlation From Myth to Reality," *Network World*, October 25, 1999, pp. 65–66.

[10] Dubie, D., "Probe Gives User Better Handle on WAN," *Network World*, October 29, 2001, pp. 2–26.

[11] Drogseth, D. "Digging for the Root Cause of Network Problems," *Network Magazine*, May 2002, pp. 96–100.

[12] Connolly, P. J., "Boost Your Bandwidth Efficiency," *Infoworld*, March 27, 2000, p. 41.

[13] McCabe, J. D., *Practical Computer Network Analysis and Design*, San Francisco, CA: Morgan Kaufmann, 1998.

[14] Davis, K., "Traffic Shapers Ease WAN Congestion," *Network World*, April 22, 2002, p. 49.

[15] Browning, T., "Planning Ahead," *Enterprise Systems Journal*, September 2000, pp. 52–55.

[16] Dubie, D., "Demystifying Capacity Planning," *Network World*, October 29, 2001, p. 28.

[17] Yasin, R., "Software that Sees the Future," *Internet Week*, No. 827, September 4, 2000, pp. 1, 60.

[18] Ashwood-Smith, P., B. Jamoussi, , and D. Fedyk, "MPLS: A Progress Report," *Network Magazine*, November 1999, pp. 96–102.

[19] Xiao, X., et al., "Traffic Engineering with MPLS," *America's Network*, November 15, 1999, pp. 32–38.

[20] Sudan, R., "Signaling in MPLS Networks with RSVP-TE," *Telecommunications*, November 2000, pp. 80, 84.

[21] Jamoussi, B., "CR-LDP for Value Added Services," *Telecommunications*, November 2000, pp. 81–82.

[22] Griffin, C., and G. Goddard, "Searching for the QoS Holy Grail," *Network World*, June 3, 2002, pp. 53–55.

[23] Bernet, Y., et al., *RFC 2998: A Framework for Integrated Services Operation over Diffserv Networks*, November 2000.

[24]   Nichols, K., V. Jacobsen, and L. Zhang, *RFC 2638:A Two-bit Differentiated Services Architecture for the Internet*, July 1999.

[25]   Miller, M. A., "How's Your Net Working?" *Infoworld—Enterprise Connections: The Networking Series*, Part 7, 1999, pp. 3–7.

[26]   Jacobson, V., "Congestion Control and Avoidance," *ACM Computer Communication Review: Proceedings of the Sigcomm '88 Symposium in Stanford, CA*, August 1988.

[27]   Khan, M., "Network Management: Don't Let Unruly Internet Apps Bring Your Network Down," *Communication News*, September 2000, pp. 30–32.

[28]   "Bandwidth Management Pays Off," *Communication News*, November 2001, pp. 54–56.

[29]   Parrish, S. J., "Regular or Premium: What Are You Buying," *Phone+*, February 2002, pp. 48, 65.

[30]   Boardman, B., "Orchestream Conducts PBNM with Precision," *Network Computing*, January 21, 2002, pp. 41–49.

[31]   Saperia, J., "IETF Wrangles over Policy Definitions," *Network Computing*, January 21, 2002, p. 38.

[32]   Conover, J., "Policy Based Network Management," *Network Computing*, November 29, 1999, pp. 44–50.

CHAPTER 13

Using Recovery Sites

Creating and maintaining an in-house data center operation requires resources, infrastructure, and staff. Operation of a mission-critical data center requires careful attention to quality and redundancy, particularly in the areas of power supply, networking, and data storage. But only so much can be invested in these areas to maximize the reliability of a single site operation. Having a second data center at another site to take over some or all of the primary functions can certainly be the next option to protect against catastrophic situations.

If a mission is time critical, quick transition of operation to an alternate data center is required in the event of a disaster. This implies having the alternate data center accessible to the constituency that relies on the operation. Depending on the type of failover required, it also implies having primary and secondary locations connected using a wide area network (WAN) for status and transaction updating. The secondary data center need not be a duplicate. It does not have to handle all functions of the primary center, only those that are essential to the mission. Prioritizing those functions that are considered to be essential is the single first step in establishing a recovery site.

Today there are organizations that internally implement and operate their own recovery sites. In recent years, recovery service providers have arisen offering ready-made business recovery centers with computing and telecommunications equipment, local area networks (LANs), personal computers (PCs), and storage that an organization can lease in the event of a significant outage at their main processing facility.

This chapter reviews some of the key considerations to be made when using recovery sites for mission-critical networking. There are various approaches as to how a secondary site should be used and how operations should transition during adverse situations. There are various ways to cutover and activate a site, many of which reflect the types of failover discussed previously in this book. The following sections focus on these issues.

13.1 Types of Sites

Redundant facilities in the same geographic location do not fulfill true, foolproof recovery. A geographically diverse alternate facility is usually a better measure, as it protects against local disasters. A second facility located in a distant metropolitan region will likely be unaffected by the same physical disaster affecting the primary location. However, depending on the nature of the facility, the secondary site may still require full or partial staffing from the main facility. In this case, a secondary site may have to be accessed remotely by staff, unless there is a means to supply staff during adverse conditions.

Many organizations will be faced with the decision of implementing a recovery site in house or outsourcing it to a service provider. An in-house implementation can provide some distinct advantages. It can be operated by staff that understands the in-house systems and applications. Another advantage is the priority that the organization will receive in the event of a crisis. Because the organization owns and operates its own facility, it has exclusive priority to use it, unlike having a recovery site service that operates on a first-come, first-served basis.

Reciprocal agreements between organizations in a consortium arrangement are a popular alternative to outsourcing. In this scenario, two or more firms establish reciprocal backup and recovery agreements with each other. They can even go so far as to use each other's systems for application processing if needed. Such arrangements should be made with firms in diverse industries or diverse geographic locations, to avoid being both subject to the same adverse event.

Constructing a redundant site from scratch can be an enormous capital expense. A better option is to establish the site elsewhere in the same enterprise. In large corporations, internal reciprocal agreements between departments or data centers are often made. Having duplicate identical servers housed in other corporate locations can make life easier to control redundancy and failover, but sweeping duplication can be cost prohibitive. Many organizations, for instance, mirror across external data or application centers for redundancy. The challenge is then to determine how systems and data should be duplicated in these redundant sites.

Failover to a redundant site is largely dependent on networking and data replication. Hot failover requires investing in expensive storage systems, mirroring capabilities, and highly reliable WAN infrastructure. The systems and network must be tested several times a year to ensure they will work when needed. For these reasons, reliance on less expensive approaches may be more viable. This includes using scale-downed servers that can process only the most critical applications and a simpler data replication mechanism, such as remote journaling, to update the redundant site.

The following sections review several types of recovery sites. One will find that hybrid approaches involving combinations of the following may work as well. The exact choice will depend mostly on the cost, function, and mission of the secondary site, and whether the site should be implemented internally or in conjunction with a recovery site provider. There should be a conscious decision made beforehand on what type of recovery site option, if any, is to be used. Deciding and implementing an approach at the time of disaster is too late.

13.1.1 Hot Sites

Hot sites are preconfigured data centers that are maintained at a separate location, internally by an organization or contracted through a service provider [1]. They are intended to permit mission-critical operations to continue uninterrupted through hot failover if an outage affects a primary site. From earlier discussion in this book, hot failover between elements, whether system components or entire data centers, offers the best approach to reduce recovery time.

A hot site must be kept constantly available to an organization for near-instantaneous recovery. In one extreme, a hot site can be equipped with exact replicas of hardware, software, and communication systems used at the primary site. In another extreme, they can be equipped with a configuration that is partially or only functionally similar to the main site and capable of processing only the most critical functions. In either case, the site is ready to go live at a moments notice or is already in a live state.

Hot site service providers offer many different types of service packages. Some provide a total solution with duplicate data centers complete with the systems needed to resume operation—the customer need only supply the software and the data. Such locations have preinstalled computers, raised flooring, air-conditioning, telecommunications equipment, networking equipment, technical support, and redundant power plant. Some are designed so that a firm's staff can readily inhabit the site in the event of an outage.

The following are some general recommendations for developing a hot site, either internally or through outsourcing:

- *Failover operations*. An organization should ensure that it has access to its standby systems at the provider's recovery facility in the event of a widespread disaster. Because the facility will most likely be shared with other firms contending for the same resources, priority should be clearly understood beforehand, to the extent that it even become one of the provider's contractual obligations.

  A hot site should have correct operating system (OS) versions for application processing. The site should be ready to accept any recent OS or application updates required for comparable operation. Equipping the site with preconfigured boot drives and system configuration files can make the transition more instantaneous. From a security perspective, a hot site must support the exact policy enforced at the primary, accomplished only when the hot site patch fix levels are the same as the primary. For critical or sensitive applications, it is best to keep the hot site as current as possible.

- *Data*. File systems and data should be replicated and transmitted to the location. A service provider should be able to implement and operate the standby application systems with real-time data replication. Critical information should be at most mirrored at the site, using some of the approaches discussed in the chapter on storage. At a minimum, daily data changes should be transferred to the site. A dedicated full backup disk should exist at the site to which incremental or differential backups can be applied.

- *Networking*. Data traffic should be redirected to the site using a variety of approaches, many of which are discussed in the chapter on network management. Internet traffic can be redirected in a number of ways, as discussed earlier in this book. They include the use of load balancing, border gateway protocol (BGP), and domain name server (DNS) changes.

  Load sharing or offloading peak processing to a hot site over a network on a normal basis not only facilitates instantaneous recovery, it helps maximize its value and recapture implementation costs. It also keeps the site alive nearly all the time. Another option is to use the site for different functions during nonessential periods, such as running less critical applications, software development, testing, and training.

Well-documented failover and recovery procedures are a must when using an internal or external hot site or any other type of recovery site, for that matter. They should illustrate how the site will operate when in use. As needs grow and change, the plans must be modified to meet those needs. Periodic testing should be conducted to process applications at the hot site to ensure the validity of the plans. Critical information should be replicated there so that nondisruptive tests can be conducted.

13.1.2 Cold Sites

A cold site, sometimes referred to as a *shell site*, is maintained space that is ready for the installation of networking and computer systems. Preinstalled wiring, raised floors, air conditioning, electric power, and fire protection characterize a shell site. A shell site is usually situated on or near a network backbone for quick and easy access. While cold sites only facilitate a delayed recovery, they provide an affordable approach for organizations that can sustain a few days of downtime before resumption. They are also useful for organizations that might lose a primary site for months following a disaster.

A cold site approach requires identifying and leasing an appropriate facility. Because a delayed recovery is involved, there is flexibility in finding a cost-effective site location that can be accessed by internal staff in a longer time interval. Once procured, the main task is then to plan on how to acquire, transport, install, configure, and activate the hardware, software, and systems at the site after an outage. It involves locating and installing the correct hardware, loading the software and data from backups, and making final network connections. Cabling and telecommunications circuits should be installed in advance to avoid turn-up delays.

Hardware and software for installation in a cold site could be pulled from a couple of sources. One approach is to transport systems from the primary site, assuming it is not damaged. Otherwise, appropriate agreements must be made with hardware vendors in advance for quick shipment of critical equipment upon notice. However, if a widespread disaster has occurred, equipment shortages and lag times in delivery are highly likely. It is important to contractually require vendors to provide priority in obtaining systems and equipment that can be functionally used as replacements, perhaps even the first systems off the production lines if their inventories are exhausted.

Service providers that operate cold sites will have multiple subscribers contracted to use a facility. In the event of an outage, one of them will move into the facility, at least temporarily. As stated earlier, if a widespread disaster has occurred, chances are that several companies in the same geographical area may be contending to use the same site. To avoid this problem, it is important to require a cold site provider to state contractual limits on the number of companies in the same area to which they can simultaneously provide service.

A cold site can be used in conjunction with a hot site. During the time failover to a temporary hot site is initiated and sustained, a semipermanent cold site could be set up. The hot site is used to bridge a service gap and buy time for setting up the cold site. Once the cold site is operational, service is transferred from the hot site to the cold site. When the permanent main facility is recovered, service is transferred from the cold site back to the main site.

13.1.3 Warm Sites

The exact definition of a warm site can vary, but fundamentally it is a facility whose functionality is midway between a cold site and hot site. Unlike a cold site, a warm site is either partially or fully equipped with some IT infrastructure, designed to facilitate a noninstantaneous, or warm, failover of an operation. The objective of a warm site is to activate operation upon an outage of the primary facility. Because failover is not immediate, a warm site must be readied in some way prior to operation. It may require additional equipment, software, state information, staff, and other resources in order to be activated. Possible disruption of service and loss of transactions during the failover period could occur, depending on the type of failover and subsequent level of service provided by the warm site.

13.1.4 Mobile Sites

A mobile recovery site is data center space on wheels, delivered via tractor-trailer to a designated location. Some mobile site providers offer sites that are mobile data centers, equipped with everything from electrical generators, environmental systems, workstations, and telecommunications systems. Mobile sites are normally used in situations where a deferred recovery approach is involved, and can be considered as a cold or warm site on wheels. The sites are usually mobilized when a catastrophic event takes place, usually within hours of a disaster.

When using mobile sites, plans should reflect how computer systems and data will be transferred to the unit and how it will be activated. Plans may have to include preconfiguration prior to mobilization to a specific location. When using a mobile site provider, they should specify how long it would take for the site to arrive and its physical requirements.

Mobile sites provide several advantages. Staff does not have to travel to another location during an adverse event. Also, depending on the type of disaster, the site can be continuously relocated to other locations. But mobile sites are also subject to some of the same caveats as other types of recovery sites. They may also not provide the same degree of physical security as a permanent site. In times of widespread crisis, mobile site availability can be limited in a particular region, so arrangements should be made with the vendor beforehand to obtain priority.

13.2 Site Services

Enterprises with limited staff, resources, or capital to invest in their own data centers are turning to service providers for recovery site options. The following considerations should be made when choosing a recovery site service:

- *How will data be stored and intermittently transferred to the site? What related technologies should the provider offer?* Mechanisms that are

compatible with an organization's current systems and technologies are preferred. The recovery point objective (RPO), the nature of the data replication,
and the recovery mechanism are critical factors that require consideration [2].

- *A state-of-the-art hardened physical facility* that can support high-
  performance server processing and network access should be used. It should
  conform to many of the facility characteristics discussed in the chapter on
  facilities.

- *How will servers running an organization's applications be maintained? Who
  is responsible for addressing application problems?* Consistent application
  and server upgrades between the main site and the recovery site need to be
  maintained.

- *Direct physical access to the recovery site* will most likely have to be provided,
  at least to an organization's essential staff. Some services provide end-user
  staff space, office services, telephone access, and other amenities.

- *Data and voice networking to the recovery site* should be made available. Network access and bandwidth should be available at the recovery site to enable
  transfer and processing of live traffic during recovery. If the site is going to be
  used as a hot site, networking should be available to allow for continuous data
  and application updates.

- *The types of recovery services the provider has available* at the site are worth
  identifying. The service provider could assist in helping an organization
  recover during adverse situations. They should have the ability to recover data
  and applications when there is a problem.

- *The types of testing services available* at the site are also worth identifying.
  Testing keeps recovery plans up to date. Testing periods should be scheduled
  with the provider as part of the service to ensure that failover can be conducted
  as required and applications process correctly at the recovery facility.

- *The available in-house technical expertise of the provider* can supplement an
  organization's staff if needed. The recovery site staff must be knowledgeable
  on the use of the facility and should be made accessible during crisis situations.

Many of these issues should be spelled out in a service agreement with the provider. In addition to standard legal and business clauses regarding costs, contract
duration, and termination, they should include specific terms that address testing,
security, system change notification, hours of service operation, specific hardware
or software requirements, circumstances constituting specific adverse events, service
priorities, guarantee of compatibility with an organization's systems, availability of
service, and use of noncomputing/networking resources.

The following sections discuss several types of services that can be utilized as a
recovery site option. There are various types of providers and arrangements [3].
Figure 13.1 illustrates the types of service providers and the levels of services offered.

13.2.1 Hosting Services

A popular solution for Web and extranet sites, a hosting service provides the servers
and associated networking infrastructure while the customer is responsible for
updating their Web content or applications [4]. Hosting providers offer a full service

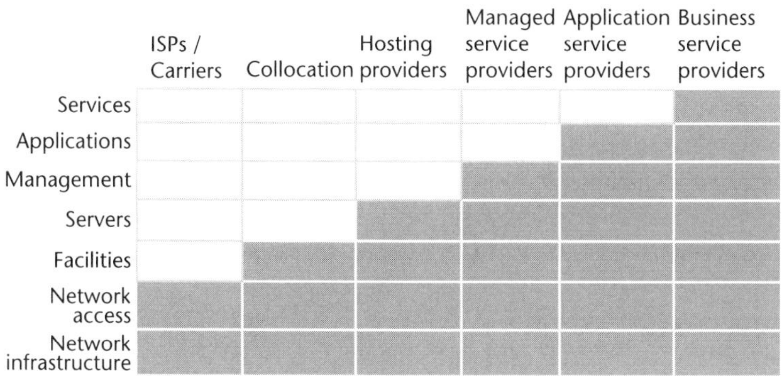

	ISPs / Carriers	Collocation	Hosting providers	Managed service providers	Application service providers	Business service providers
Services						■
Applications					■	■
Management				■	■	■
Servers			■	■	■	■
Facilities		■	■	■	■	■
Network access	■	■	■	■	■	■
Network infrastructure	■	■	■	■	■	■

Figure 13.1 Recovery site service provider options.

that includes high levels of service monitoring and recovery. However, many hosting service providers will not perform OS and application level security and administration on servers a customer provides. Many e-commerce sites place their operations in a Web hosting facility and rely on the hosting service to deliver site availability and response at prenegotiated service levels through a service level agreement (SLA). In some arrangements, the hosting provider agrees to pay penalties if the availability and response requirements are violated [5].

In a *managed hosting* environment, the hosting provider is responsible for day-to-day management of the hosting environment, including the infrastructure, platforms, and connectivity. A typical managed hosting environment consists of multiple types of application servers, firewalls, and links to customer locations. They also offer extra services such as load balancing and traffic monitoring which often better serve the hosting provider than the client. Before purchasing any extra services, it is best to scrutinize their value to daily operation.

Managed hosting has a couple variants. *Shared hosting* involves multiple customers housed on a single server and is geared toward the consumer and small business markets [6]. In contrast, *dedicated hosting* involves a client leasing a dedicated server, with the option of buying additional services. This service is popular among large and midsize enterprises. Charging elements will vary depending on the type of service, but they typically include the number of servers, disk space, number of Domains, server management costs and bandwidth consumption.

A prerequisite to using a hosting provider for mission critical services is having comfort with surrendering technology control to another party. To this end, it is important to obtain and review a hosting provider's backup and recovery plans if they encounter an outage. Hosting sites should distribute traffic across multiple Internet points of presence (POPs) or peering points to provide routing diversity. If used as a main processing site, a hosting service provider can be a single point of failure. Therefore, having multiple providers can ensure greater availability.

13.2.2 Collocation Services

A collocation site can be used to offload Web site hosting or data center applications, or it can be used as a recovery site [7]. A collocation facility is a building used to house servers in a safe, secure environment. For a fee, customers place their systems in the space and have physical access to them for servicing. The collocation provider does not care what server or software systems clients are using. Collocation providers typically provide the compartmentalized and physically secured space, power, air conditioning, and network access required for service [8]. The facilities are typically hardened with temperature and humidity controls, clean and reliable power, and access to communication backbone networks. As always, an appropriate level of geographical diversity is desirable to maintain service during widespread outage.

The concept of collocation is an old one and arose out of the telecommunication industry. After the breakup of AT&T in 1984, the regional Bell operating companies (RBOCs) were flooded with carrier requests to collocate their systems in RBOC central offices (COs) or POPs to cut down on the high cost of maintaining their own facilities and network access links [7]. Interexchange carriers (IXCs) and competitive access providers (CAPs) later seized this as an opportunity to gain competitive advantage by leasing their POP space to customers purchasing network capacity as a value-added service. Since then, new breeds of collocation providers have emerged, offering collocation to various types of network service providers and private enterprises.

Trends in computing and networking have somewhat blurred the distinction between collocation and data center space. Traditional data centers focused on application hosting equipment using mainframe and midrange systems and vast arrays of storage, too large to fit into the space of many carrier POPs. However, the trends toward rack-mounted systems, smaller hardware footprints, and growing bandwidth requirements have driven many carriers and customers closer to each other.

It is not unusual to find collocation spaces in the same building where carrier POP sites, COs, and hosting providers reside. These facilities, sometimes referred to as *carrier hotels*, are buildings where these entities can meet and easily connect with one another, avoiding the expense of lengthy network connections. However, carrier hotels are notorious single points of failure. An outage or physical disaster in a hotel can affect service for nearly all tenants [9]. The following considerations should be made when choosing a collocation site for mission-critical operations [10]:

- *Network connectivity.* Multiple broadband carriers should supply service to the facility. A properly placed collocation facility should relieve the local-access charges that can be quite significant. The savings should outweigh the costs of networking to a customer location. A synchronous optical network (SONET) access ring topology, as discussed earlier in this book, can better achieve reliable service. Physical proximity to a POP or Internet network access point (NAP) also improves reliability and reduces distance-sensitive network costs for both the client and the collocation provider.

- *Survivability.* At a minimum, a collocation site should provide reliable bandwidth, floor space, power, cooling, fire prevention, and physical security. Collocation facilities are prone to the same physical mishaps of any data center facility. Thus, the survivability features of the facility and provider recovery plans should be evaluated to ensure that they support the customer's availability criteria. Some collocation providers offer redundant facilities for their tenants as well. Facilities that serve carriers should meet carrier-grade standards for building and operating high-availability facilities. Such standards include the Telcordia Technologies G.1275 specifications for COs and POPs.

- *Tenants.* The tenants inhabiting a collocation facility each have a share in the facility's resources. Whether the shares are equitable under normal and exceptional circumstances depends on the nature of the service agreements tenants have with the collocation provider. It is most likely that different clients will have different types of agreements each with their own guarantees of resources and availability. In the event of network or facility problems or peak usage periods, chances are that priority will be given to certain tenants. It is important to know what order of precedence one has. While tenants such as application service providers (ASPs), Internet service providers (ISPs), carriers, data centers, or other companies can provide valuable resources, it is important to make sure that critical-service suppliers are not tenants. It is also important to know how many companies within a facility have contracted with the same suppliers, potentially straining their assets during a crisis.

- *Management services.* The facility provider should offer skilled staff to assist customers in the installation and maintenance of equipment. This support is most likely needed in provisioning connectivity to the local backbone. Some providers offer packages whereby their local staff is available to respond to customer equipment or network issues. Additionally, some offer various levels of hardware and software systems monitoring. Providers will likely emphasize certain brands of hardware or software, as it is impossible to maintain expertise in every brand. Furthermore, they may have already established agreements with particular vendors. This could be a problem for customers whose systems do not reflect the favored brands.

- *Customization.* Quite often, collocation sites resemble jail-like facilities with card, key code, or biometrically controlled access, secure raised or concrete floors, underfloor cooling, overhead cable trays, locking chain-link cage space, or locking cabinets of fixed size that house client systems. A collocation provider will likely maintain a lease agreement with the facility owner. The lease should be verified to see if it permits the collocation provider to build or expand any key facilities or infrastructure as business requirements change.

13.2.3 Recovery Services

Recovery service providers offer various ranges of services aimed at delivering the equivalent of a second data center to a client. Depending on the client needs, they can provide anything from a full duplicate site to hosting [11]. They typically design service solutions on a case-by-case basis from their own shared facilities. Clients need not go through the capital and operational expense of developing and customizing their own recovery site solutions, as the recovery service does this for them [12].

In recent years, many of these providers have not been able to economically offer these services. The high prices of these services attract only a limited market consisting of mostly large organizations. This, coupled with the expense of offering custom versus standard solutions, has resulted in the inability of recovery site providers to cost effectively deliver services.

13.3   Implementing and Managing Recovery Sites

The most fundamental rule in establishing a recovery site is to ensure there is no common point of failure with the primary site. Both the primary and recovery sites should be able to function independently. From a facility perspective, the site should provide redundant components as discussed in the chapter on facilities. It should have backup and dual power served by multiple separate power grids.

Other rules suggest that a recovery site should be 30 to 50 mi away from a primary site to avoid the damage disruptions that might come with a regional disaster, yet allow critical staff from the main site to reach it. Although some might say that this distance is still too close for comfort, the greater the distance, the more complex and expensive the implementation. If critical staff is unable to reach the site during a crisis, then remote computing is required. Some organizations even establish a third site to triangulate information for backup and recovery.

The essential business processes to be supported at a recovery site should be determined prior to implementation. A recovery site, at best, can have systems and connectivity identical to the primary site and support all required applications. An alternative is to use systems and connectivity that are functionally similar, but not identical, to those at the primary site. To address these issues, it is best to view the recovery site as resource, just like any other component in an IT infrastructure. From this perspective, several approaches to using the resource should be considered. Table 13.1 illustrates how these options compare with one another:

- *Dedicated resource.* A dedicated site is there for the exclusive use of the organization. If the site is dedicated, then there is a strong likelihood that an organization will want to own or lease and operate the systems. Dedicated sites are conducive to a hot site model. Using a service provider site as a possible dedicated resource can prove to be an expensive option, as most providers favor sharing their facilities. In this case, having a service provider as only a functional or partial backup may be more economical.

- *Shared resource.* A shared site involves multiple organizations sharing a single site or resource Recovery service provider sites are examples of shared sites. Sharing a facility with other parties leads to contention if more than one party declares an outage and all parties attempt to simultaneously use the same recovery resources [13]. This is a likely scenario when using a service provider.

- *Standby resource.* A standby site is one that is not used until an outage or overload situation takes place. Because the site sits idle outside these situations, it can be an expensive option. Implementation through a shared site service might be a more economical approach, unless priority and hot failover are

Table 13.1   Types of Recovery Resources

	Dedicated Resource	Shared Resource
Standby Resource	Supports warm failovers; no resource contention	Supports warm failovers; resource contention
Active Resource	Supports hot failovers; no contention	Supports hot failovers; resource contention

required. A standby site is best used as a partial recovery mechanism to support only critical business applications during critical periods.

- *Active resource.* Continuously operating a recovery site ensures the integrity of a failover. Using the recovery site as an operational resource rather that sitting idle adds both value and payback. The recovery site could operate as a live production load-shared site, supporting Web traffic, processing transactions, and keeping databases synchronized with the primary site. Another option is to use the site to provide nonessential functions during normal, off-peak periods but activate it for critical applications during peak periods or times of crisis. Technologies such as wide area clustering, load balancing, mirroring, storage area networks (SANs), and WANs are likely candidates for use with an active recovery site.

An active site can also be used in conjunction with a standby site. For instance, a standby site can be used to offload critical or noncritical processing from an active recovery site. This enables the active recovery site to scale to 100% capacity should a primary site fail.

13.3.1   Networking Recovery Sites

Data transfer during normal operation and failover between the production site and the recovery site are prime networking considerations. Failover mechanisms will involve both public and private interfaces. Figure 13.2 illustrates just one example

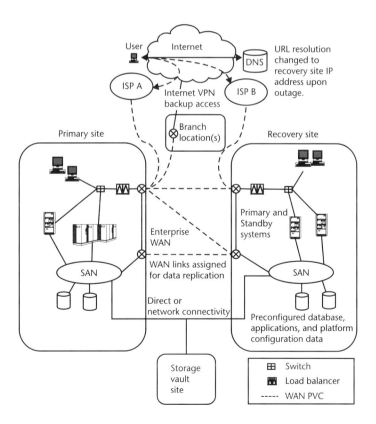

Figure 13.2   Recovery site network example.

of networking with a recovery site. In this example, the recovery site has a network presence on an enterprise WAN. In a frame relay environment, the location could be connected using permanent virtual circuits (PVCs). Routers are then configured with the recovery site address to preserve the continued use of the same domain names. This makes service seem available anywhere in the domain, regardless of the physical location. Load balancing can be used to redirect and manage traffic between the primary and recovery sites, enabling traffic sharing and automatic rerouting.

Transparent failover and traffic redirection to the recovery site can be achieved in combination with domain name and Web infrastructure to ensure continuous Web presence. Ideally, whether a user accesses a primary or recovery site should be transparent. Redirection of the domain to a different Internet protocol (IP) address can be done using some of the techniques discussed earlier in this book. Internet access to the recovery site can be realized through connectivity with the same ISPs that serve the primary site or an alternate ISP that serves the recovery site. Branch locations that normally access the primary location for transaction processing and Web access through a PVC could use virtual private network (VPN) service as an alternate means of accessing the recovery site.

Automatic failover and redirection of traffic requires that data and content between the two sites be replicated such that the recovery site is kept up to date to take over processing. This implies that data and content must be updated at either the primary or recovery site, implying bidirectional replication between the sites. Data replication should be done on an automated, predetermined schedule if possible. Data should be instantaneously backed up to off-site storage devices as well. Active and backup copies of data should exist at both sites. As already mentioned, a SAN can be good way to facilitate data transfer and backup through sharing common storage over a network. It also alleviates physical transport of tape libraries to the recovery and storage sites [14]. Regardless of what type of data replication is used, database, application software, and hardware configurations data should exist at the recovery site so that proper processing can be resumed.

The reader should keep in mind that Figure 13.2 is just one illustration of how sites can be networked. Network configurations will vary depending on costs, recovery needs, technology, and numerous other factors.

13.3.2   Recovery Operations

A failover and recovery site plan should be well defined and tested regularly. Most recoveries revive less than 40% of critical systems [15]. This is why testing is required to ensure recovery site plan requirements are achieved. This is especially true when using a hosting or recovery site provider. Many organizations mistakenly assume that using such services obviates the need for recovery planning because, after all, many providers have their own survivability plans. This is quite the contrary. Agreements with such providers should extend beyond OS, hardware, and applications. They should cover operational issues during times of immediacy and how they plan to transfer their operations to a different site. They should also define the precise definition of an outage or disaster so that precedence and support can be obtained if the site is in a shared environment.

After an outage is declared at a primary site, the time and activity to perform failover can vary depending on the type of failover involved. The operating environment at the recovery site should be the same or functionally equivalent to the primary site. This may require additional recent changes and software upgrades to the recovery site servers. Configuration settings of servers, applications, databases, and networking systems should be verified and set to the last safe settings. This is done in order to synchronize servers and applications with each other and configure routers appropriately. System reboots, database reloads, application revectoring, user access rerouting, and other steps may be required. Many recovery site providers will offer such system recovery services that assist in these activities.

If the recovery site were running less critical applications, then these must be transferred or shut down in an orderly fashion. The site is then placed in full production mode. Initial startup at the site should include those applications classified as critical. A standby backup server should be available as well, in the event the primary recovery server encounters problems. A server used for staging and testing for the primary backup server can work well. The standby should have on-line access to the same data as the primary backup server. Applications running on both servers should be connection aware, as in the case of cluster servers, so that they can automatically continue to process on the standby server if the primary server fails.

13.4   Summary and Conclusions

Implementing a recovery site, whether internally or outsourced, requires identifying those critical service applications that the site will support. The site should be geographically diverse from the primary processing site if possible, yet enable key staff to access the site either physically or remotely during an adverse situation. The site should be networked with the primary site so that data can be replicated and applications can be consistently updated.

Hot sites are intended for instantaneous failover and work best when they share normal traffic or offload peak traffic from a primary processing site. Cold sites are idle, empty sites intended for a delayed recovery and are a less expensive alternative. Warm sites are equipped sites that are activated upon an outage at a primary site. Although a more expensive option than a cold site, they result in less service disruption and transaction loss.

Outsourced recovery sites can be achieved using various types of site service providers, including hosting, collocation, and recovery site services. Regardless of the type of provider, it is important to understand the priority one has during widespread outages relative to the provider's other customers. During such circumstances, a provider's resources might strain, leading to contention among customers for scant resources.

A recovery site should be networked with a primary site to support operational failover and rerouting of traffic upon an outage. The network should also facilitate data replication and backup to the recovery site on a regular basis so that it can take over operation when needed. The recovery site should have a server and data backup of its own for added survivability.

References

[1]   Yager, T., "Hope for the Best, Plan for the Worst," *Infoworld*, October 22, 2001, pp. 44–46.

[2]   Dye, K., "Determining Business Risk for New Projects," *Disaster Recovery Journal*, Spring 2002, pp. 74–75.

[3]   Emigh, J., "Brace Yourself for Another Acronym," *Smart Partner*, November 13, 2000, p. 28.

[4]   Bannan, K. J., "What's Your Plan B?" *Internet World*, July 15, 2000, pp. 38–40.

[5]   Benck, D., "Pick Your Spot," *Hosting Tech*, November 2001, pp. 70–71.

[6]   Chamberlin, T., and J. Browning, "Hosting Services: The Price is Right for Enterprises," *Gartner Group Report*, October 17, 2001.

[7]   Payne, T., "Collocation: Never Mind the Spelling, It's How It's Delivered," *Phone Plus*, September 2001, pp. 104–106.

[8]   Facinelli, K., "Solve Bandwidth Problems," *Communications News*, April 2002, pp. 32–37.

[9]   Coffield, D., "Networks at Risk: Assessing Vulnerabilities," *Interactive Week*, September 24, 2001, pp. 11, 14–22.

[10] Henderson, K., "Neutral Colos Hawk Peace of Mind," *Phone Plus*, January 2003, pp. 30–32.

[11] Carr, J., "Girding for the Worst," *Teleconnect*, May 2001, pp. 42–51.

[12] Torode, C., "Disaster Recovery, as Needed," *Computer Reseller News*, August 13, 2001, p. 12.

[13] Walsh, B., "RFP: Heading for Disaster?" *Network Computing*, January 11, 1999, pp. 39–56.

[14] Apicella, M., "Lessons Learned from Trade Center Attack," *Infoworld*, September 24, 2001, p. 28.

[15] Berlind, D., "How Ready is Your Business for Worst Case Scenario?" October 11, 2001, www.ZDNet.com.

CHAPTER 14

Continuity Testing

Technology alone does not ensure a successful mission; rather, it is how the technology is installed, tested, and monitored. Many organizations fail to see the importance of spending more time and money than needed on a project and forego thorough, well integrated, and sophisticated testing prior to deployment of a system or network. Few phases of the system development cycle are more important than testing. Good testing results in a better return on investment, greater customer satisfaction, and, most important, systems that can fulfill their mission. Insufficient testing can leave organizations susceptible to exactly the types of failures and outages continuity planning hopes to eliminate or reduce.

Almost two-thirds of all system errors occur during the system-design phase, and system developers overlook more than half of these. The cost of not detecting system errors grows astronomically throughout a system project, as shown in Figure 14.1 [1]. Problems, gaps, and oversights in design are the biggest potential sources of error. Testing verifies that a system or network complies with requirements and validates the structure, design, and logic behind a system or network.

More than half of network outages could have been avoided with better testing. Several years ago, about two-thirds of the network problems originated in the layers 1 and 2 of the Internet protocol architecture. Growing stability in network elements such as network interface cards (NICs), switches, and routers has reduced this percentage down to one-third. Today the root cause of many network problems has moved up into the application layer, making thorough application testing essential to network testing.

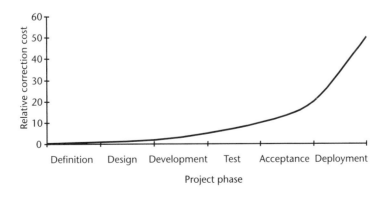

Figure 14.1 Cost of undetected errors.

Network testing is performed through either *host-based* or *outboard* testing [2]. In a host-based approach, testing functions are embedded within a network device. This can be effective as long as interoperability is supported with other vendor equipment though a standard protocol such as simple network management protocol (SNMP). Outboard testing, on the other hand, distributes testing functions among links or circuit terminations through the use of standalone devices or an alternative testing system, often available in network-management system products.

Quite often, time and cost constraints lead to piecemeal testing versus a planned, comprehensive testing approach. Some organizations will perform *throwaway tests*, which are impromptu, ad hoc tests that are neither documented nor reproducible. Quite often, organizations will let users play with a system or network that is still incomplete and undergoing construction in order to "shake it down." This is known as *blitz* or *beta testing*, which is a form of throwaway testing.

Experience has shown that relying on throwaway tests as a main form of testing can overlook large amounts of error. Instead, a methodology for testing should be adopted. The methodology should identify the requirements, benchmarks, and *satisfaction criteria* that are to be met, how testing is to be performed, how the results of tests will be processed, and what testing processes are to be iteratively applied to meet satisfaction criteria. The tests should be assessed for their completeness in evaluating systems and applications. They should be logically organized, preferably in two basic ways. *Structural tests* make use of the knowledge of a system's design to execute hypothetical test cases. *Functional tests* are those that require no design knowledge but evaluate a system's response to certain inputs, actions, or conditions [3].

Structural and functional tests can be applied at any test stage, depending on the need. This chapter reviews several stages of tests and their relevance to mission-critical operation. In each, we discuss from a fundamental standpoint, those things to consider to ensuring a thorough network or system test program.

14.1 Requirements and Testing

Requirements provide the basis for testing [4]. They are usually created in conjunction with some form of user, customer, or mission objective. Functional requirements should be played back to the intended user to avoid misunderstanding, but they do not stop at the user level. Systems specifications should also contain system-level requirements and functional requirements that a user can completely overlook, such as exception handling.

A change in requirements at a late stage can cause project delay and cost overrun. It costs about five times more if a requirement change is made during the development or implementation rather than design phases. Assigning attributes to requirements can help avert potential overruns, enhance testing, and enable quick response to problems, whether they occur in the field or in development. Requirements should be tagged with the following information:

- *Priority*. How important is the requirement to the mission?
- *Benefit*. What need does the requirement fill?

- *Difficulty*. What is the estimated level of effort in terms of time and resources?
- *Status*. What is the target completion date and current status?
- *Validation*. Has the requirement been fulfilled, and can it serve as a basis for testing?
- *History*. How has the requirement changed over time?
- *Risk*. How does failure to a meet requirement impact other critical requirements? (This helps to prioritize work and focus attention on the whole project.)
- *Dependencies*. What other requirements need to be completed prior to this one?
- *Origin*. Where did the requirement come from?

Strategic points when requirements are reviewed with systems engineering for consistency and accuracy should be identified throughout the entire systems-development cycle, particularly in the design and testing phases.

14.2 Test Planning

Testing methodologies and processes should be planned in parallel with the development cycle. The testing plan should be developed no later than the onset of the design phase. Throughout the network design phase, some indication of the appropriate test for each network requirement should be made. A properly designed testing plan should ensure that testing and acceptance are conducted expeditiously and on time and assure that requirements are met. Planning can begin as early as the analysis or definition phase by formulating a testing strategy, which can later evolve into a testing plan. The strategy should establish the backdrop of the testing process and spell out the following items [5]:

- *Requirements*. Those requirements that are most important to the mission should be tested first [6]. Each requirement should be categorized in terms of its level of impact on testing. For each, a test approach should be outlined that can check for potential problems. In addition, special adverse conditions that can impact testing strategy should be indicated. For instance, an e-commerce operation will be exposed to public Internet traffic, which can require rigorous testing of security measures.
- *Process*. The testing process for unit testing, integration testing, system testing, performance testing, and acceptance testing should be laid out. This should include major events and milestones for each stage (discussed further in this chapter). In its most fundamental form, a testing process consists of the steps shown in Figure 14.2 [7].
- *Environment*. The technologies and facilities required for the testing process should be identified, at least at a high level. Detailed arrangements for a testing environment are later made in the testing plan. Testing environments are discussed in the next section.

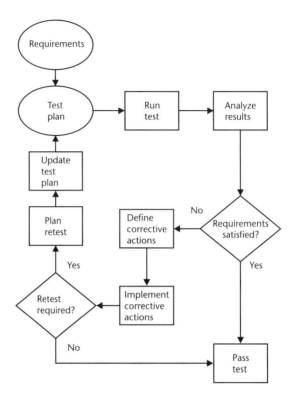

Figure 14.2   Basic steps of a testing process.

The test plan should specify how the testing strategy would be executed in each testing phase. It also should include detailed schedules for each of the tests and how they will be conducted. Tests are comprised of test cases—collections of tests grouped by requirement, functional area, or operation. Because creating test cases is inexact and is more of an art than a science, it is important to establish and follow best practices. An important best practice is to ensure that a test case addresses both anticipated as well as invalid or unexpected inputs or conditions. The test plan should include, at a minimum, the following information for each test case:

- The test case;
- The type of tests (e.g., functional or structural);
- The test environment;
- The configuration setting;
- The steps to execute each test;
- The expected results;
- How errors or problems will be corrected and retested;
- How the results will be documented and tracked.

The appropriate test methodology must be chosen to ensure a successful test execution. There are many industry-standard test methodologies available. No methodology will address all situations, and in most cases their use will require some adaptation to the test cases at hand. A walkthrough of the testing plan should be performed to plan for unforeseen situations, oversights, gaps, miscalculations, or other problems in the test planning process. The plan should be updated to correct problems identified during testing.

14.3 Test Environment

Investing in a test environment requires weighing the risks of future network problems. This is the reason that many organizations fail to maintain test facilities—or test plans, for that matter. Quite often, new applications or systems are put into production without adequate testing in a protected and isolated lab environment. Test environments should be designed for testing systems from the platform and operating system (OS) levels up through middleware and application levels. They should be able to test communication and connectivity between platforms across all relevant network layers. Furthermore, a common omission in testing is determining whether appropriate levels of auditing and logging are being performed.

In a perfect world, a test environment should identically simulate a live production network environment [8]. In doing so, one can test components planned for network deployment, different configuration scenarios, and the types of transaction activities, applications, data resources, and potential problem areas. But in reality, building a test environment that recreates a production environment at all levels is cost prohibitive, if not impossible, particularly if a production network is large and complex.

Instead, a test environment should reasonably approximate a production environment. It could mimic a scaled down production environment in its entirety or at least partially. To obtain greater value for the investment made, an enterprise should define the test facility's mission and make it part of normal operations. For instance, it can serve as a training ground or even a recovery site if further justification of the expense is required.

Network tests must be flexible so that they can be used to test a variety of situations. They should enable the execution of standard methodologies but should allow customization to particular situations, in order to hunt down problems. There are three basic approaches, which can be used in combination [9]:

- *Building-block approach*. This approach maintains building block components so that a tester can construct a piece of the overall network one block at a time, according to the situation. This approach provides flexibility and adaptability to various test scenarios.
- *Prepackaged approach*. This approach uses prepackaged tools whose purpose is to perform well-described industry-standard tests. This approach offers consistency when comparing systems from different vendors.
- *Bottom-up approach*. This approach involves developing the components from scratch to execute the test. It is usually done with respect to system software, where special routines have to be written from scratch to execute tests. Although this option can be more expensive and require more resources, it provides the best flexibility and can be of value to large enterprises with a wide variety of needs. This approach is almost always needed when custom software development by a third party is performed—it is here that thorough

testing is often neglected. A proper recourse is to ensure that a test methodology, test plan, and completed tests are included in the deliverable.

When developing a test environment for a mission-critical operation, some common pitfalls to avoid include:

- Using platform or software versions in the test lab that differ from the production environment will inhibit the ability to replicate field problems. Platforms and software in the test environment must be kept current with the production environment.

- Changing any platform or software in the field is dangerous without first assessing the potential impact by using the test environment. An unplanned change should be first made in the test environment and thoroughly tested before field deployment. It is unsafe to assume that higher layer components, such an application or a layer 4 protocol, will be impervious to a change in a lower layer.

- Assuming that OS changes are transparent to applications is dangerous. The process of maintaining OS service packages, security updates, and cumulative patches can be taxing and complicated. All too often, the patches do not work as anticipated, or they interfere with some other operation.

- Organizations will often use spare systems for testing new features, as it is economical to do so. This approach can work well because a spare system, particularly a hot spare, will most likely be configured consistently with production systems and maintain current state information. But this approach requires the ability to isolate the spare from production for testing and being able to immediately place it into service when needed.

- Quite often, organizations will rely on tests performed in a vendor's test environment. Unless the test is performed by the organization's own staff, with some vendor assistance, the results should be approached with caution. An alternative is to use an independent third-party vendor to do the testing. Organizations should insist on a test report that documents the methodology and provides copies of test software as well as the results. An unwillingness to share such information should raise concern about whether the testing was sufficient.

- Automated testing tools may enable cheaper, faster, easier, and possibly more reliable testing, but overreliance on them can mask flaws that can be otherwise detected through more thorough manual testing. This is particularly true with respect to testing for outages, which have to be carefully orchestrated and scripted.

14.4 Test Phases

The following sections briefly describe the stages of testing that a mission-critical system or network feature must undergo prior to deployment. At each stage, entrance and exit criteria should be established. It should clearly state the conditions a component must satisfy in order to enter the next phase of testing and the conditions it must satisfy to exit that phase. For instance, an *entrance criterion* may require all functionality to be implemented and errors not to exceed a certain level. An *exit criterion*, for instance, might forbid outstanding fixes or require that a component function be based on specific requirements.

Furthermore, some form of *regression testing* should be performed at each stage. Such tests verify that existing features remain functional after remediation and that nothing else was broken during development. These tests involve checking previously tested features to ensure that they still work after changes have been made elsewhere. This is done by rerunning earlier tests to prove that adding a new function has not unintentionally changed other existing capabilities.

Figure 14.3 illustrates the key stages that are involved in mission-critical network or system testing. They are discussed in further detail in the following sections.

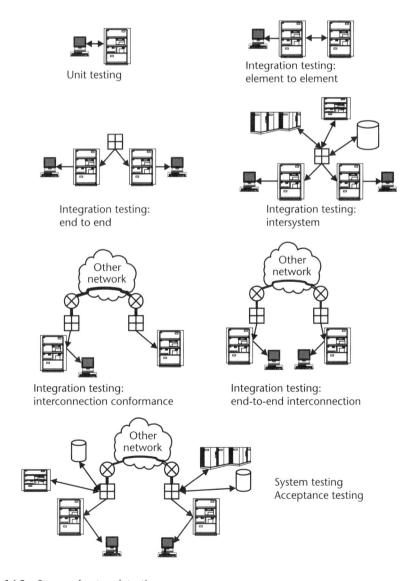

Figure 14.3 Stages of network testing.

14.4.1   Unit Testing

Unit testing involves testing a single component to verify that it functions correctly on an individual basis. It is conducted to determine the integrity of a component after it has been developed or acquired. In software development, for instance, this test takes place once code is received from the development shop. In general, there are two classes of tests that can be performed: *black-box* tests and *white-box* tests [3]. The approaches, which can be used in conjunction with each other, can yield different results.

Black-box tests are functional tests, and in this respect they measure whether a component meets functional requirements. To this end, they test for omissions in requirement compliance. Black-box tests are not necessarily concerned with the internal workings of a particular component. Instead, they focus on the behavior of a component in reaction to certain inputs or stimuli. Black-box tests are conducted once requirements have been approved or have stabilized and a system is under development. White-box tests, on the other hand, are structural tests. They rely on internal knowledge of the system to determine whether a component's internal mechanisms are faulty. White-box testing is usually more resource intensive than black-box testing and requires accurate tester knowledge of the system.

14.4.2   Integration Testing

While component testing assures that network elements function as intended, it does not ensure that multiple elements will work together cohesively once they are connected. *Integration tests* ensure that data is passed correctly between elements. In network projects, integration testing begins once IT development groups have conducted rigorous unit tests on individual systems, satisfying required criteria. Integration testing is one of the most important test phases because it is the first time that various disparate elements are connected and tested as a network.

Network integration testing entails a methodical, hierarchical bottom-up approach. It includes the following series of tests, illustrated in Figure 14.3 [10]:

- *Element-to-element testing*. This test, usually done in a test-lab environment, checks for interoperability among the different elements, such as interoperability between a switch and router device or two servers.
- *End-to-end testing*. This test is aimed at testing services end to end. It tests all relevant combinations of services and network configurations. Different types of performance and recovery tests can be applied (these are discussed in the next section).
- *Intersystem testing*. This test verifies that many other existing systems operate properly after network remediation. If done in a test environment, the production environment should be accurately simulated to the extent possible.
- *Interconnection testing*. This test verifies the integrity of interconnecting with other networks, including carrier networks. Carriers should be provided with interface conformance requirements. They will typically have their own requirements and may be subject to additional regulatory or government requirements that must be included in the testing.

- *End-to-end interconnection testing.* This test confirms end-to-end service across interconnections. It should reflect end-to-end test activities over network interconnections.

As problems are identified through the course testing, corrective actions must be taken and tests repeated to confirm repairs. This may even require repeating component tests. However, at this stage, any changes made internally to an element could have significant implications on testing, as many other elements can be affected. Furthermore, changes can invalidate previous component tests, requiring earlier tests to be repeated.

Another approach to integration testing is the top-down approach, illustrated in Figure 14.4. This involves first creating a network core or skeleton network and repeatedly adding and testing new elements. Often used in application development, this method is used for networks as well [11]. It follows a more natural progression and more accurately reflects how networks grow over time. This type of testing is best used for phasing in networks, particularly large ones, while sustaining an active production environment along the way. One will find that ultimately using both approaches in combination works best. This involves using a top-down approach with bottom-up integration tests of elements or services as they are added to the network.

14.4.3 System Testing

System testing is the practice of testing network services from an end-user perspective. Corrections made at the system-test phase can be quite costly. When conducting these tests, failure to accurately replicate the scale of how a service is utilized can lead to serious flaws in survivability and performance, particularly when the

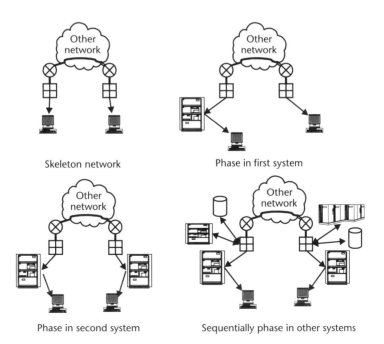

Figure 14.4 Top-down integration testing.

unforeseen arises. For instance, traffic load and distribution, database sizes, and the number of end users under normal, peak, and outage conditions must be accurately portrayed. Using nondevelopment IT staff at this stage can provide greater objectivity during testing.

It is common practice to forego thorough system testing for several reasons. As this phase occurs towards the end of the development cycle, shortcuts are often taken to meet delivery deadlines and budgets. Some even feel that it is not necessary to spend additional time and money on testing, as prior tests may have already satisfied many critical requirements to this point. But because of these practices, obvious flaws in survivability are overlooked. The following sections describe several types of system tests that should be performed.

14.4.3.1   Performance Testing

The primary goal of performance testing is to deliver comprehensive, accurate views of network stability. True network performance testing should qualify a network for production environment situations by satisfying performance requirements and ensuring the required quality of service (QoS) before unleashing users to the network. It can also avoid unnecessary countermeasures and associated product expenditures. It can also reduce the risks associated with deploying unproven or experimental technology in a network. But overall, network performance testing should address the end-user experience; the end user is ultimately the best test of network performance.

Good network performance testing should test for potential network stagnation under load and the ability to provide QoS for critical applications during these situations. Thorough network performance testing entails testing the most critical areas of a network, or the network in its entirety if possible. It should test paths and flows from user constituencies to networked applications and the resources the applications utilize—every element and link between data and user. The test methodology should address specific performance requirements and should include loading a network on an end-to-end basis to the extent possible. Various device and network configurations should be tested.

In order to completely assess an end-to-end network solution prior to deployment, a methodology should be defined that outlines those metrics that will be used to qualify each performance requirement. Testing should, at a minimum, involve the performance metrics that were discussed in the chapter on metrics. Traffic should be simulated by proprietary means or through the use of automated tools. Testing should be planned during the design stage. As service features are designed, performance tests should be planned that involve relevant transactions. Transactions should be tested from start to finish from the component to the end-to-end level to verify that portions of the network are performing adequately at all levels [12].

Network performance testing requires having some fundamental test capabilities that address identifying and classifying traffic flows. Traffic flows can be identified at the transaction, packet, frame, and even physical layer. These capabilities include [8]:

- The ability to generate, receive, and analyze specially marked or time stamped units of traffic (e.g., packets, frames, or cells) at line speed concurrently on multiple ports;

- The ability to correlate specially marked traffic with external events and to detect failed streams;
- The ability to synchronize transmitting and receiving of traffic to ensure accurate measurement;
- The ability to check for any data corruption in traffic elements as they traverse the network;
- The ability to place special counters in traffic elements to check for skipped, missing, or repeated elements;
- The ability to arrange traffic priority and timing to challenge the network's ability to prioritize flows.

When testing across network links that use different transport technologies, care must be taken to understand how overhead at each layer, particularly layer 2, will be changed, as this can impact performance metrics such as throughput. For example, traffic traversing from an Ethernet link to a serial T1 link will likely replace media access control (MAC) information in the Ethernet frame, changing the overall frame size.

Furthermore, link overflow buffers inside devices can inflate true performance when they accumulate frames or packets during a test, giving the perception of higher throughput. Varying the range of source network addresses can test a device's ability to internally cache for efficiency but inflate true link performance. It is important to vary frame or packet sizes as well. Higher volumes of smaller packets can cause devices to work harder to inspect every packet.

The following are examples of the various types of network performance tests that can be performed, depending on their relevancy [13, 14]. Many of the related metrics were discussed in the chapter on metrics:

- *Throughput* measures the maximum capacity per connection with no loss, at various frame or packet sizes, and flow rates for the network resource under test. It entails monitoring traffic volumes in and out of a device or port.
- *Loss* measures the performance of a network resource under a heavy load, by measuring the percentage of traffic not forwarded due to lack of resources.
- *Stability over time* measures the loss or percentage of line capacity for a connection over a prolonged period of time.
- *Buffering* measures the buffer capacity of a network resource by issuing traffic bursts at the maximum rate and measuring the largest burst at which no traffic is dropped.
- *Integrity* measures the accuracy of traffic transferred through various network elements by evaluating incorrectly received or out-of-sequence frames or packets.
- *End-to-end performance* measures the speed and capacity of an element or network to forward traffic between two or more end points.
- *Response* measures the time interval for an application to respond to a request. It is performed through active or passive transaction monitoring.
- *Availability* can be measured by polling a circuit or element for continued response. For a service application, it can be tested through active transaction monitoring.

- *Latency* measures one-way latency for a traffic stream received at an element, with no end-station delay. These tests should also include the latency distribution across various time intervals, yielding the minimum and maximum latency in each interval.

In each test, the maximum load levels should be measured in which respective requirements can be satisfied. An efficient way to do this is based on a binary search strategy presented in Request for Comment (RFC) 1242 and similarly in RFC 2544. Once a reading is taken at a particular rate for which a requirement is violated, the rate is halved, and the process is repeated. If at any time during this process the requirement is satisfied, the next rate is chosen as half the distance between the previous rate and the next higher rate at which the requirement was violated.

Passing such performance tests with flying colors does not ensure true QoS. Testing should be done to determine how well systems and networks optimize, prioritize, and segment individual user traffic streams and flows. Network optimization tests should be conducted with changes in network parameters, conditions, or traffic shaping policies, such as differential services (DiffServ), resource reservation protocol (RSVP), multipath label switching (MPLS), or IEEE 802.1 virtual local area network (VLAN). By tagging test transactions with the appropriate traffic priorities under each condition, the QoS offered to classes of users can be evaluated to determine in which scenarios packets or frames are dropped. Bandwidth for high-priority traffic should be preserved in cases where bandwidth is oversubscribed [15].

Performance validation should enable the abstraction of benchmarks from the test data. Benchmarks can be used to compare different traffic streams with each other under different scenarios. For each traffic stream or class, the frequency or percentage of occurrence of observed requests should be classified according to the performance metric or QoS criteria observed during the test, resulting in a probability distribution (Figure 14.5). These distributions can be used to compare different traffic streams and determine how performance would be affected if test parameters or conditions are changed.

14.4.3.2   Load Testing

Load testing is sometimes referred to as *stress testing* or *pressure testing*. Load tests verify that a system or network performs stably during high-volume situations, especially those where design volumes are far exceeded. It is a special case of

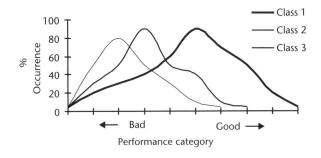

Figure 14.5   Classification of observed test traffic.

performance testing that identifies which elements operationally fail while stressing a network or the element to its maximum. Load testing can serve several purposes:

- As a scalability test to develop safeguards against further damage caused by errors or misbehavior when a network or element is stressed;
- To identify the inflection points where traffic volumes or traffic mixes severely degrade service quality;
- To test for connectivity with a carrier, partner, or customer network whose behavior is unknown, by first emulating the unknown network based on its observed performance;
- To identify *factors of safety* for a network. These are measures of how conservatively a network is designed relative to those loads or conditions that can cause failure.

Load testing can be done using several approaches. Unless testing is done in a real-world production environment, traffic load must be simulated. *Synthetic load testing* is a common approach whereby a network or element is inundated with synthetic traffic load that emulates actual usage. The traffic must be easily controlled so that different levels can be incrementally generated. Various loads are simulated that are far in excess of typical loads. The number of users and types of transactions should be changed.

Another type of load test is *accelerated stress testing*. This type of test subjects elements to harsh conditions, and then failure analysis is performed to determine the failure mechanisms, with the aim of identifying an appropriate reliability model. Depending upon the model, the failure rate could decrease, increase, or remain constant, as in the case of the exponential life model discussed in the chapter on metrics. Conditions must be chosen in such a way such that the model and failure mechanisms remain the same under all conditions. This last point is key to load testing. Any load testing is of no use unless the precise failure mechanisms can be identified so that measures can be devised to protect the functional integrity of the network under stress.

14.4.3.3 Backup/Recovery Testing

Backup systems and recovery sites cannot be trusted unless they are periodically tested to find any potential problems before an outage occurs. After all, it is much better to discover problems during a test than in an outage situation. These tests should ensure that a backup system or recovery site could activate and continue to perform upon an outage. Failover mechanisms should be tested as well to ensure their integrity.

This type of testing should be ongoing to ensure that recovery and backup plans address changes in technology and in business requirements. Tests should be conducted at least once or twice a year to be sure that the backup and recovery resources are still functioning. They should also be conducted upon any significant changes introduced into a network environment. These tests should demonstrate and verify recovery procedures and test the feasibility of different recovery approaches. The intent is to identify weaknesses and evaluate the adequacy of resources so that they can be corrected.

Recovery test data should be evaluated to estimate, at a minimum, the observed mean time to recover (MTTR). As shown in Figure 14.6, a cumulative distribution of test results can be created to estimate the MTTR for a recovery operation, which in this example is about 85 minutes. It can also convey the recovery time that can be achieved 95% of the time. This value should be compared with the desired recovery time objective (RTO) to ensure that it is satisfied.

Recovery site providers, as discussed in a previous chapter, can provide staff to work with an organization's own staff to assist in recovery testing, depending on the service agreement. If the site is remote from the main site, it will be necessary to use the service provider's local staff. Some providers perform an entire recovery test as a turnkey service, from loading and starting systems to performing network tests. In a shared warm-site environment, it is most likely that recovery will involve working with unloaded systems, requiring installation and startup of system software, OSs, and databases. In a cold-site scenario, some providers physically transport backup servers, storage devices, and other items to the site. They connect them to the network, load the OS and related data, begin recovery, and perform testing.

14.4.3.4 Tactical Testing

Tactical testing, sometimes referred to as *risk-based testing*, allocates testing priority to those requirements that are most important, those elements that have the greatest likelihood of failure, or those elements whose failure poses the greatest consequence. The most critical items are tested thoroughly to assure stable performance, while lower priority ones receive less testing. For each requirement or resource to be tested, two factors must be determined—the importance of each and the risk of failure. They are then ranked accordingly to determine the level of effort to devote in the testing process.

The goal of the ranking is sequencing the testing schedule by priority, thoroughly testing the most critical requirements or resources first. The process assumes that once critical functions are tested, the risk to an organization will rapidly decline. Lower priority requirements or resources are then dropped if a project runs out of time, resources, or budget. Examples of tactical tests include restoration of critical

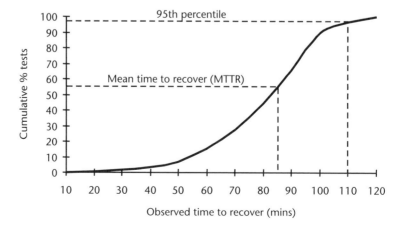

Figure 14.6 Evaluation of recovery test data.

data, testing of critical applications, testing of unprotected network resources, and mock outages.

14.4.4   Acceptance Testing

The purpose of acceptance testing is to ensure that a system, application, or network is ready for productive service. Acceptance testing is a demonstration of a finished product. In procurement situations, a customer should sign off on the acceptance test, signifying that the system or network satisfies their requirements. Even though it may be costly to correct errors found during acceptance testing, it is still better to find them prior to going live. A client can still find major problems beyond acceptance, which is quite common. However, depending on the service agreement, a supplier may not necessarily be held liable for fixing the problems.

When procuring a system or network product from a supplier, clauses should be placed in the request for proposal (RFP) or service agreement stipulating that it must pass specific tests in lab, field trial, and live field operation before the vendor is paid. Many agreements will require that specifications presented in the RFP are binding as well. These measures ensure technical support from the vendor and force the vendor to deliver a quality product. Acceptance testing should be conducted on internal developments as well. They should involve acceptance from the internal user organization. All system upgrades and application updates should undergo acceptance testing.

Unfortunately, acceptance testing is not widely practiced. One reason is that many organizations feel that prior system tests will have already uncovered and corrected most major problems. Waiting until the acceptance test stage would only delay final delivery or deployment. This belief is invalid because one of the goals of acceptance testing is to verify that a system or network has passed all of the prior system tests. Another misconception is that acceptance tests should be done only for large complex systems or those where there is a high level of customer involvement in system testing. If the customers are satisfied with the overall testing strategy, they do not need to undergo acceptance testing. In fact, quite the opposite is true. More emphasis may have to be place on acceptance testing when customers are not actively engaged to prevent problems down the road.

Acceptance testing should enable a user or customer to see the system or network operate in an error-free environment and recover from an error state back to an error-free environment. It should provide both the customer and provider the opportunity to see the system, network, or application run end to end as one consolidated entity, versus a piece part system test view. It is an opportunity for the user or client to accept ownership for the entity and close out the project. Ideally, the users or clients should be given the opportunity to validate the system or network on their own. Once accepted, the burden of responsibility lies in their hands to have problems addressed by the IT team.

There are many ways to structure acceptance tests. Approaches will vary depending on the client needs and the nature of the system, application, or network under development. Installing a new resource, turning it on, and using it is simply too risky. A more common approach is *parallel testing*, which involves installing and testing a new system or network while the older production version remains in operation. Some approaches entail having the two completely separate, while others have the two interoperate. Upgrades or changes are then performed on the test entity while the consistency between the two is evaluated for acceptance.

An acceptance test period is conducted over a period of time, sometimes referred to as a *burn-in* period, a term adapted from hardware testing that refers to electronic components getting warm when they are turned on. A new system is considered accepted after the burn-in period plus a period of additional use by the user or client. Although these periods typically last from 30 to 90 days, this author has seen systems installed that have not been accepted for close to a year due to persistent problems.

Quite often, controversy can arise between a user or client and an IT development team regarding whether a requirement was truly satisfied. To avoid these situations, it is important that acceptance criteria are clarified upfront in a project. Acceptance criteria should not be subjective or vague—it should be quantified and measurable to the extent possible. It should be presented as criteria for success versus failure.

14.4.4.1   Cutover Testing

Once acceptance testing is complete, the test system is put into production mode and the old production system is taken off-line for upgrade or decommissioned as required. This involves the process of *cutover testing* [16]. Placing the new resource into production while the old one is left in service is the safest way to handle cutover. This is preferred over a *flash cut* approach that involves turning on the new resource and turning off the old. Once users are asked to start using the new resource, the old resource can remain in operation as a backup. If major problems are encountered at this stage, a *back-out strategy* can be implemented. This allows users to go back to using the old resource until the new one is corrected.

Once the new resource is proven in, the old resource is discontinued. Hopefully, if an organization did its job correctly, there will be no major errors or problems during and after acceptance testing and cutover. Problems that do arise can signify serious flaws in the system development cycle, particularly in the testing phases. In this respect, both an IT development organization and user can benefit from acceptance testing.

14.4.5   Troubleshooting Testing

Troubleshooting testing is done to find the location of an error (i.e., what is wrong). It is somewhat distinguished from root cause analysis, which is intended to identify the source of an error (i.e., why is it wrong). It can be conducted during any phase of the development cycle. Although newer capabilities such as load balancers, virtual private network (VPN) appliances, caching servers, and higher layer switching, to name a few, mandate new types of troubleshooting tools, the fundamental tasks in troubleshooting a network problem do not necessarily change. Some of these tasks were alluded to in the chapter on network management. Relying solely on troubleshooting tools and products can be ineffective unless a logical procedure is followed to guide use of the tools.

Comprehensive troubleshooting requires detecting and fixing problems at multiple layers of the Internet protocol architecture. Cable checks, network protocol issues, and application problems, among others, may all come into play when troubleshooting a network problem. Cable and fiber testing can address physical layer media and protocols using tools such as the time domain reflectometers (TDR), which were discussed earlier in the chapter on facilities. Troubleshooting high rate 100BaseT and Gigabit Ethernet over copper can be more demanding to test, as they are less forgiving than 10BaseT.

Data link, network, and transport layer troubleshooting rely on the use of *protocol analyzers*. Protocol analysis involves capturing packets or frames and placing them into a special buffer for scrutiny. For high-speed networks, such as Gigabit Ethernet, this may require substantially large buffers to avoid buffer overflow. Because these tools typically connect using standard NICs, they can overlook some types of data link layer error conditions. They can also be oblivious to physical layer and cabling problems as well. On the other hand, some protocol analyzers can test for some application-level capabilities, such as database transactions. Many can be programmed to trap particular types of data, look for different types of errors, perform certain measurements, or convey operating status.

While protocol analyzers are typically portable devices, *probes* or *monitors* connect permanently to a network or device. As discussed in the chapter on network management, probes are devices that passively collect measurement data across network links or segments. Some devices are equipped with memory to store data for analysis. Many are based on the remote monitoring (RMON) protocol, which enables the collection of traffic statistics and errors. With the introduction of RMON II, layers 2 through 7 can be supported. Many network managers find it desirable to use tools compliant with the RMON standard versus a proprietary implementation.

Whether using an analyzer, probe, or some other approach, network troubleshooting is conducted through a process of elimination. The following are some basic steps:

1. Rule out the obvious first, eliminating rudimentary causes such as power, OS, and configuration errors [17]. More often than not, configuration problems are the leading cause of many errors. It is also important to determine whether the problem stems from misuse by a user.

2. Check first to see if the application is faulty. Attempting to reproduce the problem at each layer can help narrow down the cause. This means entering the same inputs that created the initial problem to see if it reoccurs.

3. If the application is not faulty, connectivity must then be checked. At layer 3, this typically involves checking IP addressing and the ability to reach the server hosting the application. At layer 2, this typically involves checking NICs, network adapters, hubs, switches, or modem devices. Some practices that should be considered include:

   • Turning off switches or nodal elements one by one to isolate a problem is common practice, but can force unexpected traffic to other nodes, causing them to stagnate as they recompute their traffic algorithms [18].

   • Temporarily forcing traffic onto predefined *safe paths* from every source to every destination is an exercise that can buy troubleshooting time, as long as users can tolerate some congestion. A safer approach is to have an

> alternative service, network, or backup network links available for users to temporarily utilize.

- Removing redundant links is one way to help identify unnecessary spanning tree traffic violations. These can cause traffic to indefinitely loop around a network, creating unnecessary congestion to the point where the network is inoperable.
- Another commonly used approach in smaller networks is to replace a questionable switch device with a shared hub while the device is being tested [19]. This may slow traffic somewhat, but it can keep a network operational until a problem is fixed.
- Converting a layer 2 switched network to a routing network can buy time to isolate layer 2 problems. This is particularly useful for switched backbone networks.
- Troubleshooting a device through an incumbent network that is undergoing problems can be useless. It is important to have back door access to a device either through a direct serial or dial-up connection.
- Troubleshooting a switch port can be cumbersome with an outboard probe or analyzer because it can only see one port segment versus an entire shared network, as in the case of a hub. A host-based troubleshooting capability may be a better option.
- *Port mirroring*, discussed earlier in this book, is a technique where the traffic on a troubled port can be duplicated or mirrored on an unused port. This enables a network-monitoring device to connect into the unused port for nondisruptive testing on the troubled port.
- At the physical layer, cabling and connections, particularly jumper cables, need to be checked. These are notorious for causing physical communication problems.

4. Once a problem is found and is fixed, it is important to attempt to reproduce the error condition again to see if it reoccurs or if the fix indeed corrected the problem.

5. Before returning to a production mode, traffic should be gradually migrated back onto the repaired network to avoid surging the network and to determine if the problem arises again during normal traffic conditions.

14.5   Summary and Conclusions

The cost of fixing errors grows significantly throughout the course of a development project. Next to requirements, testing is one of the most important phases of the network development cycle. Requirements should drive testing and should serve as the basis for a well-structured and logical testing plan, which is required for all new developments or upgrades. Testing should be conducted in a background that emulates the production environment as much as possible, but does not interfere with or disrupt daily operations.

There are several stages of tests, including unit tests, integration tests, system tests, and acceptance tests. Unit testing should test the integrity of an individual element, while integration testing verifies how elements work together. System tests should be conducted from the user perspective, and should include tests of performance and recovery. Acceptance testing demonstrates the finished product and should be based on clear and, if possible, quantified requirements.

Regression testing should be conducted at each stage to ensure that legacy functions were not inadvertently affected during remediation. Troubleshooting tests are also done along the way. Although there is a vast array of test tools available, the process of logically using the tools to find and fix problems is most important. A top-down approach beginning at the application level is preferred.

Although some may view testing as a troublesome activity for fear of exposing errors and flaws, its benefits are far reaching. It can lead to improved network performance, enhanced survivability, a higher quality IT environment, a better development and delivery process, and, most importantly, satisfied clients and users. One can never do enough testing—ideally, it should be ongoing. There is an old saying: "You either test it now or test it later."

References

[1] Wilkins, M. J., "Managing Batch Control Projects: The ABC's of Sidestepping Pitfalls," *Control Solutions*, November 2002, p. 26.

[2] Green, J. H., *The Irwin Handbook of Telecommunications Management*, New York: McGraw–Hill, 2001, pp. 698–703.

[3] Fielden, T., "Beat Biz Rivals by Testing," *Infoworld*, September 18, 2000, pp. 57–60.

[4] Abbott, B., "Requirements Set the Mark," *Infoworld*, March 5, 2001, pp. 45–46.

[5] Mochal, T., "Hammer Out Your Tactical Testing Decisions with a Testing Plan," *Tech Republic*, August 28, 2001, www.techrepublic.com.

[6] Wakin, E., "Testing Prevents Problems," *Beyond Computing*, July/August, 1999, pp. 52–53.

[7] Blanchard, B. S., *Logistics Engineering and Management*, Englewood Cliffs, NJ: Prentice Hall, 1981, pp. 254–255.

[8] Carr, J., "Blueprints for Building a Network Test Lab," *Network Magazine*, April 2002, pp. 54–58.

[9] Schaefer, D., "Taking Stock of Premises-Network Performance," *Lightwave*, April 2001, pp. 70–75.

[10] Parker, R.,"A Systematic Approach to Service Quality," *Internet Telephony*, June 2000, pp. 70–78.

[11] Metzger, P. W., *Managing a Programming Project*, Englewood Cliffs, NJ: Prentice Hall, 1981, pp.74–79.

[12] MacDonald, T., "Site Survival: Testing Is Your Best Bet for Curbing Web Failure," *Infoworld*, April 3, 2000, p. 61.

[13] Ma, A., "How to Test a Multilayer Switch," *Network Reliability—Supplement to America's Network*, June 1999, pp. S13–S20.

[14] Karoly, E., "DSL's Rapid Growth Demands Network Expansion," *Communications News*, March 2000, pp. 68–70.

[15] Morrissey, P., "Life in the Really Fast Lane," *Network Computing*, January 23, 2003, pp. 58–68.

[16] "System Cutover: Nail-biting Time for Any Installation," *Cabling Installation & Maintenance*, May 2001, pp. 108–110.

[17] Davis, J., "Three Rules for Faster Troubleshooting," *Tech Republic*, October 2, 2001, *www.techrepublic.com*.

[18] Berinato, S., "All Systems Down," *CIO*, February 15, 2003, pp. 46–53.

[19] Snyder, J., "High Availability's Dark Side," *Network World*, December 11, 2000, p. 84.

CHAPTER 15

Summary and Conclusions

During these critical times, network planners and operators more than ever before are under pressure to create reliable and survivable networking infrastructure for their businesses. Whether a terrorist attack, fiber cut, security breach, natural disaster, or traffic overload, networks must be self-healing. They should be designed to withstand adverse conditions and provide continuous service. Network continuity is a discipline that blends IT with reliability engineering, network planning, performance management, facility design, and recovery planning. It concentrates on how to achieve continuity by design using preventive approaches, instead of relying solely on disaster recovery procedures.

We presented an "art of war" approach to network continuity. We covered some basic principles that center upon a reference model describing the mechanics of responding to network faults. We discussed several approaches to redundancy, along with their merits and caveats. We reviewed the concept of tolerance and showed how it relates to availability and transaction loss. The chapter on metrics reviewed how availability is estimated and how it differs from reliability—although the two are often used interchangeably. Because availability can mask outage frequency, it must be used in conjunction with other metrics to characterize service.

Much of this book focused on how to embed survivability and performance within various areas that encompass most network operations: network topology, protocols, and technologies; processing and load control; network access; platforms; applications; storage; facilities; and network management. In each area, we reviewed different technologies and practices that can be used to instill continuity and discussed the precautions in using them. We covered techniques and strategies designed to keep enterprise data and voice networks in service under critical circumstances. We presented approaches on how to minimize single points of failure through redundancy and elimination of serial paths. We also showed how to choose various networking technologies and services to improve performance and survivability.

Network continuity boils down to some simple basics: applying sound processes and procedures; employing thorough performance and survivability planning; and maintaining consistent control over network architecture, design, and operation. Sense and sensibility should prevail when planning continuity. The following are some guiding principles that we would like to leave with the reader:

- *Keep it simple.* Simplicity is a fundamental rule that should be applied across all aspects of network operation. It should apply to network architecture, protocols, platforms, procurement, processes, and suppliers. The more complexity, the greater chance something will go wrong.

- *Avoid sweeping changes.* Changes should be made gradually, one step at a time. New systems, procedures, and suppliers should be phased in. The results of each phase should be checked to see that there are no adverse effects on existing operations.
- *Protect the most important things.* Priority is another prevailing rule. From an economic standpoint, not everything can be protected—this is a fundamental law of continuity. Thus, the objective is to identify the most important services and protect the IT resources that support those services. These should also include resources that enable customer and supplier access. The potential business impacts of a service disruption should be understood from the customer perspective.
- *Look beyond disasters.* Recent world events have triggered an emphasis on disaster and security. Facing a volatile economy, many firms are striving to protect their IT operations against such occurrences without blowing their budgets. However, risk, as discussed earlier in this book, is driven by the likelihood of an occurrence and its ability to cause actual damage. For most enterprises, the likelihood of such major occurrences pales to those frequent little mishaps and bottlenecks that regularly plague operations. The cumulative effect of many short but frequent disruptions and slow time can be more costly than a single major outage.
- *Know your thresholds.* At the outset, there should be well-defined performance thresholds that, if violated, constitute an outage. An envelope of performance should be constructed for each critical service. Each envelope should be comprised of service metrics that are tied to the service's objectives.
- *Focus on effect.* There are infinite ways a system could fail. To a user, a downed server is of little use, regardless of the cause. That is why it is important to first focus on effect versus cause when planning for continuity. The ramifications of not having a critical resource should be considered first, followed by precautions that ensure its survivability and performance.
- *Restore service first.* In a recovery, priority should be placed on restoring and continuing service, rather than finding and repairing the root cause. It can take some time to identify and fix the root cause of a problem. Placing users or customers at the mercy of lengthy troubleshooting will only make things worse. A contingency plan for continuing service after an outage should buy the time to fix problems.
- *Strike a happy medium.* Putting all of your eggs in too many baskets can be just as dangerous as putting them all in one basket. A happy medium should be sought when allocating and diversifying resources for continuity. Network and system architectures should be neither too centralized nor too decentralized. The number of features in any one system should be neither too many nor too few.
- *Keep things modular.* We discussed the merits of modular design. Using a zonal, compartmental, or modular approach in the design of a network or system not only aids recovery, it also enhances manageability. Modules should be defined based on the desired network control granularity, failure group size, and level of recovery. Failure of an individual module should minimally affect others and help facilitate repair.

- *Put redundancy in the right places.* Much discussion in this book was dedicated to redundancy and eliminating single points of failure. But we cannot overemphasize the importance of placing redundancy and protective capacity in the right places. Indiscriminate placement can lead to a false sense of security and wasted dollars. A backup component, for instance, must have adequate capacity to satisfy the required service levels during an outage. Another example is paying for expensive high-availability systems over a serial network that itself can pose a single point of failure.

- *Seek scalability.* Scalability is a feature desired of any mission-critical solution. The better solutions are those that not only exhibit economies of scale with growth, but also produce double value. These include intangible benefits or solutions to problems that extend beyond those of continuity. This strengthens the business case for the solution.

- *Don't rely on SLAs.* IT operations often rely on contracts and agreements with system suppliers and service providers. Service level agreements (SLAs) have been regarded as a mechanism for assuring reliability from service providers. But merely drawing up an SLA with a service provider does not ensure available or reliable service. SLAs present punitive measures that are intended as provider incentives to supply better service. SLAs do not necessarily prevent outages.

Finally, as we have implied throughout this book, the most important characteristic of a mission-critical network is predictability—being able to predict response to certain stimuli under various conditions. Predictable and consistent behavior is indicative of a reliable network developed by those having a true understanding of mission needs. If you can achieve this, then you know you have done your job.

About the Author

Matthew Liotine, Ph.D., is vice president of BLR Research, located in Naperville, Illinois. He is a renowned speaker, educator, and consultant in information technology. With over 25 years of industry experience, he provides enterprises with management and technical expertise in business continuity, wireless issues, and telecommunications planning. His clients have included financial institutions, property developers, broadcasters, public agencies, insurance firms, convention centers, and foreign governments. During his years at AT&T Bell Laboratories, he received the AT&T Network Architecture Award for the development of new network switching and routing features. He holds a doctorate in engineering from Princeton University. Dr. Liotine can be contacted at mliotine@blr-services.com.

Index

Digital Clocks for Synchronization and Communications, Masami Kihara,
   Sadayasu Ono, and Pekka Eskelinen

Digital Modulation Techniques, Fuqin Xiong

E-Commerce Systems Architecture and Applications, Wasim E. Rajput

Engineering Internet QoS, Sanjay Jha and Mahbub Hassan

Error-Control Block Codes for Communications Engineers,
   L. H. Charles Lee

FAX: Facsimile Technology and Systems, Third Edition,
   Kenneth R. McConnell, Dennis Bodson, and Stephen Urban

Fundamentals of Network Security, John E. Canavan

Gigabit Ethernet Technology and Applications, Mark Norris

Guide to ATM Systems and Technology, Mohammad A. Rahman

A Guide to the TCP/IP Protocol Suite, Floyd Wilder

Home Networking Technologies and Standards, Theodore B. Zahariadis

Information Superhighways Revisited: The Economics of Multimedia,
   Bruce Egan

*Installation and Maintenance of SDH/SONET, ATM, xDSL, and
   Synchronization Networks*, José M. Caballero et al.

*Integrated Broadband Networks: TCP/IP, ATM, SDH/SONET, and
   WDM/Optics*, Byeong Gi Lee and Woojune Kim

Internet E-mail: Protocols, Standards, and Implementation,
   Lawrence Hughes

*Introduction to Telecommunications Network Engineering,
   Second Edition,* Tarmo Anttalainen

Introduction to Telephones and Telephone Systems, Third Edition,
   A. Michael Noll

An Introduction to U.S. Telecommunications Law, Second Edition,
   Charles H. Kennedy

IP Convergence: The Next Revolution in Telecommunications,
   Nathan J. Muller

LANs to WANs: The Complete Management Guide, Nathan J. Muller

The Law and Regulation of Telecommunications Carriers,
   Henk Brands and Evan T. Leo

Managing Internet-Driven Change in International Telecommunications,
   Rob Frieden

For further information on these and other Artech House titles,
including previously considered out-of-print books now available through our
In-Print-Forever® (IPF®) program, contact:

Artech House
685 Canton Street
Norwood, MA 02062
Phone: 781-769-9750
Fax: 781-769-6334
e-mail: artech@artechhouse.com

Artech House
46 Gillingham Street
London SW1V 1AH UK
Phone: +44 (0)20 7596-8750
Fax: +44 (0)20 7630-0166
e-mail: artech-uk@artechhouse.com

Find us on the World Wide Web at:
www.artechhouse.com